T0341052

PORTFOLIO MANAGEMENT

A STRATEGIC APPROACH

Best Practices and Advances in Program Management Series

Series Editor
Ginger Levin

RECENTLY PUBLISHED TITLES

Portfolio Management: A Strategic Approach
Ginger Levin and John Wyzalek

Program Governance
Muhammad Ehsan Khan

Project Management for Research and Development: Guiding Innovation for Positive R&D Outcomes
Lory Mitchell Wingate

The Influential Project Manager: Winning Over Team Members and Stakeholders
Alfonso Bucero

Project Management for Research and Development: Guiding Innovation for Positive R&D Outcomes
Lory Mitchell Wingate

PfMP® Exam Practice Tests and Study Guide
Ginger Levin

Program Management Leadership: Creating Successful Team Dynamics
Mark C. Bojeun

Successful Program Management: Complexity Theory, Communication, and Leadership
Wanda Curlee and Robert Lee Gordon

From Projects to Programs: A Project Manager's Journey
Samir Penkar

Sustainable Program Management
Gregory T. Haugan

Leading Virtual Project Teams: Adapting Leadership Theories and Communications Techniques to 21st Century Organizations
Margaret R. Lee

Applying Guiding Principles of Effective Program Delivery
Kerry R. Wills

PORTFOLIO MANAGEMENT

A STRATEGIC APPROACH

Edited by

DR. GINGER LEVIN, PMP, PgMP

JOHN WYZALEK, PfMP

CRC Press is an imprint of the
Taylor & Francis Group, an **Informa** business
AN AUERBACH BOOK

CRC Press
Taylor & Francis Group
6000 Broken Sound Parkway NW, Suite 300
Boca Raton, FL 33487-2742

Printed on acid-free paper
Version Date: 20141113

International Standard Book Number-13: 978-1-4822-5104-3 (Hardback)

Visit the Taylor & Francis Web site at
http://www.taylorandfrancis.com

and the CRC Press Web site at
http://www.crcpress.com

To the eighteen experts who diligently shared their knowledge in the chapters they contributed to this book; to Rich O'Hanley and John Wyzalek, who approved my proposal for this book; to John Wyzalek, who in addition to writing a chapter in this book then supported me as a co-editor; and to my husband, Morris Levin, for his continued support and love.

Ginger Levin, Lighthouse Point, Florida

Contents

Preface

While portfolio management has been applied in the financial industry since the early 1950s, it is only within the past two to three decades that academic research, plus guidelines for practitioners, had been conducted and made available. Although some organizations used portfolio management techniques to select and prioritize programs and projects to pursue since the 1960s, these organizations rarely discussed its use recognizing it was a competitive advantage for them to do so. In the late 1970s and 1980s, software to assist in prioritizing programs and projects and to allocate resources became available, and there was increased interest in organizations to adopt the software and then recognition that tools alone were insufficient to manage a portfolio.

Portfolio management requires a culture change, with processes and procedures in place that are consistently followed at all levels to support organizational strategies and promote organizational success. It requires strategic goals to ensure the work being done, whether a program, project, or an operational activity, supports these goals; having an inventory of existing work in progress available to determine if it supports organizational strategy and should be continued; and business cases, which are prepared and approved for proposed work to undertake. Such a culture change takes time and dedication to implement, but increasingly, organizational leaders are doing so recognizing its necessity especially in terms of the complexity of work under way and the often lack of qualified and available resources to do this work effectively. It also requires some type of governance structure, such as a Portfolio Oversight Group or Review Board, to meet regularly and make decisions as to work to continue and work to pursue recognizing continual change affecting its overall vision and strategic goals.

While each organization has a portfolio, as it represents the overall organization's strategic intent, the extent of implementation of portfolio management varies significantly. In some organizations, portfolio management may lack visibility or may not exist in formal way, while in others portfolios exist at the highest level, in a department or business unit, and in individual programs with detailed processes to govern how work will be selected and prioritized and how decisions are communicated to stakeholders at all levels.

Some organizations have a dedicated person at the highest level as the portfolio manager with a Portfolio Management Office. However, the portfolio manager faces far more difficult challenges than those of a program, project, or operations manager. These challenges are due to the increased number of stakeholders, the ability to communicate effectively with stakeholders at all levels, the need to make difficult decisions in order that the work being done is continually aligned with the strategic goals and objectives, and to manage risks as they occur and hopefully turn them into opportunities to be pursed and embraced. As well, he or she is responsible for preparing the processes to follow and the tools and techniques to use and ideally adopts an approach that is flexible and is not overly bureaucratic so it is useful to people at all levels and not viewed as extra meaningless work.

Recognizing the importance of selecting and pursuing programs, projects, and operational work that adds business value with benefits that are transitioned to end users and customers and can be sustained, in 2006 the Project Management Institute (PMI®) issued its first *Standard on Portfolio Management.* PMI issued a second edition in 2008, followed by a third edition in 2013. Also in 2014, PMI launched a Portfolio Management Professional (PfMP®) credential in which several of the authors who contributed chapters to this book participated, to recognize the advanced expertise of practitioners in the field. The United Kingdom also recognized the increased interest in portfolio management and established Management of Portfolios (MoP®) guidelines in 2010, with a guide for senior executives followed by one for practitioners in 2011. In January 2014, a joint venture was established, AXELOS, to handle a number of certifications for practitioners including one for the MoP®.

As portfolio management is beginning to be more formally recognized as a profession of choice, increasingly peer-reviewed journal articles and books have been and are being published, doctoral dissertations and theses have been and are written, and universities and training providers are offering courses in it. However, much remains to be learned and shared.

This book attempts to help fill this void. First, the chapters are written by thought leaders in academia and business from around the world. Several of the chapters are written by PfMPs, and the authors have varied experience working with different organizations across the globe in portfolio management. The people who contributed chapters for this book also represent each region of the world, the Americas, Europe, the Middle East, Africa, and Australia. The intent of the chapters is to provide different perspectives as to why portfolio management is essential to all organizations of various sizes and types.

Following the PMI standard, this book is organized according to its five domains: strategic alignment, governance, portfolio performance management, portfolio risk management, and portfolio communications management. The authors were invited to select one of these five areas for their chapters, and the area of communications management received the greatest amount of interest recognizing the importance of communications for success in portfolio management. Each chapter, however, presents key insight as to why portfolio management is essential to organizational success with guidelines, examples, and models to consider along with a discussion and analysis of the relevant literature in the field. This approach can provide guidance whether one is a portfolio manager; a member of a portfolio governance group; a program, project or operations manager; or a team member, targeting a broad audience.

The book is organized so that it can be read in its entirety, or specific chapters of interest can be read separately and accompanying chapters at a later time. The majority of the chapters contain references for further reading on the specific topic. Overall, the book is a reference for the five portfolio management themes. It also serves then as a reference for individuals who desire to attain the PfMP® credential. Many chapters reference PMI standards, complement their concepts, and specifically expand on the concepts and issues that the standards only mention in passing or not at all.

It is our wish that this book enhances your own work in portfolio management at any level whether you are just beginning or are a seasoned professional as you work to achieve business value and excellence in your work in your organization. Although the book provides a variety of examples, guidelines, and techniques to consider, all portfolios are different and complex—there is no 'one best size fits all' approach in this profession. We hope this book assists you in managing the complexities and changes that affect your work and enhances your portfolio management practice. We also hope the research and experiences covered in these upcoming chapters guide you in your work in portfolio management.

Editor Bios

Dr. Ginger Levin is a Senior Consultant and Educator in project management. Her specialty areas are portfolio management, program management, the Project Management Office, metrics, and maturity assessments. She is certified as a PMP, PgMP and as an *OPM3* Certified Professional. She was the second person in the world to receive the PgMP. As an *OPM3* Certified Professional, she has conducted many maturity assessments using the *OPM3* Product Suite tool.

In addition, Dr. Levin is an Adjunct Professor for the University of Wisconsin-Platteville where she teaches in its M.S. in Project Management Program, for SKEMA (formerly Esc Lille) University, France, and RMIT in Melbourne, Australia in their doctoral programs in project management.

In consulting, she has served as product or project manager in numerous efforts for Fortune 500 and public sector clients, including Genentech, Cargill, Abbott Vascular, UPS, Citibank, the Food and Drug Administration, General Electric, SAP, EADS, John Deere, Schreiber Foods, TRW, New York City Transit Authority, the U.S. Joint Forces Command, and the U.S. Department of Agriculture. Prior to her work in consulting, she held positions of increasing responsibility with the U.S. Government, including the Federal Aviation Administration, Office of Personnel Management, and the General Accounting Office.

Dr. Levin is the editor of *Program Management: A Life Cycle Approach* (2012), author of *Interpersonal Skills for Portfolio, Program, and Project Managers,* published in 2010. She is the co-author of *Program Management Complexity: A Competency Model (2011), Implementing Program Management: Forms and Templates Aligned with the Standard for Program Management* Second Edition (2008) and aligned with the Third Edition (2013), *Project Portfolio Management, Metrics for Project Management, Achieving Project Management Success with Virtual Teams, Advanced Project Management Office A Comprehensive Look at Function and Implementation, People Skills for Project Managers,*

Essential People Skills for Project Manager, The Business Development Capability Maturity Model, and the *PMP Challenge!* (now in its sixth edition) *PMP Practice Test and Study Guide* (now in its ninth edition), the *PgMP Study Guide* (now in its fourth edition), and the *PgMP Challenge!*. In 2014, she authored the *PfMP Practice Tests and Study Guide.*

Dr. Levin received her doctorate in Information Systems Technology and Public Administration from The George Washington University, and received the *Outstanding Dissertation Award* for her research on large organizations.

John Wyzalek, PfMP uses portfolio management in his position as Senior Acquisitions Editor at Taylor & Francis, where he acquires, edits, and markets lines of books on project, program, and portfolio management, as well as information technology and software engineering. His publication lines include books for professionals, textbooks, certification guides, and encyclopedias. He is also editor and Web master of the on-line magazine IT Performance Improvement. He published *Green Project Management,* which won the David I. Cleland Project Management Literature Award in 2011. Prior to Taylor & Francis, he was a book editor at Artech House and Computing McGraw-Hill. His career in publishing began as a reporter for the *Today Newspaper,* which serves communities in Northern New Jersey. Subsequently, he edited the magazines *Computer Products* and *Fiberoptic Product News.* He was also the editor of the journal *Information Systems Management* for Warren Gorham & Lamont and Taylor & Francis.

Mr. Wyzalek is currently pursuing a doctoral degree in project management at SKEMA University, Lille, France. He earned the PfMP certification from the Project Management Institute during pilot program for the certificate. He holds a Master of Computer Science Degree from the New Jersey Institute of Technology and a Bachelor of Arts Honors Degree from McGill University, where he studied German language and literature, as well as computer science.

Mr. Wyzalek resides in Weehawken, NJ, where manages his residence's condominium homeowners association and building. In his spare time he enjoys reading, running, and training with kettlebells. He has travelled frequently to Mexico, Colombia, and Quebec and speaks both French and Spanish.

Contributor Bios

Amaury Aubrée-Dauchez is a global leader with a 21+ year proven career track record in delivering businesscentric solutions that achieve organization strategic needs. In 2014, he joined the UBS Group Strategy to augment and evolve its portfolio management framework. In 2010, he established a management consulting firm, Webbed Star, providing strategic advisory services to prestigious clients such as Nestlé for achieving excellence and adopting industry standards. From 1997–2010, he worked in a number of senior management roles at the United Nations, the North Atlantic Treaty Organization and the European Union. He holds multiple certifications in programme, portfolio and project management such as the PgMP, PMP, PRINCE2, PPSO and ASAP. In 2007, he was nominated for the best project manager by the Belgian Chapter of the Project Management Institute. From 1994 to 1997, he controlled a 400 million euros portfolio and managed the bank's relationship with regulatory bodies. In addition to his role at Kookmin Bank, was Lecturer at the Luxembourg Institute for Training in Banking.

He has an M.Sc. from Lille University of Science and Technology (France) strengthened by advanced university studies in business administration from the European Bank Academy (Luxembourg) and attended La Sapienza University (Italy) as a recipient of an Erasmus scholarship.

Jennifer Baker, PgMP, PMP, MBB, ITIL and PfMP is a certified manager with more than 25 years of experience in delivering both complex technical infrastructure projects and strategic business development programs on time and within budget. Her experience includes many business sectors including finance, government, hospitality, education and energy. She is currently an Enterprise Project Portfolio Governance Manager at Duke Energy and an Adjunct Professor at Northeast University in the MSPM Program. She is also very active with several non-profit organizations including serving on the Board of Directors for her local PMI chapter. She was both an Alpha and Beta Contributor for PMI's "Navigating Complexity" guide and was featured in a portfolio management article called "Going the Distance" in the *PM Network* magazine.

Dr Lynda Bourne, FAIM, FACS, PMP is a senior management consultant, professional speaker, trainer and an award winning project manager with 30+ years professional industry experience. She is the CEO and Managing Director of Stakeholder Management Pty Ltd and Director of Professional Services with Mosaic Project Services Pty Ltd focussing on the delivery of stakeholder management and other project and organisation related consultancy, mentoring, and training for clients worldwide.

She is a member of the International Faculty at EAN University, Columbia, teaching in the Masters of Project Management program. She is also a visiting International Professor in the Master's program "Innovation" (MSc) at the Faculty of Exact Sciences and Innovative Technologies, Sholokhov Moscow State University for the Humanities. The modules Lynda teaches are focused on stakeholder management, communication, and leadership in project management.

Dr Bourne is a Fellow of the Australian Institute of Management and a Fellow of the Australian Computer Society. She was awarded PMI Australia's 'Project Manager of the Year', and was included in PMI's inaugural list of '25 Influential Women in Project Management'.

Lynda is a recognised international author, seminar leader and speaker on the topic of stakeholder management and the *Stakeholder Circle* visualisation tool. Her book *Stakeholder Relationship Management: A Maturity Model for Organisational Implementation* (Gower, 2009, 2011) defines the SRMM® model for stakeholder relationship management maturity. She has presented at conferences and seminars in Europe, Russia, Asia, New Zealand, Australia, and the Middle East to audiences in the IT, construction, defence, and mining industries and has been the key speaker on stakeholder engagement practices at meetings, workshops, and conferences. She was editor of the book *Advising Upwards* (Gower, 2011) selecting a group of experts in their field to provide practical advice for those seeking to influence their senior stakeholder and help their managers help them. She presents workshops regularly in the Government sector on stakeholder engagement and project governance. She is currently working on a book: *Making Projects Work: Effective Stakeholder and Communication Management* for publication early 2015.

Lynda's career has combined practical project experience with business management roles and academic research to deliver successful projects that meet stakeholders' expectations. She has worked as a Senior Management Consultant with various organisations in Australia and South East Asia including senior roles with telecommunication companies. Other industry-related roles include strategic planning, account management within the IT industry, business process re-engineering (BPR) and business development in both private enterprise and Government bodies in Australasia and South East Asia.

Dr. Wanda Curlee has been married to her husband, Steve, for almost 34-year. She has three children and is proud that all three serve or will serve in the US Military. Her eldest, Paul, is a Captain in the Army, soon to transition to the civilian world. Sam, her second son, serves in the US Navy as a nuclear reactor office on the USS R. Regan. Finally, her daughter, Tiffany, is in Army ROTC as a freshman at Virginia Tech. Dr. Curlee is a veteran of establishing and running virtual projects and program management offices, tackling complex, international programs and projects by bringing diverse parties together to attain shared goals. Wanda is a program management leader experienced in global markets, government projects and highly complex IT, information systems, and managed services projects. She is also an adjunct professor at several universities. Her present employment is with a Fortune 500 as an ePMO lead. She is active on several Project Management Institute's (PMI) committees and has spoken at several congresses. Most notably is she serves on the core team for the new Requirements Standard and she sat on the team that developed the latest PMI certification, Portfolio Management Certification (PfMP).

Kari Dakakni is an international risk management professional consultant and researcher in the field of project management with experience in different countries. She has experience in multiple roles as a risk engineer, account manager, business analyst, actuary, underwriter, and manager. She has conducted site assessments for the insurance industry to measure risk exposure in manufacturing plants in the food industry. She worked for consulting companies in Italy, where she also gained a strong IT background working at Ericsson and other telecommunications companies. She was involved with originating billing systems and data filtering activities.

Kari is an engineer by background with a foundation in the mechanical technology, robotics and biomedics. As an engineer and project manager, she was responsible for technological processes in the machine tool industry, foundry design, and software development. She started her career as assistantship assignment at the University of Rome "La Sapienza" after her degree in engineering, where she completed the study gaining the title of "*dottore in ingegneria*" from the same University. Then she moved into the coaching and training field to conduct courses in engineering design for consulting companies.

In her academic life, Kari received the nomination as Baccalaureate Board's Professor as an external membership professor at the Graduate Studies Committee directly from the Education and Skills Department for high school proficiency assessment test. She also gained other certifications included a master in robotics from Iowa State University. Kari was involved in research studies at ETH–Zürich and other consulting companies, where she worked in model developing about perception-cognition-action, skill-based behavior in risk management, and developing conceptual frameworks. She presents at conferences where the last one was at the Australian risk engineering society a research in fracking technology and manmade activities developing a risk management stream for the risk appraisal. She has interests in philosophy and ethics in general and how to fit ethical values into modern organizational applicability. In addition, she is also interested in analyzing safety aspects of the food industry, their impacts on the population, and the mapping of disease development. Kari is certified engineer with the State Exam since 2006, from the National Alliance of Italian Engineers.

Robert Joslin, PhDc, BSc (Durham), PgMP, PMP, CEng, MIEEE, MBCS, is a project/program management consultant and academic researcher. He has considerable experience in designing, initiating and program management delivery of large scale business transformation, reengineering, infrastructure, strategy development initiatives including winning awards for ideas and product innovation. Previously, he has been a consultant in a wide range of industries including telecom, banking, insurance, manufacturing and direct marketing whilst working for McKinsey & Co, Logica and his own consulting company. Robert is currently studying for a PhD in 'Strategy, Programme & Project Management' at SKEMA Business School in France. During this time he has published book chapters, research papers and is in the process of authoring a book on portfolio, program and project success factors.

Dr. Muhammad Ehsan Khan is an entrepreneur and an internationally acknowledged professional on the subject of Governance and Management of Strategic Initiatives. An award-winning strategist with over a decade of leadership success, and a career that has driven multi-million dollar projects for various clients in Middle East and Pakistan, Ehsan is a founding member, and presently serves as a Partner/VP Operations and Service delivery, in a UAE based firm "Inseyab Consulting & Information Solutions LLC".

Ehsan is a PhD in Strategic, Programme and Project Management (Major de Promotion/Valedictorian) from SKEMA Business School, France, and a certified Program (PgMP) and Project Management Professional (PMP). Ehsan is a recipient of the PMI James R. Snyder Award for the year 2012 and was awarded the Young Researcher of the year award by IPMA in 2013. He is the author of first every book on the subject of Program Governance.

With a special inclination towards strategic planning and governance of projects and programs, Ehsan has provided management, consulting and mentoring services in the Middle East Region. He has been involved in establishment of PMOs, implementation of management/governance frameworks, and related practices and tools, in order to create an environment of project management excellence. He has also been managing medium-large scale ICT programs and projects for various customers, especially in the government sector.

Dr. Catherine Killen is the coordinator for innovation programs in the Faculty of Engineering and IT at the University of Technology, Sydney (UTS), Australia. Catherine conducts research on innovation processes with a focus on project portfolio management and has published more than 50 journal articles and conference papers in the area. Her current research themes include the relationship between strategy and the project portfolio, organisational capabilities for survival in dynamic environments, and the management of project interdependencies within a project portfolio. Recent work includes 'best practice' survey-based research, qualitative analyses drawing upon strategic management theories, and industry-based and classroom-based 'experiments' where new methods are employed and evaluated.

Catherine maintains strong links with industry and convenes a project portfolio management special interest group with more than 80 industry-based members. She is regularly invited to speak about innovation and project portfolio management at conferences, seminars, and industry events. She also delivers corporate workshops on technology management tools and assists organisations in benchmarking and improving their innovation processes.

As the coordinator for innovation programs, Catherine develops and teaches courses and programs on technological innovation to undergraduate and postgraduate engineering students. She also supervises research students ranging from doctoral candidates to undergraduate student completing final year projects.

Catherine has a Bachelor of Science in Mechanical Engineering from the University of Virginia (USA) with high distinction, a Master of Engineering Management from the University of Technology, Sydney (awarded the MEM prize), and a PhD from the Macquarie Graduate School of Management (MGSM) in Sydney, Australia.

The five most recent journal publications are:

1. Killen, C. P. (2013), "Evaluation of project interdependency visualizations through decision scenario experimentation", *International Journal of Project Management*, Volume 31, Issue 6, pp 804–816.
2. Killen, C. P., & Hunt, R. A. (2013), " Robust project portfolio management: Capability evolution and maturity", *International Journal of Managing Projects in Business*, Volume 6, Issue 1, pp 121–151.
3. du Plessis, M., & Killen, C. P. (2013). "Valuing water industry R&D: A framework for valuing water research and development investments in financial and non-financial terms". *Water* 2013 (September), 63–67.
4. Killen, C. P., Jugdev, K., Drouin, N., & Petit, Y. (2012), "Advancing project and portfolio management research: Applying strategic management theories", *International Journal of Project Management*, Volume 30, Issue 5, pp 525–538.
5. Killen, C. P., & Kjaer, C. (2012), Understanding project interdependencies: The role of visual representation, culture and process, *International Journal of Project Management*, Volume 30, Issue 5, pp 554–566.

Professor Laurence Lecoeuvre was formerly an International Director within the industrial sector and automotive industry (1984–2001). She integrated SKEMA's Business School in 2001. After few years as the Business Programs Director, she is today Director of Project Management Department, Member of the Board of Directors, Associate Dean in charge of the coordination of the doctoral programmes of the Group. She leads the PhD and the DBA in Programme & Project Management.

Laurence is mainly teaching research methodology to MBA and PhD/DBA students; especially qualitative research and data interpretation, and system modelling; she also teaches project management fundamentals, project marketing and management of stakeholders in the masters' programmes. Laurence is also involved in Executive Education.

Her PhD (2005), at Ecole Centrale Paris, focused on the links between project marketing and project management, in the sector of Business to Business; she continues to develop her research on the topic of project marketing; but also on governance. She received the "Habilitation to Direct Research" (HDR) award in 2009 at Lille University.

She publishes in international journals, publishes books and chapters with renowned colleagues in the area of project management, in particular on the topics of project marketing, sustainable project performance, and project governance. Her recent publications are in the *International Journal of Project Management* and *Global Business Perspectives.* In 2014 she was a presenter at the 13th International Marketing Trends Conference and has also presented at the European Academy of Management Conference.

Professor Carl Marnewick's academic career started in 2007, when he joined the University of Johannesburg. He traded his professional career as a senior Information Technology (IT) project manager for that of an academic career. The career change provided him the opportunity to emerge himself in the question why IT/IS-related projects are not always successful and do not provide the intended benefits that were originally anticipated. This is currently a problem internationally as valuable resources are wasted on projects and programmes that do not add value to the strategic objectives of the organisation. It is an international problem where there is a gap between theory and practice, and he is in an ideal position to address this problem.

The focus of his research is the overarching topic and special interest of the strategic alignment of projects to the vision of the organisations. This alignment is from the initiation of a project to the realisation of benefits. He developed a framework (Vision-to-Project i.e. V2P) that ensures that projects within an organisation are linked to the vision. Within this framework, a natural outflow of research is the realisation of benefits to the organisation through the implementation of IT/IS systems. Benefits realisation is part of a complex system, and his research to date has identified the following impediments in the realisation of benefits: (i) IT project success rates as well as IT project management maturity levels did not improve over the last decade, and these results are in line with similar international research; (ii) IT project managers are not necessarily following best practices and industry standards; (iii) Governance and auditing structures are not in place, and (iv) IT project managers' training and required skills are not aligned. If these four aspects are addressed through research and practice, then benefits realisation can occur.

His research has given him national and international presence. He is a regular reviewer for national and international journals. He is actively involved in the development of new international project management standards ISO21500 and ISO21502 (portfolio management). Project Management SA awarded him the Excellence in Research Award as recognition for his active contribution to the local and global body of knowledge by conducting and publishing scientific research in portfolio, programme, and project management.

David Maynard, BSEE, MBA, PMP is a native New Yorker who after graduation from engineering school from the State University of New York, traveled to Houston, Texas to work for NASA at the Johnson Spacecraft Center at a very exciting time. The Apollo program was wrapping up with the Apollo-Soyuz Test Project under way and the Space Shuttle design, which had started long before but still had a long way to go before it was mature.

Starting as an engineering aide, he gained responsibilities and widely participated in the Shuttle avionics architecture design and in the "glass cockpit" development. Working closely with crew members and scientists, full-sized Shuttle simulators were used as an engineering tool helping design and implement and evaluate extremely complex test cases and designs of the Shuttle.

Incrementally David's level of responsibility increased, and he became a Senior Engineer, Project Engineer, a Project Manager, Program Manager, and Section Chief. David oversaw many varied crew-training and Shuttle-related engineering challenges at the Johnson Space Center in Houston and elsewhere.

After the Challenger disaster, David's life-time focus shifted from solving technical problems to diagnosing troubled projects and assisting in their turn-around. David's MBA thesis was "The Accuracy of Stochastic Methods of Risk Assessment."

Leaving NASA, David was asked to become the General Manager of Systems Management Inc. (SMI) in Orlando, Florida whose mission was to turnaround troubled projects, programs or operations. At SMI projects with values up to $46 million USD were acquired and turned around. Customers included the U.S. Navy, U.S. Coast Guard, Alaska Airlines, and many other major corporations. After seven years, Mr. Maynard left SMI for another attempt at retirement.

David now teaches project management and risk management related topics at Indiana University and Purdue University and is the manager of PMI's Risk Community of Practice, which currently has over 23,000 members.

His interests remain in examining the accuracy of estimating risks.

Werner G. Meyer, PhD, has been a consultant in the field of project management for 19 years. He led software development projects for a number of years before getting involved in mining, construction, and engineering projects. His speciality areas are portfolio management, project management, project offices, and methodology development. He is certified as a PMP®, an *OPM3*® consultant and is a Certified Cost Professional (CCP™).

Werner has developed and presented a number of project management courses ranging from the fundamentals of project management, earned value management, program and portfolio management, cost engineering, Monte Carlo simulations, and project office establishment.

In consulting, he is the founder and managing director of ProjectLink Consulting, a South Africa-based company, but he has consulted and trained in the Middle East, Australia, and Russia. One of his focus areas is the development of scalable project, program, and portfolio management methodologies. He created the CENTRIX methodology, which is a generic and iterative project management methodology.

He received his PhD from SKEMA Business School in France, with his dissertation on the effect of optimism bias on decision makers who are faced with failing projects. His current research is on the use of computer simulations to investigate the behaviour of decision makers.

Dr. Subbu Murthy is CEO of UGovernIT, the only product in the market that provides an analytics-based IT management solution on a common integrated technology platform that works in customer data centers or in the cloud as a SaaS solution. UGovernIT provides service management, project management, resource management, IT analytics, and dashboards that provide a 360-degree view of budgets, IT spend, resources, services delivered, projects implemented, and value created. Using configurable workflows, intelligent agents, analytics and dashboards, UGovernIT is a complete solution to run the business of IT driving efficiency and fostering innovation.

Dr. Murthy has C-Level expertise in creating and marketing software products. He developed the global delivery models for IT services in 1985, SaaS based ERP, and CRM systems for claims processing in 2003, web-based B2B outsourcing portal, and the first SaaS fuzzy logic system in 2006.

He developed an On-Demand CIO retainer model to help SMBs obtain IT strategy expertise at a fraction of the cost of hiring a full-time CIO. He serves as the CIO at the Braille Institute driving technology initiatives to ensure that the high-tech/high-touch strategy reaches the visually-impaired community.

Previously he was CIO at QTC Management, Inc. QTC is the government outsourced occupational health, injury evaluations, and disability examinations in the country. The web foot print also helped QTC move from a $30M to $100M in revenues.

Prior to that he was a senior partner at Relsys, which was acquired by Oracle. He developed the SQA framework for the medical device and drug industry. He helped launch this business line with leading medical device makers such as Alcon, Allergan, Baxter, Boston Scientific, and J & J. He designed the EasyTrak product line, which provided effective tracking of user complaints.

Subbu earned his Bachelors degree in Electronics and his Masters in Computer Science from the University of Southern California. He earned his Doctorate in Information Systems from Claremont Graduate University, where he pioneered the use of metrics and analytics in managing information technology. His interests include value-based technology, IT investments, and pragmatic outsourcing which embrace quality principles drawn from his expertise in FDA GMP practices, CCE (Malcolm Baldrige criteria for excellence), ISO-9001, CMM and Six Sigma. Subbu is considered an expert in Indian Classical Music, and enjoys solving Sudoku puzzles, playing bridge, and keeping up with his twin son and daughter.

David Tennant draws on more than 25 years experience as a successful engineer, manager, and executive. He has led numerous successful organizational change initiatives and also offers expertise in project management consulting and training, business re-engineering, rescue of troubled projects, and company turnarounds.

Mr. Tennant has directed over $3.5-billion in projects and resources and served as the COO of a publicly held company. He holds a BS in Mechanical Engineering from Florida Atlantic University, a MS in Technology and Science Policy from Georgia Tech, and an Executive MBA from Kennesaw State University. Additionally, he is a registered Professional Engineer (PE) and a certified Project Management Professional (PMP).

His firm, Windward Consulting Group, received the "Project of the Year" award (2011) for the development of a $200-million project from the Project Management Institute, Atlanta Chapter.

David is currently the Chairman of the Board of Directors of Cobb EMC, a $1-billion corporation and also serves as 2014 Chairman Emeritus for the Atlanta chapter of the Project Management Institute.

He has had engagements in Finland, Germany, New Zealand, the Caribbean basin, Canada, Czech Republic, and the United Arab Emirates.

David is the President and founder of Windward Consulting Group, located in the metro Atlanta area, which has been in business for 12 years. He has authored over 30 papers and presentations on technical and managerial topics.

Rodney Turner is a Professor of Project Management at Kingston Business School and at SKEMA Business School, in Lille, France, where he is Scientific Director for the PhD in Project and Programme Management at. He is an Adjunct Professor at the University of Technology Sydney.

Rodney was educated at Auckland University where he did a Bachelor of Engineering and Oxford University where he received an MSc in Industrial Mathematics and a DPhil in Engineering Science.

Rodney was introduced to project management working for ICI as a mechanical engineer and project manager in the petrochemical industry. He then worked for Coopers and Lybrand as a management consultant, working in shipbuilding, manufacturing, telecommunications, computing, finance, government, and other areas. He was then director of Project Management at Henley Management College and Professor of Project Management at Erasmus University Rotterdam before joining the Lille School of Management (now SKEMA Business School) in 2004.

Rodney is the author or editor of 16 books, including *The Handbook of Project-based Management*, the best selling book published by McGraw-Hill, and the *Gower Handbook of Project Management*. He is editor of *The International Journal of Project Management*. His research areas cover project management in small to medium-enterprises, the management of complex projects, the governance of project management, including ethics and trust, project leadership and human resource management in the project-oriented firms.

Rodney is Vice President, Honorary Fellow and former chairman of the UK's Association for Project Management, and also an Honorary Fellow and former President and Chairman of the International Project Management Association. In 2004 he received a life-time research achievement award from the Project Management Institute, and in 2012 from the International Project Management Association. From 1997 to 2005, he returned to the oil, gas and petrochemical industry as Operations Director of the Benelux Region of the European Construction Institute. He is a member of the Institute of Directors and a Fellow of the Institution of Mechanical Engineers.

Manuel Vara is Project Portfolio Manager at Nartex Software with over 15 years of professional IT and management experience.

He holds a Bachelor of Business Administration from Universidad Autonoma de Madrid, a Master in Financial and Management Control at School of Industrial Organization (Madrid), and a Stanford Certified Project Manager (SCPM) and Project Management Professional (PMP) certifications.

His expertise is managing strategic initiatives and technology programs and projects. Currently, he collaborates with *Proeictus* a Spanish project management magazine.

J. LeRoy Ward, President of Ward Associates, is a seasoned global executive with 38+ years of progressive experience in project, program, and portfolio management in all S&P industry sectors and the U.S. Federal Government. Formerly, Mr. Ward was Executive Vice President at ESI International responsible for global product strategy and consulting in the areas of project, program, and portfolio management, business analysis, contract management and leadership/business skills. He has broad and deep international business experience, including negotiating licensing partnerships in North America, Europe, Asia and Australia.

He has authored a number of publications and articles including *Project Management Dictionary* (3rd ed); with Ginger Levin, PMP® Exam Practice Test and Study Guide, PMP® Exam Challenge, *PgMP® Exam Practice Test and Study Guide*, *PgMP® Exam Challenge*, and, *Program Management Complexity, A Competency Model*; and, with Carl Pritchard, a collection of audio CDs entitled Conversations on Passing the PMP® Exam (4th ed), which has helped more than 20,000 professionals earn the credential. He also authors his popular blog WardWired.com.

His articles have appeared in *Chief Learning Officer*, *PM Network* and *Project Manager Today*, and he is frequently quoted in key industry print and online publications. He is a dynamic and popular speaker on a wide range of topics in the project management arena. For his extraordinary contributions to the project management profession Mr. Ward was named the distinguished Winner of the Project Management Institute's (PMI) 2013 Eric Jenett Project Management Excellence Award, one of PMI's highest accolades.

Mr. Ward was a member of the adjunct faculties of The George Washington University and the American University presenting courses in remote sensing and information systems. He holds a BS and MS in geography/cartography from Southern Connecticut State University and an MSTM in Information Systems from the American University. He is also a graduate of the Federal Executive Institute, the U.S. Government's premiere executive leadership program.

He is certified by PMI as a PMP (No. 431), a PgMP (one of the first to earn it) and a PfMP (again, one of the first to earn it); and is certified by the Scrum Alliance as a CSM (Certified Scrum Master). Mr. Ward is also a licensed New York City taxicab driver (Taxi & Limousine Commission Hack License No. 5433772) that he earned just for the fun of it.

Ward Associates offers specialty advisory services in project, program and portfolio management practice and process. Mr. Ward can be reached at jleroyward@gmail.com.

Acknowledgments

Many thanks to the editorial, marketing, and production team at CRC Press, with special recognition to: Randy Burling, Jessica Vakili, and Rebecca Rothschild.

A great team!

1

Organizational Agility through Project Portfolio Management

Catherine P. Killen, PhD

Abstract

In dynamic environments, organizational agility is essential for survival; organizations must be able to adapt to change in order to succeed. In project-based organizations, a dynamic project portfolio management (PPM) capability can enhance organizational agility. PPM is an important organizational capability that enables organizations to manage and balance the portfolio holistically, to align projects with strategy, and to ensure adequate resourcing for projects in order to maximize the benefits from project investments. A dynamic PPM capability enables organizations to be agile and flexible by facilitating adjustments to the project portfolio and reallocating resources in response to the changes in the environment. In order for the PPM capability to remain relevant, it must evolve to reflect changes in the environment. Examples of aspects of PPM that enhance organizational agility are outlined in this paper to provide guidance for practitioners.

Introduction

During the past two decades, PPM has become established as a discipline, and organizations have been increasingly turning to PPM to help them manage their portfolios of projects and improve their competitive position (Wideman, 2004; Levine, 2005; Kester, Griffin, Hutlink & Lauche, 2011). Primary goals for the adoption of PPM are to effectively implement the organizational strategy through the portfolio of projects and to enhance the long-term value of the portfolio as a whole. As part of these aims, PPM assists with the management of resources across the portfolio to avoid a 'resource crunch' where the organization attempts too many projects (Cooper & Edgett, 2003). PPM methods also provide the holistic oversight required to ensure that there is balance in the portfolio. The use of formal and mature PPM approaches have been linked with higher success levels in research studies (Cooper, Edgett & Kleinschmidt, 2001; Killen, Hunt, & Kleinschmidt, 2008) prompting organizations to focus on the establishment and development of PPM.

Until recently, PPM has been presented as a series of processes and procedures that organizations tailor to suit their environment. The common refrain has been that, once tailored appropriately, the PPM process will assist an organization to achieve a competitive advantage by implementing strategy, balancing the portfolio, maximizing the value, and ensuring resource adequacy for projects. However recent research highlights many other aspects of PPM that paint a picture of increased complexity and dynamism and offers insight into additional ways that PPM can create value for an organization (Killen & Hunt 2010; Petit, 2012). PPM is now seen as more than a process—PPM is an organizational capability that also includes the organizational structure, the people, and the culture. These elements must work together for effective PPM, and top management support is an important factor in PPM capability success. Recent studies also indicate that PPM has an important role to play in helping organizations achieve advantages in dynamic environments, and the PPM capability itself needs to evolve and adjust to enhance organizational agility and contribute to sustainable competitive advantage (Killen & Hunt, 2010).

This paper first introduces PPM concepts and outlines typical processes before discussing the additional challenges for PPM in dynamic environments. To guide practitioners, several examples are presented to illustrate aspects of PPM that enhance organizational agility in dynamic environments.

PPM Concepts

As many organizations shift to 'management by projects', projects are often the main vehicle for delivering organizational strategy. Definitions of PPM have been evolving as the discipline has become established. A widely accepted and often referred to definition of PPM developed by Cooper et al. (2001, p. 3) is that "Portfolio management... is a dynamic decision process wherein the list of... projects is constantly revised. In this process, new projects are evaluated, selected, and prioritized. Existing projects may be accelerated, killed, or deprioritized and resources are allocated and reallocated". McDonough and Spital (2003 p. 40) point out that PPM is more than project portfolio selection as it also involves the "day to day management of the portfolio including the policies, practices, procedures, tools and actions that managers take to manage resources, make allocation decisions and ensure that the portfolio is balanced in such a way to ensure successful portfolio-wide new product performance". Levine (2005 p. 22) offers a broad definition of PPM: "Project portfolio management is the management of the project portfolio so as to maximize the contribution of projects to the overall welfare and success of the enterprise". Recent research highlights the fact that an organization's capability to manage the project portfolio encompasses much more than the processes and methods identified for PPM; it also requires the people and a culture that supports information transparency and portfolio level perspectives, and it requires organizational structures that provide appropriate levels of visibility and responsibility to support the PPM capability (Killen & Hunt, 2010).

Although PPM is tailored for each organization, there are many common elements and approaches to PPM. In its most simple form, PPM facilitates decisions across the entire portfolio of projects by (1) collecting information from all projects (existing and proposed projects), (2) collating and organizing the information, (3) presenting information to a carefully selected decision-making team for portfolio-level reviews, and (4) providing a structure for communicating and implementing decisions. These four steps are explained with extensions for dynamic environments in the section labeled ***Outline of a Dynamic PPM Approach***.

Figure 1.1 illustrates a range of common methods and tools for organizing and presenting portfolio data for decision meetings. Portfolio mapping is a common method to provide a central view of all projects in the portfolio. Portfolio maps plot projects on two axes and can be used to assist with the selection of a balanced portfolio of projects. Commonly used portfolio maps balance aspects such as risk versus return and can also display other information through the size, color, patterns, or notes associated with the symbol for each project. Scoring models use weightings and ratings to compare projects based on multiple criteria. Many software applications for PPM offer 'dashboard' displays that show the status of projects on dials and graphs; stoplight reporting uses the red and amber colors to highlight trouble areas and green to show the 'all clear'. Pie charts are often used to communicate the balance in the portfolio; for example by displaying the breakdown of funding across types of projects in a portfolio. All of these methods and others must be customized for each environment to best support decision making.

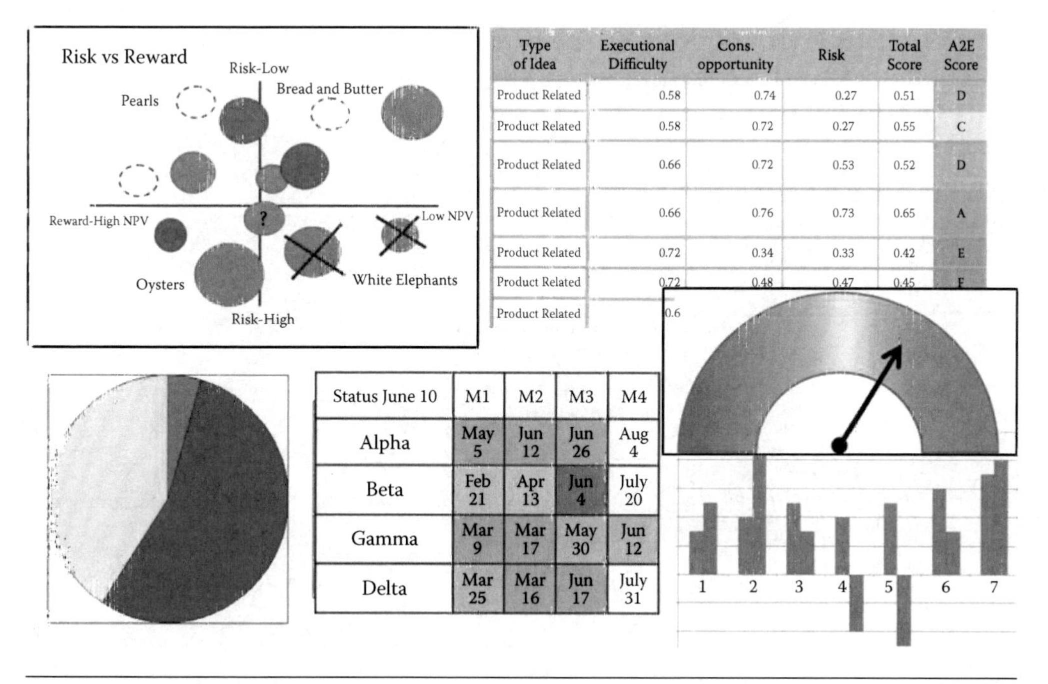

Figure 1.1 Typical methods for organizing and presenting PPM data.

While PPM capabilities often have common elements, they must be developed over time and adjusted to the environment. There is an order of implementation to many aspects of a PPM capability (Eisenhardt & Martin, 2000; Cooper, et al., 2001). For example, establishing a foundational capability such as a gated project management process is an antecedent to the development of an effective PPM capability, and data gathering capabilities must be developed before the capability to evaluate and adjust the portfolio mix can be established (Martinsuo & Lehtonen, 2007).

As shown in Figure 1.2, PPM capabilities generally include a gated project management process integrated with a portfolio-level review process at one or more of the gates or decision points. In addition the figure also reflects the fact that many

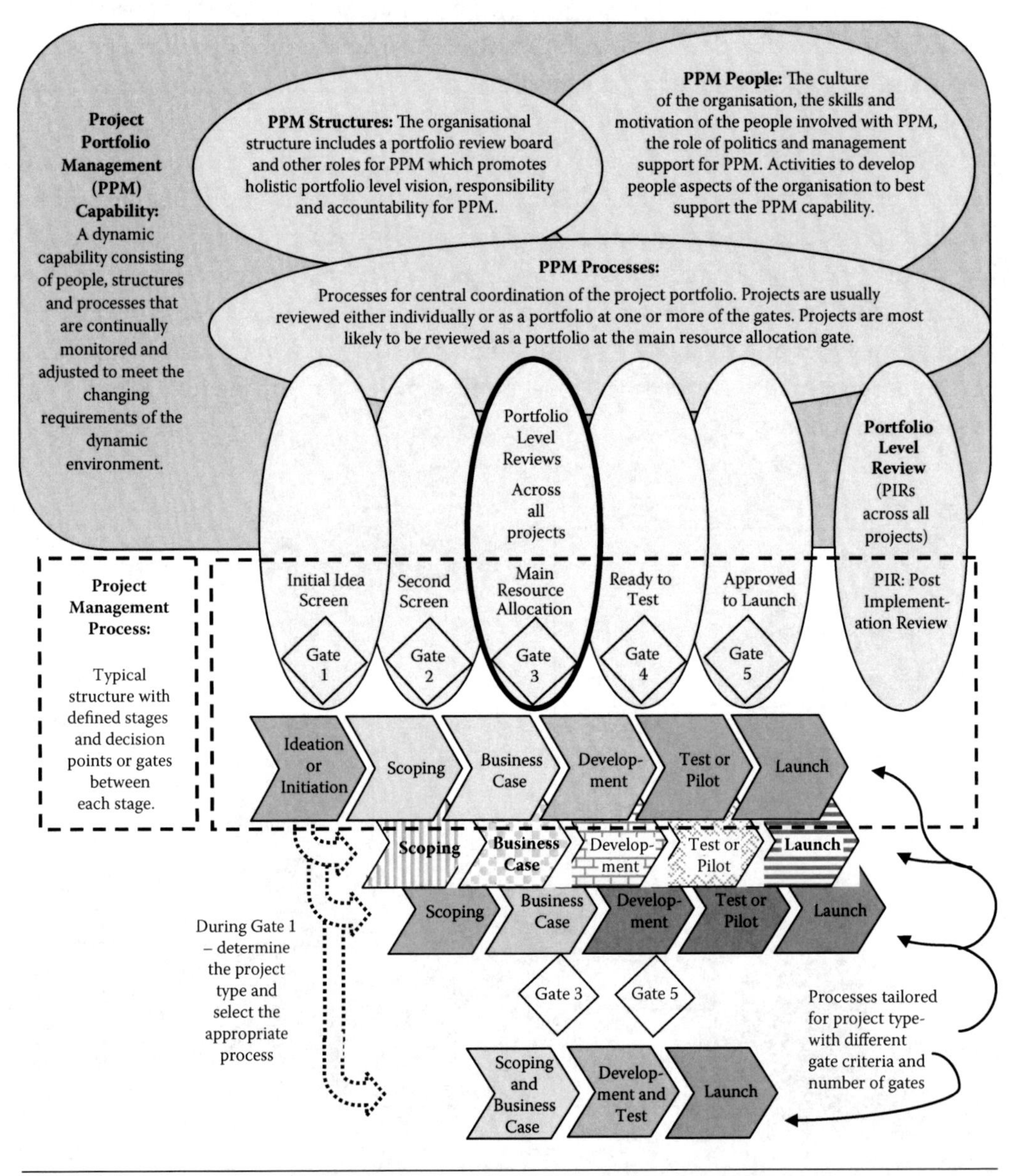

Figure 1.2 Three dimensions of PPM integrated with tailored gated project management processes.

organizations develop more than one version of a project management process to cater to different project types. The main differences between versions of the gated project management processes are in the number of stages and gates and in the types of criteria used to evaluate projects at the gates. The three main dimensions of a PPM capability are also illustrated: 'process' dimensions, 'structure' dimensions, and 'people and culture' dimensions.

Figure 1.2 also depicts the post implementation review (PIR) as part of the process. The PIR is an important stage of the process because the feedback enables the review, evaluation, and improvement of the project management and PPM processes. However, research indicates that this is a weak area in many organizations; it is common for managers to recognize the importance of PIRs, but many find it difficult to allocate resources or gain support for such tasks.

Recent research shows how PPM capabilities can improve organizational flexibility and performance by providing a holistic and responsive decision-making environment in dynamic environments. The role of the project portfolio manager is becoming formalized as organizations aim to gain the best results from PPM (Jonas, 2010). In addition to the challenge of multi-project management, organizations must address the challenges of an increasingly competitive, globalized, and deregulated environment characterized by shortening life cycles and dynamic markets. Organizational agility, the ability to adapt and respond to change, is essential in such dynamic environments (Killen & Hunt, 2010).

The focus on 'organizational agility' in this paper should not be confused with agile project management approaches. Agile project management approaches offer an incremental and responsive approach to the management of projects and are becoming adopted in an increasing range of environments; however such approaches are not the topic of this paper. This paper focuses on organizational agility from a strategic portfolio perspective. From this perspective, PPM can provide organizational agility by allowing an organization to identify changes in the environment and to evaluate, analyze, and adjust the portfolio to respond to changes in the environment. In order to observe changes in the environment, PPM requires a 'sensing' capability that involves scanning the environment and re-visiting assumptions regularly (Teece, 2007). The PPM capability is responsible for configuring the organization's efforts by building and allocating resources. A PPM capability that is able to do this in a timely fashion to respond to the environment provides organizational advantages in dynamic environments: it is a dynamic capability.

Dynamic Capabilities and Competitive Advantage through PPM

Dynamic capabilities are a special type of capability that enables an organization to respond to changes in the environment. Frameworks to identify and understand dynamic capabilities have emerged from research on strategy and competitive advantage. One of the goals of strategy research is to determine why some organizations are more successful than others and to understand the mechanisms that help some organizations achieve a competitive advantage. PPM has been identified as one of these mechanisms (Killen, et al., 2007; Killen & Hunt, 2010). Competitive advantage is the ability of an organization to create more value than its rivals, and therefore, achieve superior return on investment (Barney & Hesterly, 2012). One of the streams of strategy research is the resource-based view; the resource-based view proposes that the differences in the levels and types of resources between competing organizations can be used to explain differences in organizational success rates. An extension or offshoot of the resource-based view is the identification of a special class of organizational capabilities that enable organizations to effectively respond to changes in the dynamic environments in which they compete (Teece, Pisano & Shuen, 1997). "'Dynamic capabilities' do this by providing a capacity for 'an organization to purposefully create, extend, or modify its resource base'" (Helfat, et al., 2007 p. 4).

An organization's PPM capability is one of the internal organizational capabilities or resources that an organization uses to gain competitive advantage. In a dynamic environment, a PPM capability that acts as a dynamic capability can enable an organization to be agile and respond to change in the environment. Although dynamic capabilities are a type of resource-based capability, they do not have the ability to create value independently. Dynamic capacities add value by working with the existing resource-base (Eisenhardt & Martin, 2000) and therefore can be considered 'enabling resources' (Smith, Vasudevan & Tanniru, 1996). It is also important that supporting capabilities are established before a dynamic capability can be effective (Eisenhardt & Martin, 2000). Therefore, a dynamic capability such as PPM must be accompanied by underlying resources and capabilities such as the project management capability in order to provide long-term competitive advantage in dynamic environments. Dynamic capabilities play an important role in allocating resources, as well as in identifying the desired development and direction of resources and capabilities in line with strategy (Wang & Ahmed, 2007). As a dynamic capability, PPM can improve an organization's "ability to integrate, build, and reconfigure internal and external competencies to address rapidly changing environments" (Teece, et al., 1997, p. 516) and through these mechanisms improve the competitive advantage in dynamic environments.

PPM in Dynamic Environments

Learning and change are an important part of PPM's ability to provide advantages in dynamic environments. Figure 1.3 illustrates the effect of learning and change on the PPM capability in order that it evolves to meet the requirements of a dynamic environment. With learning and change, PPM can be a dynamic capability and enhance competitive advantage. Organizational learning is embedded in PPM capabilities through mechanisms for tacit and explicit learning. For example, tacit learning—the type of learning that is difficult to document or codify and is best transferred through experience or observation—is achieved through the interaction of experienced managers in

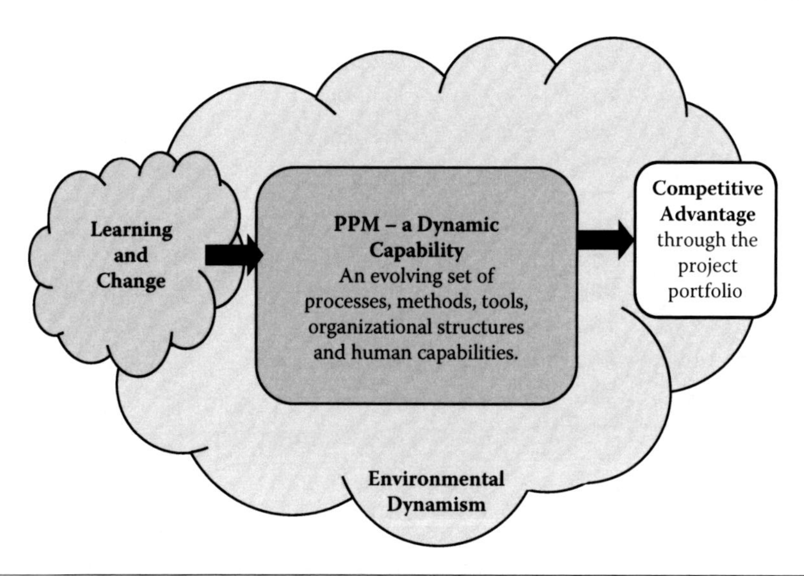

Figure 1.3 Learning and change: Competitive advantage through the evolution of PPM in dynamic environments.

PPM meetings and through the ability of PPM to act as a focus point for decision experiences to be shared and for learning to accumulate. On the other hand, explicit learning—the type of learning that can be codified and documented—is incorporated in PPM through aspects such as standard templates, data bases, and defined and documented methods and routines. Both types of learning inform the evolution of the PPM capability and ensure that it remains up to date and relevant in a changing environment. Through this learning, the PPM process is thus able to deliver a competitive advantage in dynamic environments.

Outline of a Dynamic PPM Approach

A typical portfolio-level review process includes the four steps outlined in Figure 1.4 and explained below. The general aspects of the four steps are outlined first followed by specific *aspects of PPM for dynamic environments in italics.*

1. **Single project data collection.** Data are collected for new project proposals and on existing project status in order to inform decision making (Kester, et al., 2011). The data are generally collected from all relevant projects in a standard form that defines the types of data required to facilitate evaluation. Project data may be obtained from a computer system or through templates or proposal documents. Templates often include a one-page executive summary that highlights the main criteria for the decisions makers to consider (for example risk, reward, investment, skills and resources required, benefits, and aims). *Dynamic environments may require more frequent refreshing of project data. The relevant types of data must be kept up to date; the templates for data collection may change periodically in response to capability reviews. In addition, beyond simply collecting project data, a dynamic PPM capability may promote or*

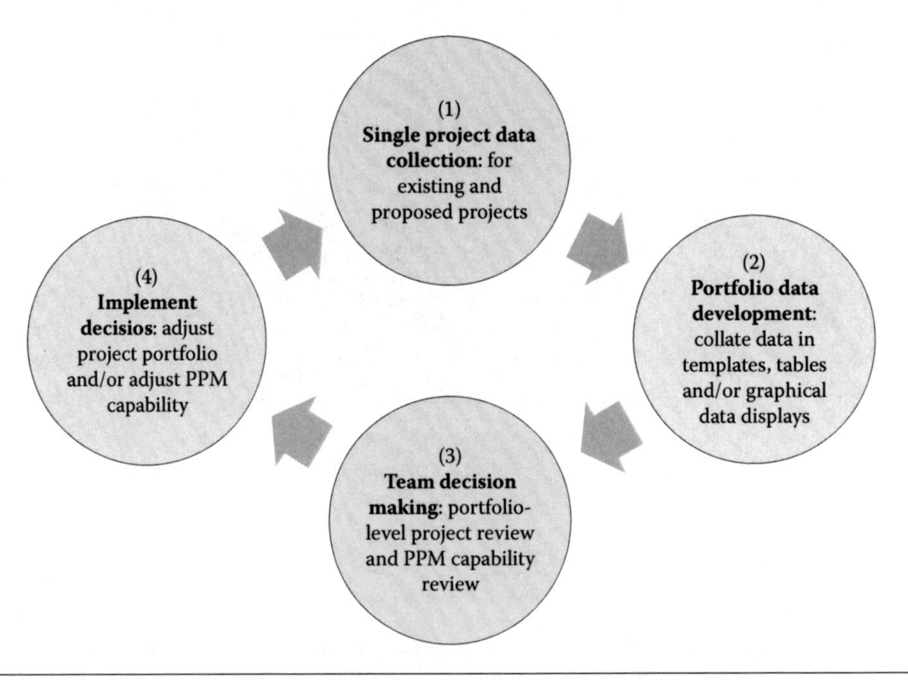

Figure 1.4 Outline of a dynamic PPM approach including evolution of processes through capability reviews.

encourage project ideas that will support the organizational strategy. Idea management portals and collaborative tools can be used to assist in the idea and project proposal development process.

2. **Portfolio data development.** Drawing upon the single project information for all projects in the portfolio, the data are collated or 'rolled up' to provide portfolio-level summaries. The data are arranged to assist decision makers with the comparison and evaluation of portfolio data.

   Research indicates that 'best practice' organizations create graphical and visual information displays such as portfolio maps, to facilitate group decision making (Cooper, et al., 2001; Mikkola 2001; Killen, et al., 2008; de Oliveira Lacerda, Ensslin & Ensslin, 2011). Figure 1.1 shows some common portfolio-level data displays, including portfolio maps that are developed in this stage of the process. Portfolio maps display projects and the strategic options they represent on two axes, augmented with additional data to provide a visual representation that incorporates information such as strategic alignment, risk, return, and competitive advantage. Because of the multiple types of data represented, these types of visual displays are often called two-and-a-half dimensional (2½-D) displays (Warglien, 2010). *Such displays and all portfolio-level summaries must be kept up to date in dynamic environments. Some tools and techniques may be better suited to dynamic environments, and new tools and techniques are regularly being developed and tested to meet current challenges. For example, network mapping approaches may help identify flow-on effects among interdependent projects arising from changes in the portfolio (Killen & Kjaer, 2012).*

3. **Team decision making.** In many organizations, a portfolio review board meets periodically to discuss the options available and to make project decisions in the context of the entire portfolio of projects (including ongoing projects as well as new proposals). The portfolio review board generally consists of five to ten experienced executives or managers that represent diverse organizational perspectives and responsibilities. There are many approaches to the timing of portfolio level reviews—for example, some organizations create an annual portfolio plan, while others meet to refine the portfolio every week or two. The timing depends on the organization's environment—timing is influenced by aspects such as complexity, dynamism in the market, levels of technological change, and project duration. Meetings often employ graphical data representations to inform group discussions and negotiations (Mikkola, 2001; Killen, et al., 2008). Decisions are made with the entire portfolio in mind and will consider resourcing, strategic alignment, and other aspects. Typical decisions on new project proposals at a portfolio review meeting range from approval, hold for a later date, rejection, or requests for more information. The decisions relate to new projects as well as existing projects through mid-stream reviews (Rad & Levin, 2008); for example ongoing projects can be cancelled, delayed, accelerated, or left unchanged. *In dynamic environments enhanced 'sensing' capabilities need to be incorporated to detect changes in the environment; the time between decision meetings may need to be shorter; and/or special mechanisms may be required to enable agile response to unanticipated changes in the environment.*

   In addition, in dynamic environments the regular review of the PPM capability is particularly important. The portfolio review board and/or other executives must also review the processes used and their outcomes. The reviews are done to keep track of the results of the process, and if necessary, recommendations for adjustments to the process will result from the review. In a dynamic environment, such adjustments are periodically required so that the portfolio outcomes will continue to reflect the desired strategy and balance.

4. **Implement decisions:** The outcomes of the portfolio review decision meetings are implemented in this step. For example, new projects may be initiated, some existing projects may be cancelled and resources reallocated, and other existing projects may be accelerated to beat the competition; these changes flow from the decisions made by the portfolio review board. Through this process, there is continual adjustment to the portfolio of projects. *In dynamic environments, the adjustments to the portfolio may be more frequent. In addition, the suggestions arising from the reviews of the PPM capability are implemented, and the cycle continues with evaluation and adjustment of the capability as required. In dynamic environments the decisions and suggestions will drive the continual evolution of the processes for managing the portfolio.*

Examples of PPM in Dynamic Environments

What does a dynamic PPM capability look like in practice? The following examples are taken from a study of the PPM approaches used by successful innovators. The examples have been selected to illustrate practical examples of aspects of PPM capability that can improve organizational agility—an organization's ability to adjust to changes in the environment.

'Sensing' the Environment 'Sensing' changes in the environment is necessary for an organization to start the process of evaluating and adapting to the changes (Teece, 2007). A medical devices company recognized the importance of keeping abreast of developments in medical treatments that have the potential to influence its product development directions. The medical specialists employed by the organization have always played an important role in the 'sensing' of the environment, however their time is limited, and their expertise is focused in specific areas. In recognition of the importance of 'sensing' the environment, the organization developed several strategies to enhance their ability to keep track of trends and developments in the field. One of these strategies was the development of a medical review board consisting of external advisors and specialists from a range of related professions. This initiative greatly extends the available expertise and provides a diversity of perspectives.

Similarly, an approach employed by a telecommunication company is to encourage and facilitate employee involvement in specialist communities through conferences and professional associations. Through these contacts and conference presentations, the employees are better able to contribute to the organization's ability to 'sense' the environment.

Reallocating Resources An important aspect of PPM in any environment, and especially in dynamic environments, is the ability to stop poor projects and reallocate the resources to other projects. This ability is the key to organizational agility through PPM: the organization must be able to ensure that the current set of projects represents the best overall mix at the current time. Often a project that was strongly supported when initiated becomes less desirable as the environment changes. Changes such as the emergence of a new technology or a competitive product, changes in demographics or foreign exchange rates, or changes in commodity or property prices can radically alter a project's prospects for success. However many organizations find it difficult to cancel a project, and often the people involved resist changes to the project. One manufacturing organization felt that a culture that supported information and decision transparency and communication is the key. They implemented steps to ensure that the criteria, data, and methods for evaluation are openly shared and discussed. In addition all levels of management visibly supported and participated in the PPM processes. Through these measures, the organization gained strong buy in and support for the process. With such support, the organization felt that decisions to cancel a project and reallocate resources were understood and supported—making it easier to make the difficult decisions to cancel projects when necessary.

Ensuring Ambidexterity In many industries, it is important that an organization is able to successfully 'exploit' and 'explore' at the same time—this is sometimes called 'organizational ambidexterity' (Tushman & O'Reilly, 1996; Tushman, Smith, Wood, Westerman & O'Reilly, 2002). Exploitation projects are generally short-term, incremental, or low-risk undertakings that are relied on for day-to-day improvements in existing offerings or operations. In contrast, exploration projects are longer-term, higher-risk, radical, or breakthrough initiatives that aim to create innovative new capabilities and offerings to bring the organization to the next level. Collating data across the portfolio of projects through PPM can provide an organization with the ability to determine the current balance of project types. This is often done using graphical data displays such as portfolio maps or pie charts.

If an imbalance is found, PPM processes can be used to help redress the balance. For example, a digital services organization introduced targeted idea generation activities to increase the number of radical ideas when it realized that its portfolio was skewed toward 'exploitation' over 'exploration'. This type of skewing is common and has been named the 'success trap' because accumulated decision-making experiences can reinforce the support for short-term 'exploitation' projects at the expense of the longer-term 'exploration' projects that organizations believe are essential for long-term success (March, 1991). As one manager in a financial services organization commented during an interview "Short versus long-term is most difficult to balance, especially with pressure to turn around in a shorter term. Longer term no one gives you any credit for and it is harder to get justification."

To address this problem an industrial machinery manufacturer allocates a set percentage of its budget for each type of project to ensure the appropriate balance. Another approach that is commonly used is to develop a separate tailored process with appropriate evaluation criteria to be used with the longer-term explorative projects as illustrated in the model in Figure 1.2. This approach ensures that good ideas and projects are not disadvantaged by having to meet rigid criteria that are not appropriate for 'exploration' projects.

Adjusting the Portfolio Review Board The membership of the portfolio review board is an important part of a PPM capability. In a dynamic environment, the profile of portfolio review board members may need to be adjusted as the environment changes. For example, one successful manufacturer traditionally had a strong engineering and technical influence on the review board. This served the organization well during its early stages of developing a best-in-class technology and enabled it to extend its market internationally. However, as the international competitive environment evolved, the portfolio review decisions failed to incorporate marketing and customer-related input and instead resulted in a number of technologically driven projects that failed to find a market. Upon review of the situation, the organization decided to radically change the membership of the portfolio review board to include marketing experience across the main regions. This change allowed the portfolio to better reflect marketing requirements in the regions.

Reviewing and Developing PPM Methods and Tools Dynamic project environments are often characterized by complexity, interdependency between projects, and constraints in the availability of skills and resources. In such environments, PPM is a complex multi-dimensional challenge, and the PPM capability must evolve to stay relevant. The challenge is amplified by the presence of interdependencies as PPM is more than an extension or scaled-up version of project management; the inter-project effects are more complex and difficult to predict (Aritua, et al., 2009). The management of interdependences is an area of weakness for PPM (Elonen & Artto, 2003); this is one of many areas where new tools are being tested. Practitioners and researchers continually refine existing methods and tools and also develop and test new methods and tools. For example, new methods to manage project interdependencies have been proposed (Rungi 2007; Killen & Kjaer, 2012). Recent research and trials in defense and telecommunication industries suggest that new network mapping methods for visualizing projects and their interdependencies may support PPM decision making (Killen & Kjaer, 2012).

Conclusion

PPM is an important organizational capability that enables organizations to manage and balance the portfolio holistically, to align projects with strategy, and to ensure adequate resourcing for projects in order to maximize the benefits from project investments. In dynamic environments, PPM is also the key to developing organizational agility to respond to changes in the environment.

A PPM capability requires more than tools and methods for evaluating and making decisions on project portfolio data; it also requires appropriate organizational structures, a supportive culture, and top management support. One of the major challenges facing organizations is implementing a PPM capability that is flexible and responsive to changes in the environment. Although there are many common elements identified in PPM processes, there is evidence that each organization must tailor its PPM process to suit the individual environment, and that the PPM capability must be able to adapt and adjust to reflect changes in the environment.

A dynamic PPM capability can help project-based organizations respond to change in the environment and improve their organizational agility. Learning and change have been shown to be an important component of a dynamic PPM capability, and several examples of PPM capability aspects that enhance agility are outlined. Practitioners can draw upon these examples to stimulate ideas on improving their PPM capability and evolving the capability to enhance their organizational agility.

Note

A presentation and an earlier version of this paper were included in the PMI Annual conference in Lima, Peru, Tour Cono Sur, 30 November 2012.

References

Aritua, B., Smith, N.J., & Bower, D. (2009). Construction client multi-projects—A complex adaptive systems perspective. *International Journal of Project Management. 27*, 72–79.

Barney, J. B. & Hesterly, W.S. (2012). *Strategic management and competitive advantage: Concepts and cases.* Upper Saddle River, New Jersey: Pearson, Prentice Hall.

Cooper, R. G. & Edgett, S.J. (2003). Overcoming the crunch in resources for new product development" *Research Technology Management. 46*(3), 48–58.

Cooper, R. G., Edgett, S. J. & Kleinschmidt, (2001). *Portfolio management for new products.* Cambridge, MA: Perseus.

de Oliveira Lacerda, R. T., Ensslin, L. & Ensslin, S. R., (2011). A performance measurement framework in portfolio management: A constructivist case. *Management Decision. 49*(4), 648–668.

Eisenhardt, K. M. & Martin, J.A. (2000). Dynamic capabilities: What are they? *Strategic Management Journal. 21*(10/11), 1105–1121.

Elonen, S. & Artto, K.A. (2003). Problems in managing internal development projects in multi-project environments. *International Journal of Project Management. 21*(6), 395–402.

Helfat, C. E., Finkelstein, S., Mitchell, W., Peteraf, M. A., Singh, H., Teece, D. J., & Winter, S. G., (2007). *Dynamic capabilities : understanding strategic change in organizations.* Malden, MA: Blackwell Publishing.

Jonas, D. (2010). Empowering project portfolio managers: How management involvement impacts project portfolio management performance. *International Journal of Project Management. 28*, 818–831.

Kester, L., Griffin, A., Hultink, E. J., & Lauche, K., (2011). Exploring portfolio decision-making processes. *Journal of Product Innovation Management. 28*(5), 641–661.

Killen, C. P. & Hunt, R.A. (2010). Dynamic capability through project portfolio management in service and manufacturing industries. *International Journal of Managing Projects in Business. 3*(1), 157–169.

Killen, C. P., Hunt, R.A., & Kleinschmidt, E. J., (2007). Dynamic capabilities: Innovation project portfolio management. *Proceedings of ANZAM 2007,* Sydney, Australia, Australia and New Zealand Academy of Management.

Killen, C. P., Hunt, R.A. & Kleinschmidt, E. J., (2008). Project portfolio management for product innovation. *International Journal of Quality and Reliability Management. 25(*1), 24–38.

Killen, C. P. & Kjaer, C. (2012). Understanding project interdependencies: The role of visual representation, culture and process. *International Journal of Project Management. 30(*5), 554–566.

Levine, H. A. (2005). *Project portfolio management : a practical guide to selecting projects, managing portfolios, and maximizing benefits.* San Francisco, CA. Chichester, Jossey-Bass; John Wiley distributor.

March, J. G. (1991). Exploration and exploitation in organizational learning. *Organization Science. 2*(1), 71–87.

Martinsuo, M. & Lehtonen, P. (2007). Role of single-project management in achieving portfolio management efficiency. *International Journal of Project Management. 25*(1), 56–65.

McDonough III, E. F. & Spital, F.C. (2003). Managing project portfolios. *Research Technology Management. 46*(3), 40–46.

Mikkola, J. H. (2001). Portfolio management of R&D projects: implications for innovation management. *Technovation. 21*(7), 423–435.

Petit, Y. (2012). Project portfolios in dynamic environments: Organizing for uncertainty. *International Journal of Project Management. 30*(5), 539–553.

Rad, P. F. & Levin, G. (2008). What is project portfolio management? *AACE International Transactions:* TC31.

Rungi, M. (2007). Visual representation of interdependencies between projects. *37th International Conference on Computers and Industrial Engineering*, Alexandria, Egypt.

Smith, K. A., Vasudevan, S.P. & Tanniru, M.R., (1996). Organizational learning and resource-based theory: An integrative model. *Journal of Organizational Change Management. 9*(6), 41–53.

Teece, D. J. (2007). Explicating dynamic capabilities: the nature and microfoundations of (sustainable) enterprise performance. *Strategic Management Journal. 28*(13), 1319–1350.

Teece, D. J., Pisano, G. & Shuen, A., (1997). Dynamic capabilities and strategic management. *Strategic Management Journal. 18*(7), 509–533.

Tushman, M. & O'Reilly, C., (1996). Ambidextrous organizations: managing evolutionary and revolutionary change. *California Management Review, 38*(4), 8–30.

Tushman, M., Smith, W., Wood, R., Westerman, G. & O'Reilly, C., (2002). *Innovation streams and ambidextrous organizational designs: On building dynamic capabilities.* Organisational Studies Group Seminar, Massachusetts Institute of Technology, Cambridge, MA.

Wang, C. L. and P. K. Ahmed (2007). Dynamic capabilities: A review and research agenda. *International Journal of Management Reviews. 9*(1), 31–51.

Warglien, M. (2010). Seeing, thinking and deciding: some research questions on strategy and vision. Workshop on the Power of Representations: From Visualization, Maps and Categories to Dynamic Tools, Academy of Management Meeting, August 6th, 2010, Montreal.

Wideman, R. M. (2004). *A management framework for project, program and portfolio management.* Victoria B.C.: Trafford Publishing.

2
Portfolio Selection and Termination

WERNER G. MEYER, PhD

Introduction

Portfolio selection is the process of selecting the components (projects, programs, and other work) that make up the portfolio mix of the organization (Project Management Institute (PMI), 2013b). There are many ways in which this can be done. At the one extreme, projects are selected because an executive in the organization feels that the project is important. On the other end of the scale there are complicated mathematical methods that can be used to select the portfolio. Whichever method is used, the goal of the organization is to select those portfolio components that will deliver the best possible business results for the organization.

A fully optimized portfolio has a strong balance between the optimal use of organizational resources, technical feasibility, and the achievement of the organization's strategic objectives (Lee & Kim, 2000). The interaction between these three aspects can be seen as a triangle, where the business benefit is the area of the triangle, see Figure 2.1. A change in any one of the sides of the triangle will change the area and therefore the business benefit.

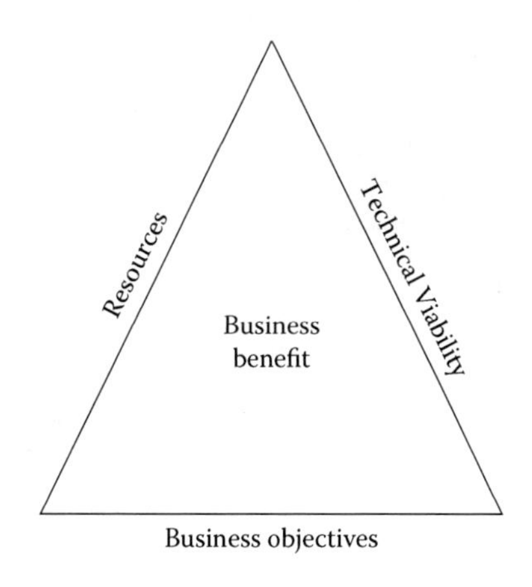

Figure 2.1 Portfolio optimization.

Portfolio optimization should consider whether new organizational initiatives should be included in the portfolio but should also evaluate whether portfolio components that are being implemented should remain in the portfolio, based on their performance (Archer & Ghasemzadeh, 1999; Archer & Ghasemzadeh, 2004).

Portfolio Selection Process

Figure 2.2 illustrates major aspects that should be considered when defining and optimizing a portfolio.

Component Analysis

Component analysis is the process of analyzing each identified component to collect information about the component. Before the component is analyzed, the criteria against which the component will be evaluated must be known. Component analyses fall into two categories i.e., numerical analysis, and non-numerical analysis. A combination of both types of data may be considered during the evaluation process.

Non-numeric Analysis Non-numeric analysis investigates attributes of a component that cannot be represented by discrete measurements. These attributes include aspects such as brand awareness, employee morale, customer perception, user friendliness, and buy in. The assessment of these attributes cannot be calculated with formulas and is often done subjectively by portfolio stakeholders. Likert scales are sometimes developed to guide the analysis, and each component is scored according to a predefined value statement (Meredith & Mantel, 2012).

Non-numerical attributes are not very useful for determining the value of a component. They are, however, valuable for comparing components against each other provided that all the components are scored against the same set of attributes.

Table 2.1 shows an example of the Likert scale for the non-numeric attribute *Employee Morale.*

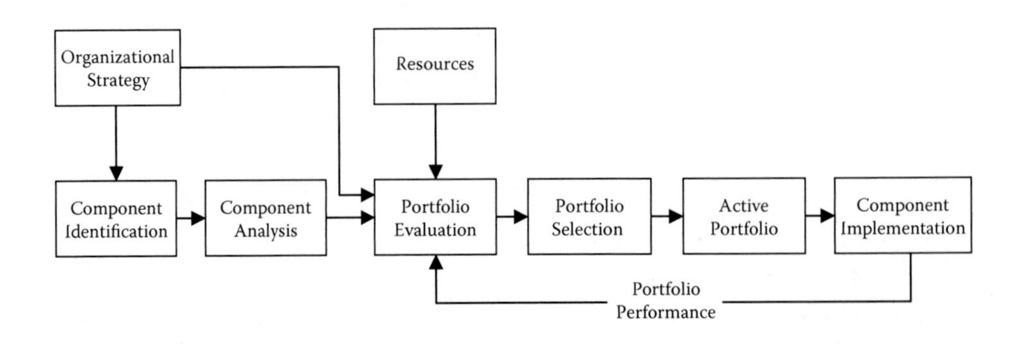

Figure 2.2 Portfolio optimization process.

Table 2.1 Non-numerical scoring attributes.

	LIKERT VALUE				
ATTRIBUTE	1	2	3	4	5
Employee Morale	The project may have a negative effect on employee morale.	The project will have an insignificant effect on employee morale.	The project will have a positive effect on less than a third of the employees' morale.	The project will have a positive effect on more than half of the employees' morale.	The project will have a positive effect on more than two thirds of the employees' morale.

Numeric Analysis Numeric analysis looks at component attributes that can be represented with a calculated numerical value. These are most often financial attributes, but other attributes, such as performance specifications of a component, can also be presented numerically. Three financial attributes that are most often measured are Net Present Value (NPV), Internal Rate of Return (IRR), and Payback Period (Sullivan, Wicks, & Luxhoj, 2006).

Net Present Value NPV is a measure of the financial return of an investment. It takes the time value of money into consideration to adjust the calculation for factors such as inflation, investment risk, and the cost of borrowing money.

Internal Rate of Return The internal rate of return is the discount rate the will yield a zero NPV. A higher IRR means that the investment is less exposed to risks that could erode the value of the investment. The IRR is found though trial and error.

Payback Period The payback period is time it will take for the returns on the investment to pay the invested amount back. It does not consider the time value of money. When comparing two investments of the same investment value, the one with the shorter payback period may not necessarily have a higher NPV.

One of the advantages of numeric models is that the result can be used as an indicator of the feasibility of a single component as well as a comparative measure against other components in the portfolio.

Portfolio Evaluation

Portfolio evaluation methods are used to help decision makers select the optimal portfolio. Before portfolio evaluation can commence, a number of inputs must be in place:

- The organization has identified a mix of existing and new components from which an optimized portfolio must be selected;
- Each component that must be evaluated was analyzed to determine the values of the attributes that will be used evaluate it;

- The selection and approval of the optimal portfolio is done by an individual or a group within the organizations, which we will call the Portfolio Selection Committee (PSC);
- The aim of portfolio optimization is to select a mix of components that will give an optimal business benefit under the constraints of risk, resource usage, and strategic fit; and;
- This benefit gained from a component is not necessarily monetary and may be a combination of numeric and non-numeric attributes.

Component Selection

Every component in a portfolio has an uncertain future outcome. Predicting the duration, cost, and business benefit of a component can only be done as a range of values with varying levels of certainty. This uncertainty is reflected in the contingency allowances during the estimation process in projects.

Economists differentiate between decisions under risk and decisions under uncertainty (Fox & Poldrack, 2009; Knight, 1921). When making a decision under risk, the decision maker knows with certainty what the probability distribution of possible outcomes is (e.g., when throwing a dice). When making a decision under uncertainty the decision maker does not know with certainty what the probabilities or the outcomes are and must therefore consider other factors to motivate the decision e.g., investing money in the stock market. Decision makers who select projects deal with both of these types of decisions; in some cases the probabilities are known, but in other cases the decision maker has to deal with much ambiguity.

The evaluation of components depends on the information that is available for each component at the time of the evaluation. Two components may, for example, have the exact same future outcome, but one may be presented in a way that appears to be more attractive, which may lead the PSC to select this component. It is therefore crucial for the PSC to employ evaluation methods that will give a fair assessment of each component. The following example highlights a problem that the members of the PSC often face, that of framing (Tversky & Kahneman, 1981). Framing refers to the tendency of decision makers to draw conclusions based on how data are delivered. The narrative and visual presentation of a particular component may influence the final decision to include or exclude a component from the portfolio.

In the following two statements, which is the better project to choose?

Project A has a 20% chance of failure.
Project B has an 80% chance of success.

Most people will consider project B to be a better option, even though the two statements say exactly the same thing. It is clear that the way in which the information is presented affects one's decision.

In summary, the selection of portfolio components is a trade-off between the physical attributes of the component (scope, time, cost, quality etc.), its organizational impact (resource requirements), and its strategic fit (achievement of business goals and benefits).

Portfolio Evaluation Models

Portfolio evaluation models present the information available for components in a way that allows the PSC to evaluate components for an optimal portfolio. The type of model used for this evaluation is determined by the information that is available for each component.

Non-numeric Evaluation Methods

Sacred Cow A project that is suggested or forced into the portfolio by an influential person inside or outside the organization. These projects consume valuable resources and can usually only be stopped by the person who proposed the project. To project team members and managers it is often clear that the project should not be done and that it holds very little or no benefit (Meredith & Mantel, 2012).

Operating Necessity The survival or the organization depends on these projects, and the organization has no choice but to do them. The projects could be the results of natural disasters. A flood that is threatening the existence of factory will require immediate action from the organization and does not need to be evaluated against other portfolio components. Legal compliance projects also fall in this category. Financial institutions, for example, must often conduct projects to comply with governmental regulations. If they do not, they may lose their license to operate.

In recent years sustainability has become an important aspect for organizations to remain in business. In some countries companies are forced to show that they have sustainable business models. Many organizations have embarked on sustainability projects to improve the way they manufacture and sell their products (Meredith & Mantel, 2012).

Competitive Necessity Competitive necessity projects are similar to Operating Necessity but may be less critical. These projects ensure that the business remains competitive and are usually performed in response to a competitor's actions or are due to a change in technology or market. For example, changing technology has forced television manufacturers to upgrade manufacturing facilities that were designed to build CRT (cathode ray tube) TVs to keep up with the fast-changing display technologies such as plasma, LCD (liquid crystal display), and OLED (organic light emitting diode) technology.

Product Line Extensions Product line extension projects are the result of new products developed or distributed by an organization. These extensions require the organization to invest capital and require careful analysis of cost, benefits, and payback periods.

Numeric Evaluation Models Evaluation models comprise a set of attributes that are measured for each component under evaluation. Each attribute represents a unique component characteristic. Attributes are broadly categorized into two categories, numerical and non-numerical.

Non-numerical attributes do not have a value that can be directly measured and are often a subjective view of a particular component (e.g., employee morale, brand awareness, user friendliness, etc.). Numerical attributes can be represented and have a precise value, or a numerical value with a variance (e.g., cost, duration, resources required, NPV, IRR, etc.). The success of a selection model lies in its ability to present the attributes of components in an unbiased way.

Unweighted Pass-Fail Factor Model A list of attributes for each component is scored on a binary scale (e.g., pass or fail, yes or no, exists or does not exist, etc.). This is the simplest form of evaluation and can be done individually, by a number of reviewers, or collectively by a group. When it is performed by individuals, the average of the scores is taken for each component and compared. The components with the highest number of positive scores are selected into the portfolio.

In Table 2.2 it appears that Project A is the better choice.

Table 2.2 Pass–Fail Factor Table

	PROJECT A		PROJECT B	
ATTRIBUTE	PASS	FAIL	PASS	FAIL
TECHNICAL ATTRIBUTES				
Increase productivity by more than 10%	✓			✓
Cloud based	✓			✓
Works on mobile devices		✓	✓	
Backward compatibility with legacy systems	✓		✓	
Integrates with existing accounting system	✓			✓
COMMERCIAL ATTRIBUTES				
Increase in brand awareness		✓		✓
Increase gross earnings by more than 5.5%	✓		✓	
Local training/e-training	✓			✓
Payback period is less than two years		✓		✓
Implementation in less than three months	✓		✓	
Monthly cost is less than $50 per user per month	✓			✓
Internal Rate of Return exceeds 20%		✓	✓	
TOTALS	8	4	5	7

The advantages of this model are that it is easy to develop and use, and often gives a large variance between projects that meet the criteria and those that do not, making a decision easier. The main disadvantage is that criteria do not allow for any intermediate values. In the above example, the monthly cost may be $55 for Project A, $40 for Project B, and a third project, Project C, may be at $48. For the particular attribute, Project C is clearly much better than Project B, but there is no way for the model to deal with these values.

Unweighted Factor Scoring Model A list of attributes is developed in the same way as before. Each attribute is scored according to a predefined scale, usually 1 to 5, or 1 to 10, to indicate to what extent the criteria for the attribute are met. The scale can be relative where a score of 1 shows that the component does not meet the criterion, and a score of 5 shows that the component completely meets the criterion. Scores from 2 to 4 show increasing levels of compliance relative to the upper and lower limits.

Alternatively a unique descriptor can be given to each value of the scale. The scores for the attributes are added, and the components with the highest scores are selected into the portfolio. Scoring can be done in a group or by individuals. In the latter case the average scores for each component are calculated to arrive a final score.

In the following table, the two projects are scored on a scale of 1 to 5. The rules of scoring for each attribute are defined as shown in the following example:

Productivity
1 – Decrease in productivity
2 – Increase of 5% or less
3 – Increase of 10% or less, but more than 5%
4 – Increase of 15% or less, but more than 10%
5 – Increase of 15% or more

Table 2.3 Unweighted Factor Scoring Table

ATTRIBUTES	PROJECT A	PROJECT B
TECHNICAL ATTRIBUTES		
Productivity	5	2
Cloud based	5	1
Works on mobile devices	1	4
Backward compatibility with legacy systems	4	3
Integrates with existing accounting system	3	2
COMMERCIAL ATTRIBUTES		
Increase in brand awareness	1	2
Increase gross earnings	4	5
Local training/e-training	3	2
Payback period	2	5
Implementation period	5	4
Monthly cost	5	3
Internal Rate of Return	2	5
TOTALS	40	38

For some attributes a range of values would not make sense, and scores are coded as discrete values. For example:

Cloud Based
1 – System is not Cloud Based
5 – System is Cloud Based

Weighted Factor Scoring Model This model is similar to the previous model, but weights of relative importance are assigned to each attribute. In this way some attributes will contribute more to the final score than others. The weights are converted to factors, which are multiplied with the score given for the attribute to get a final score for each component.

From this example it can be clearly seen that Project A is preferred over Project B when the un-weighted model is used, however, Project B is preferred over Project A when the weighted model is used as shown in Table 2.4.

The attributes used in the above examples work well when the projects that are evaluated are similar. It is, however, often the case that organizations have to choose dissimilar projects e.g., a marketing campaign, an infrastructure project, and an IT project. In these cases the evaluation attributes that apply to all the projects must be used; these would typically be commercial attributes.

Table 2.4 Weighted Factor Scoring Table

			PROJECT A		PROJECT B	
ATTRIBUTES	RELATIVE IMPORTANCE	FACTOR	SCORE	WEIGHTED SCORE	SCORE	WEIGHTED SCORE
TECHNICAL ATTRIBUTES						
Productivity increase	5	0.078	5	0.391	2	0.156
Cloud based	3	0.047	5	0.234	1	0.047
Works on mobile devices	2	0.031	1	0.031	4	0.125
Backward compatibility with legacy systems	6	0.094	4	0.375	3	0.281
Integrates with existing accounting system	4	0.063	3	0.188	2	0.125
COMMERCIAL ATTRIBUTES						
Increase in brand awareness	1	0.016	1	0.016	2	0.031
Increase gross earnings	8	0.125	4	0.500	5	0.625
Local training/e-training	3	0.047	3	0.141	2	0.094
Payback period	7	0.109	2	0.219	5	0.547
Implementation period	9	0.141	5	0.703	4	0.563
Monthly cost	6	0.094	5	0.469	3	0.281
Internal Rate of Return	10	0.156	2	0.313	5	0.781
TOTALS	64	1.000	40	3.578	38	3.656

Group vs. Individual Scoring

All the methods described above require some form of component scoring to be done. We find that in many organizations, a group of people is involved in the selection process (Meyer, 2012). Scoring can therefore be done in a group context or in an individual context. Whether a decision is taken by the group or by group members has little influence on numerical attributes, since these component attributes are measured and are not subject to bias. Non-numeric attributes are usually evaluated subjectively, and the outcome of the selection process can be influenced by individual or group scoring.

In the group context each attribute is discussed by the group, and the group decides on a score for each attribute per component. Individual scoring involves each group member scoring each component without the input from the group. The average of all the individual scores is then used as the component score. The advantage of scoring in a group context is that group members with specific information about a component can discuss it with the rest of the group. The disadvantage is that group members who are passionate about their project may sketch an unrealistically rosy picture of their project and hence influence group members to give a high score for a component, which may in fact not be a very good investment.

The advantage of individual scoring making is that all the group members have an equal vote. The disadvantage is that group members may have to take decisions without having all the information about a component.

Optimization Models

Where evaluation models seek to select components in a portfolio that have a best fit of a set of attributes, optimization models are used to select components in a portfolio that minimize or maximize a specific attribute such as risk, ROI, production throughput, resource usage, budget, etc. Optimization models use mathematical and graphical methods to find the components that give the optimal performance for a particular attribute.

Zero-One Integer Linear Programming

Zero-One Integer Linear Programming (ILP) is a linear programming method that allows one to select a subset of projects from a portfolio that will maximize or minimize some target value. This method is useful to solve complicated scenarios where the best possible projects in a set of multiple projects must be selected (Ghasemzadeh, Archer, & Iyogun, 1999; Santhanam, Muralidhar, & Schniederjans, 1989).

The model has a single decision variable, which determines whether a project is selected or not.

$$X_i = \begin{cases} 0, & \text{the project is not selected} \\ 1, & \text{the project is selected} \end{cases}$$

Where $i = 1,\ldots,N$, and N is the total number of projects under evaluation.

The aim of the method is to maximize (or minimize) a specific parameter, which is defined by the objective function:

$$\max Z = \sum_{i}^{N} a_i X_i$$

Where Z is the value to be maximized, and a_i is an attribute of a particular project X_i, for example NPV.

Constraints can be established to reflect resource limitations such as money, time, and skills. Resource constraints are defined as follows:

$$\sum_{i}^{N} c_{ij} X_i \leq v_j \quad \text{for } j = 1, \ldots, \mathrm{K}$$

Where c_{ij} is the jth constraint applied to project X_i, and the sum of the jth constraints is less than or equal to a defined constraint value v_j. The constraint value can be set as a minimum (≥) or a maximum (≤) threshold, or as an exact (=) value.

Constraints can also be set to define the relationship between projects. The constraint for a mandatory project, which influences the availability of resources to other project is defined as:

$$X_i = 1$$

Organizations sometimes have to select between projects that have overlapping deliverables. In this case either the one project or the other can be selected but not both. This relationship constraint is defined as:

$$X_1 + X_2 \leq 1$$

Example: An organization has a portfolio of five projects from which it wants to select the projects that will give the maximum NPV. The organization is, however, constrained by a budget of $1 million.

Doing all the projects would cost $1,680,000, which is not possible for the organization; see Table 2.5.

Since we want to maximize NPV, the objective function is (all values are shown in $ millions):

$$\max Z = 1.3P_1 + 3.0PL_2 + 0.78PL_3 + 3.75PL_4 + 0.28PL_5$$

Table 2.5 Zero-one ILP Data Table

	PROJECT A	PROJECT B	PROJECT C	PROJECT D	PROJECT E
Budget	$200,000	$400,000	$350,000	$130,000	$600,000
NPV	$1,300,000	$3,000,000	$780,000	$3,750,000	$280,000

For the limited resource, budget, we have (all values are shown $ in millions):

$$0.20P_1 + 0.40PL_2 + 0.35PL_3 + 0.13PL_4 + 0.60PL_5 \leq 1.00$$

The equations can be solved manually using the Simplex method (Dantzig, Orden, & Wolfe, 1955), but this is beyond the scope of this discussion. Instead we will use the solver function in Microsoft® Excel, and get the following solution:

Table 2.6 indicates the selected projects to optimize the NPV to $8.05 million, within a budget of $1 million. Note that the actual budget used is only $0.73 million, and the inclusion of any of the other projects would exceed the $1 million budget.

Zero-One ILP is a useful technique, but it has number of shortcomings. In the above example, we may have wanted to maximize the NPV and minimize the risk exposure. This is not possible using the Zero-One ILP since only a single attribute can be optimized. To solve multiple attributes we have to use Goal Programming.

Goal Programming

Zero-One Goal Programming (GP) is an extension of linear programming that uses multiple, conflicting goals and multiple resource constraints to select an optimal set of components. Zero-One GP also allows for goal prioritization, and goals can be measured in incommensurable units. Both tangible and intangible benefits can be integrated in the model (Santhanam et al., 1989).

Analytic Hierarchy Process

A further enhancement of portfolio optimization models is the use of Zero-One Linear Integer Programming with the Analytic Hierarchy Process (AHP). This approach considers interdependencies between projects, handles both quantitative and qualitative attributes, does schedule optimization, deals with multiple conflicting attributes, and integrates non-uniform resource availability. The model helps to select the optimal portfolio and schedules the projects based on the resources that are available during particular stages of the project (Ghasemzadeh et al., 1999). A similar approach using the analytic network process is proposed by Lee and Kim (2000).

Table 2.6 Zero-one ILP Solution

	PROJECT A	PROJECT B	PROJECT C	PROJECT D	PROJECT E	TOTAL	MAXIMUM
Selected	1	1	0	1	0		
Budget	$200,000	$400,000	$350,000	$130,000	$600,000	$730,000	$1,000,000
NPV	$1,300,000	$3,000,000	$780,000	$3,750,000	$280,000	$8,050,000	

Evaluation of Active Projects

In the evaluation of portfolios organizations should consider new components as well as components that are active. Active components that are not performing as planned or that will yield less benefits than some new components should ideally be cancelled.

History has shown that the cancellation of active components, specifically projects, is extremely problematic, and very few organizations will actually stop a failing project. As a matter of fact, past research has shown that decision makers are more likely to invest more resources in a failing project than they are to stop the project.

Project Success

Delivering successful projects is the primary aim of any organization, but the definition of project success is often a topic of debate. Cooke-Davies (2001) defines 12 factors that affect the success of a project. In a later contribution, Cooke-Davies defines three views of project success (Cooke-Davies, 2004).

The first is the successful delivery of the project as far as its budget, schedule, and quality are concerned (i.e., was the project done right?) (Pinto & Slevin, 1988). This definition does, however, imply that a project that holds no value for the organization but could be delivered successfully, which is sometimes the case (Shenhar, Dvir, Levy, & Maltz, 2001; Turner, 1999).

The second view of project success is that of benefits realization (Pinto & Slevin, 1988). This is the view that executives would have of a project since it supports the organization's strategic objectives (i.e., was the right project done?). This definition does, however, imply that a project could be delivered over budget, late, and with less than the desired quality and still deliver value to the organization.

The third view is whether the right projects were consistently done right.

Baccarini (1999) offers an alternative approach using the Logical Framework Method, but arrives at the same conclusion, which highlights the difference between project and product success. When executives consider project success, there is clearly a trade-off between the success of the project and that of the delivered product.

Termination of Failing Projects

Existing literature suggests that decision makers will resist the termination of a non-performing project and will escalate the commitment of resources to a failing course of action (i.e., a failing project) for a variety of mostly psychological reasons (Brockner et al., 1986; Conlon & Garland, 1993; Garland & Conlon, 1998; Kahneman & Tversky, 1979; Moon et al., 2003; Staw, 1997; Whyte, 1986). Most of initial research on Escalation of Commitment (EoC) was done by Barry Staw (Staw, 1976). Staw and Ross (1989) suggest that decision makers are influenced by five classes of determinants when they have to make a decision about a failing project, Figure 2.3.

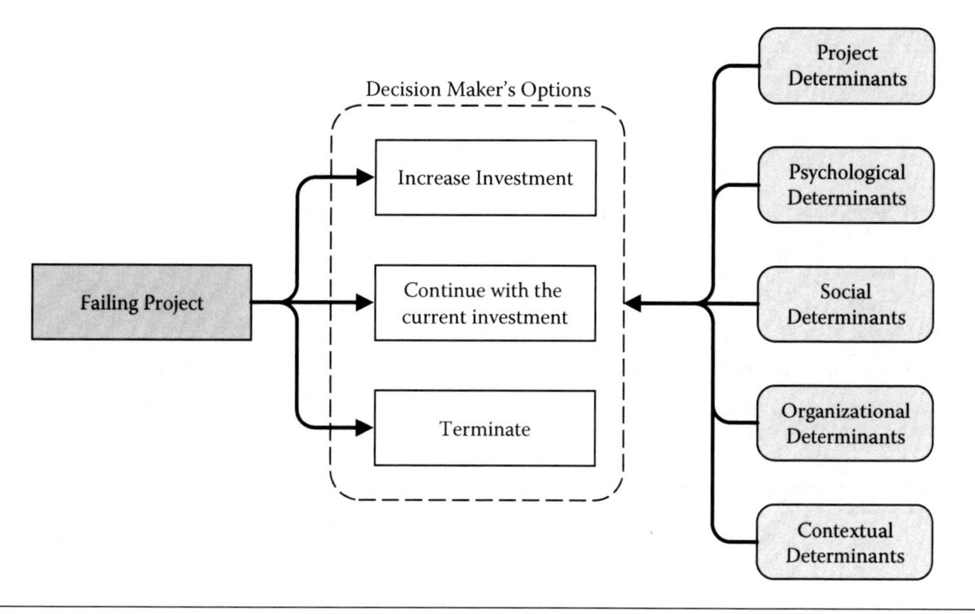

Figure 2.3 Influence of determinants on the decision about a failing project.

Project Determinants. Project-related factors are the most obvious causes for EoC and are persistence with chosen courses of action. Decision makers will consider these factors when evaluating the usefulness of further investment/actions in turning a losing situation around. Staw proposes the following factors that could lead to persistence on a project: temporary vs. permanent losses, the efficacy of further investments/actions in turning a losing situation around, the size of the project's goal or eventual payoff, availability of feasible alternatives to a course of action, and the salvage value or closing costs for ending a project.

Psychological Determinants. Staw's research has shown that people often do not turn away when they realize that their investment is not performing the way they expected and that it is unlikely that they will reap the anticipated benefits. Staw proposes four specific determinants (i.e., optimism and illusions of control, self-justification, framing effects, and sunk cost effects).

Social Determinants. Social determinants are external to the decision-maker, and Staw proposes two determinants (i.e., external justification and binding and leadership norms).

Organizational Determinants. Organizational determinants are macro-level variables that could play a role in EoC. These determinants relate to the way in which an organization establishes organizational policies and procedures to govern its decision making. These factors are not specifically related to individuals in the organization but rather to the system of the organization.

Contextual Determinants. Contextual determinants are forces that are greater than the organization itself, such as government intervention. Governments bailing out companies that are of national interest are an example of a contextual determinant that could hold an organization to a failing course of action.

EoC Determinants

Four determinants have been proposed by various researchers as major contributors to EoC:

Sunk Cost Effect Sunk cost in the context of projects is defined as: "a cost that has already been incurred and which should not be considered in making a new investment decision" (Amos, 2007, p. A.11). Northcraft and Wolf (1984) suggest that most psychological research into sunk cost focuses on situations in which explicit information about the sunk costs and negative financial situation of the project is available, but revenue information of the project is not. What appears to be throwing good money after bad could in fact be a good investment.

Arkes and Blumer (1985) present a number of experiments to show the propensity of subjects to favor investments with a high sunk cost compared to investments with a lower sunk cost. This notion is supported by Whyte (1986, 1993) who specifically notes the roles of decision framing in escalation.

From this research we find strong evidence that there is a correlation between the willingness to spend more on a project that will in all likelihood not make any money for the organization as the sunk cost increases.

Self-Justification Staw (1976) investigates the effect of personal responsibility on EoC in situations where a decision maker had a choice between alternatives, and has, at a later stage, the opportunity to make further decisions to influence the situation i.e., the decision maker can allocate additional resources or withhold resources. He argues that a decision maker may, instead of changing his or her behavior, cognitively distort the negative consequences of a prior decision to make it appear to be a rational and favorable decision.

The decision maker should, however, have not committed to a situation that cannot be easily changed (Brehm & Cohen, 1962) and feel that he or she would be held personally responsible for the negative consequences of the decision (Carlsmith & Freedman, 1968; Cooper, 1971).

In an analysis of research results of EoC prior to 1992, Brockner (1992) concludes that self-justification is one of the primary reasons for EoC. In the existing research there is a strong argument that decision makers will primarily consider their own prior decisions when deciding to invest more resources in a failing project.

Project Completion Conlon and Garland (1993) argue that project completion and sunk cost are cofounded, and prior research may have over emphasized the effect of sunk cost as a reason for EoC. Research by Garland, Conlon, and Rogers (1990) shows that the opposite of the sunk cost effect may occur (i.e., less is invested as the incremental cost of the project increases). On a typical project the amount of money spent is correlated to the progression of the available time to complete the project.

This relationship is not linear and graphing the incremental project cost over time typically takes the form of an S-curve (PMI, 2013a). A decision maker would usually be aware of the time progress of the project when decisions are made and the approaching end of the project. The approaching end of the project could entrap the decision maker and may lead to goal substitution, whereby decision makers shift their focus away from the goal the project has to achieve in terms of its deliverables to the goal of just completing the project.

Jensen, Conlon, Humphrey, and Moon (2011) suggest that decision makers will increasingly conceal negative information about a project as the project approaches completion.

Optimism Bias Optimism bias refers to the tendency of people to believe that they are less likely to experience negative events and are more likely to experience positive events, than other people. Optimism bias is a relatively new term for this phenomena, but it has been studied for many years as belief and desire (Lund, 1925), unwarranted optimism (Tversky & Kahneman, 1974), unwarranted certainty (Fischhoff, Slovic, & Lichtenstein, 1977), unrealistic optimism (Weinstein, 1980), and comparative optimism (Shepperd, Carroll, Grace, & Terry, 2002).

One of the earliest studies show that the desire of decision makers for a particular outcome is strongly correlated with their belief that the outcome will be achieved (Lund, 1925).

Humans have a built-in bias to be optimistic about future events. This bias is difficult to control, and, even when decision makers are aware of this bias, they are highly unlikely to control or alter their behavior (Lovallo & Kahneman, 2003; Sharot et al., 2012). Optimism bias has been observed in nearly every human endeavor that involves the prediction of future events, and psychologists argue that it is a major mechanism for survival in humans (Sharot et al., 2012). Sharot (2012) found that optimism is seated in the inferior frontal gyrus (IFG) part of the brain. Interfering with the physical functioning of the IFG through Transcranial Magnetic Stimulation decreases optimism bias, suggesting that optimism bias is a biological attribute of humans. Meyer (2013) defines two types of optimism bias i.e., in-project and post-project optimism bias, to explain the behavior of decision makers who believe a failing project can be salvaged, or a project will yield better returns than what was originally planned.

Preventing Escalation of Commitment

A number of strategies have been proposed to defuse escalation situations. Strategies with a high level of success in studied escalation situations are briefly described below:

Administrative Turnover Staw and Ross (1987) suggest replacing decision makers in the decision-making situation to break escalation behavior. Removing a decision maker would remove the internal justification psychological determinants, but would not remove the project determinants, and would provide only a partial remedy for social and structural determinants (Keil, 1995; Ross & Staw, 1993).

Support for Failure Staw and Ross (1987) suggest that organizations provide limited support for managers who have had failed projects. Managers are not at risk of losing their jobs, but they could be restricted from running larger projects for a period of time, following a project failure. This would reduce the high stakes a decision maker faces for making mistakes, which in turn would reduce commitment.

Bringing Phase-out Costs Forward Staw and Ross (1987) suggest that presenting the cost of stopping a project to decision makers, even before the project is started, will create consciousness of the withdrawal cost. It would furthermore be beneficial if withdrawal costs are shown at significant project review points throughout the life of the project.

Early and Frequent Risk Assessment Assessment of project risks should start at the early stages of a project and should be continued through the life of the project. The project team should specifically question the continuation of the project at frequent intervals (Keil, 1995).

Minimum Goal Setting Decision makers have to set minimum target levels of project performance, which, if not achieved, would lead to a change in policy, and possible withdrawal. Minimum goal setting resulted in the biggest reduction in EoC of the factors tested by Simonson and Staw (1992).

Summary

Selecting an optimal portfolio of components requires careful planning and analysis of the benefits and constraints of each component. Many companies still use unscientific approaches for portfolio evaluation and selection. This approach inevitably leads to wasted resources. Formal methods for portfolio evaluation and selection exist which can be implemented by organizations. These methods significantly improve the ability of organizations to ensure they select components that will give the best business benefits.

Portfolio selection does not only involve the evaluation of new components. It must consider the evaluation (and possible termination) of projects that have already started but that may be failing.

References

Amos, S. (Ed.). (2007). *Skills and knowledge of cost engineering* (5th ed.). Morgantown, WV: AACE International.

Archer, N. P., & Ghasemzadeh, F. (1999). An integrated framework for project portfolio selection. *International Journal of Project Management, 17*(4), 207–216.

Archer, N. P., & Ghasemzadeh, F. (2004). Project portfolio selection and management. In Morris, P.W.G & Pinto, J.K. (Eds.). *The Wiley guide to managing projects* (pp. 237–255). Hoboken, NJ: John Wiley & Sons, Inc.

Arkes, H. R. & Blumer, C. (1985). The psychology of sunk cost. *Organizational Behavior and Human Decision Making, 35*(1), 124–140.

Baccarini, D. (1999). The logical framework method for defining project success. *Project Management Journal, 30*(4), 25–32.

Brehm, J. W. & Cohen, A. R. (1962). *Explorations in cognitive dissonance.* Hoboken, NJ: John Wiley & Sons Inc.

Brockner, J. (1992). The escalation of commitment to a failing course of action: Toward theoretical progress. *Academy of Management Review, 17*(1), 39–61.

Brockner, J., Houser, R., Birnbaum, G., Lloyd, K., Deitcher, J., Nathanson, S., & Rubin, J. Z. (1986). Escalation of commitment to an ineffective course of action: The effect of feedback having negative implications for self-identity. *Administrative Science Quarterly, 31*(1), 109–126.

Carlsmith, J. M. & Freedman, J. L. (Eds.). (1968). *Bad decisions and dissonance: Nobody's perfect.* Chicago, IL: Rand McNally.

Conlon, D. E. & Garland, H. (1993). The role of project completion information in resource allocation decisions. *Academy of Management Journal, 36*(2), 402–413.

Cooke-Davies, T. (2001). The "real" success factors on projects. *International Journal of Project Management, 20*(3), 185–190.

Cooke-Davies, T. (2004). Project success. In Morris, P.W.G. & Pinto, J.K. (Eds.). *The Wiley guide to managing projects* (pp. 99–122). Hoboken, NJ: John Wiley & Sons, Inc.

Cooper, J. (1971). Personal responsibility and dissonance: The role of foreseen consequences. *Journal of Personality and Social Psychology, 18*(3).

Dantzig, G. B., Orden, A., & Wolfe, P. (1955). The generalized simplex method for minimizing a linear form under linear inequality restraints. *Pacific Journal of Mathematics, 5*(2), 183–195.

Fischhoff, B., Slovic, P., & Lichtenstein, S. (1977). Knowing with certainty: The appropriateness of extreme confidence. *Journal of Experimental Psychology: Human Perception and Performance, 3*(4), 552–564.

Fox, C. R. & Poldrack, R. A. (2009). Prospect theory and the brain. In Glimcher, P.W. (Ed.), *Neuroeconomics: Decision making and the brain.* New York, NY: Elsevier.

Garland, H. & Conlon, D. E. (1998). Too close to quit: The role of project completion in maintaining commitment. *Journal of Applied Social Psychology, 28*(22), 2025–2048.

Garland, H., Sandefur, C. A., & Rogers, A. C. (1990). De-escalation of commitment in oil exploration: When sunk costs and negative feedback coincide. *Journal of Applied Psychology, 75*(6), 721–727.

Ghasemzadeh, F., Archer, N. P., & Iyogun, P. (1999). A zero-one model for project portfolio selection and scheduling. *The Journal of the Operational Research Society, 50*(7), 745–755.

Jensen, J. M., Conlon, D. E., Humphrey, S. E., & Moon, H. (2011). The consequences of completion: How level of completion influences information concealment by decision makers. *Journal of Applied Social Psychology, 41*(2), 401–428.

Kahneman, D. & Tversky, A. (1979). Prospect theory: An analysis of decision under risk. *Econometrica, 47*(2), 263–291.

Keil, M. (1995). Pulling the plug: Software project management and the problem of project escalation. *MIS Quarterly, 19*(4), 421–447.

Knight, F. (1921). *Risk, uncertainty, and profit.* Boston, MA: Houghton Mifflin Company.

Lee, J. W. & Kim, S. H. (2000). Using analytic network process and goal programming for interdependent information system project selection. *Computers & Operations Research, 27*(4), 367–382.

Lund, F. H. (1925). The psychology of belief. *The Journal of Abnormal and Social Psychology, 20*(1), 63.

Meredith, J. R. & Mantel, S. J. (2012). *Project management: A managerial approach* (8th ed.). Hoboken, NJ: John Wiley & Sons, Inc.

Meyer, W. G. (2012). *Project selection and termination How executives get trapped.* Paper presented at the PMI Global Congress October 2012, Vancouver, Canada. Newtown Square: Project Management Institute.

Meyer, W. G. (2013). *The effect of optimism bias on the decision to terminate failing projects.* Paper presented at the IRNOP XI June 2013, Oslo, Norway. The International Research Network on Organizing by Projects.

Moon, H., Conlon, D. E., Humphrey, S. E., Quigley, N., Devers, C. E., & Nowakowski, J. M. (2003). Group decision process and incrementalism in organizational decision making. *Organizational Behavior and Human Decision Processes, 92*(1), 67–79.

Northcraft, G. B. & Wolf, G. (1984). Dollars, sense, and sunk costs: A life-cycle model for resource allocation decisions. *Academy of Management Review, 9*(2), 225–234.

Pinto, J. K. & Slevin, D. P. (1988). Project success: Definitions and measurement techniques. *Project Management Journal, 19*(1), 67–72.

Project Management Institute. (2013a). *A guide to the project management body of knowledge (PMBOK® Guide).* Fifth Edition. Newtown Square, PA: Project Management Institute.

Project Management Institute. (2013b). *The standard for portfolio management.* Third Edition. Newtown Square, PA: Project Management Institute.

Ross, J. & Staw, B. M. (1993). Organizational escalation and exit: The case of the Shoreham nuclear power plant. *Academy of Management Journal, 36*(4), 701–732.

Santhanam, R., Muralidhar, K., & Schniederjans, M. (1989). A zero-one goal programming approach for information system project selection. *Omega, 17*(6), 583–593.

Sharot, T., Kanai, R., Marston, D., Korn, C. W., Rees, G., & Dolan, R. J. (2012). Selectively altering belief formation in the human brain. *Proceedings of the National Academy of Sciences of the United States, 109*(42), 17058–17062.

Shenhar, A. J., Dvir, D., Levy, O., & Maltz, A. C. (2001). Project success: A multidimensional strategic concept. *Long Range Planning, 34*(6), 699–725.

Shepperd, J. A., Carroll, P., Grace, J., & Terry, M. (2002). Exploring the causes of comparative optimism. *Psychologica Belgica, 42*(1/2), 65–98.

Simonson, I. & Staw, B. M. (1992). Deescalation strategies: A comparison of techniques for reducing commitment to losing courses of action. *Journal of Applied Psychology, 77*(4), 419–426.

Staw, B. M. (1976). Knee-deep in the big muddy: A study of escalating commitment to a chosen course of action. *Organizational Behavior and Human Resourses, 16*(1), 27–44.

Staw, B. M. (1997). The escalation of commitment: An update and appraisal. In Shapira, Z. (Ed.), *Organizational Decision Making* (pp. 191–215). Cambridge: Cambridge University Press.

Staw, B. M. & Ross, J. (1987). Behavior in escalation situations: Antecedents, prototypes and solutions. *Research in Organizational Behavior, 9*, 39–78.

Staw, B. M. & Ross, J. (1989). Understanding behavior in escalation situations. *Science, 246*(4927), 216–246.

Sullivan, W. G., Wicks, E. M., & Luxhoj, J. T. (2006). *Engineering Economy.* Upper Saddle River, NJ: Pearson Education.

Turner, J. R. (1999). *The handbook of project-based management* (2nd ed.). London: McGraw-Hill.

Tversky, A. & Kahneman, D. (1974). Judgment under uncertainty: Heuristics and biases. *Science, 185*(4157), 1124–1131.

Tversky, A. & Kahneman, D. (1981). Framing of decisions and the psychology of choice. *Science, 211*(4481), 453–458.

Weinstein, N. D. (1980). Unrealistic optimism about future life events. *Journal of Personality and Social Psychology, 39*(5), 806.

Whyte, G. (1986). Escalating commitment to a course of action: A reinterpretation. *Academy of Management Review, 11*(2), 311–321.

Whyte, G. (1993). Escalating commitment in individual and group decision making: A prospect theory approach. *Organizational Behavior and Human Decision Processes, 54*(3), 430–455.

3

Corporate Strategy—Converting Thoughts and Concepts into Action

DAVID V. TENNANT, PE, PMP, MBA

High-Level Thinking

What is strategic planning? Why is it important? Strategy has different meanings to different people. To many, a strategy represents a guiding philosophy. To others, it may be a roadmap to take a company from point A to point B. To others still, it is a plan for transition or changing the corporate culture.

In a traditional sense, strategic planning is concerned with the future and with changes in capabilities, products, or services to meet future market conditions. However, it has been the author's experience that tactical plans—that is operational, one-year advance planning—are sometimes thought of as strategic planning. There are a number of challenges that companies have with strategic planning.

From a practical standpoint, how does a strategic plan differ from a business plan? That is a fair question, and the answer is that they are closely related. A business plan generally has a short-time frame, but should have some of the same components as a strategic plan such as expected costs, resources required, objectives, etc.

However, for large corporations, the marketing group should play a large role in the strategic planning process. This is because marketing is concerned with industry trends, customer preferences, competitor analysis, and with developing promotions to keep the company's

PROBLEMS WITH STRATEGIC PLANNING

- Strategic planning typically occurs in the summer so that budgets, developed in the fall are used for the operational budgets and plans for next year. This is how strategic plans change into operational (i.e. tactical) plans with a 12-month short-term view.
- Managers are generally so focused on the day-to-day operations that anticipating the long-term view is hard to visualize.
- Sometimes, markets or customer preferences change so quickly, that chaos occurs. Here, all previous planning is null and void.

brand alive and relevant. Therefore it should be clear that a professional marketing organization is needed to drive parts of the strategic plan.

To develop an effective strategic plan, there will be a number of "inputs" needed.

- Purpose of the company (mission)
- Objectives
- Where do we want to be in five years (or 10 years)?
- Industry trends
- Key stakeholders (current and future)
- Competitor analyses
- Strategic initiatives
    1.
    2.
    3.
- Costs to develop products or services
- Schedule (when will it happen?)
- Risks associated with the initiatives
- Key resources needed (dollars, people, equipment, etc.)
- Getting employee buy in
- Metrics (How will we know when we get there?)

TRUE QUICK CASE

Your firm, a major manufacturing firm in the mid-west develops equipment for home, farm, and commercial use. The firm is well established, has decent products, and a full network of dealerships with parts support and customer service.

The company is always looking over its shoulder to determine what its competitor is developing next. The competitor is equally established but always seems to be further "out front" in launching new products.

Your firm is always playing "catch up" to launch similar products just to maintain market share and prove that your products are just as good.

Key discussion points:

1. Why is your firm always behind the competitor?
2. What can you do differently to be ahead of the curve?

In the above case, it should be apparent that Firm A is simply reacting and responding to whatever Firm B does. With this thinking, the company will never be an industry leader but simply a follower. A true industry leader needs to anticipate the market and plan accordingly.

Signs of market leadership include:

- Your competitors consider your firm to be innovative, a rule maker (or breaker), and nimble in coming to market
- Top managers are not happy with the status quo but are always seeking to do things differently
- The firm is focused on innovation rather than operational efficiency as a secondary priority (this may require investment in research and development [R&D])
- Senior management is focused on strategy; middle management is focused on operations and process
- Your customers look forward to your new products, and you have a reputation for quality (Key point: people will pay more for higher quality or even the perception of higher quality)

If you are successful in developing a strategic plan, how do we get the company to embrace it? How will we communicate the plan?

In putting together the strategic plan, it is necessary to have a number of inputs from a variety of departments, see Figure 3.1.

It is key to note that buy in can be obtained with a strategy session that seeks to build consensus. The exception to this is where a company is in chaos, and the Chief Executive Officer (CEO) must drive the planning process toward his or her vision.

A subset of the strategy should be a plan to communicate the strategy. Depending on the sensitivity of the contents, one must carefully think through what can be publicly divulged and what must be kept under wraps. Reasons for secrecy include

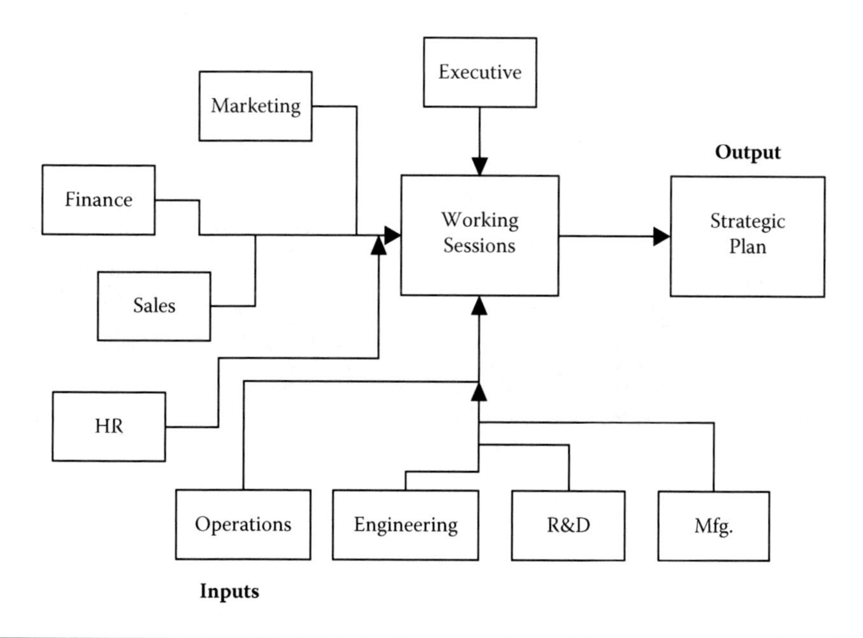

Figure 3.1 Inputs in developing the strategic plan.

proprietary processes, entrance into new markets, surprising competitors, or launching of a new product.

A communications plan can be a major part of the strategy's implementation.

Implementing the Strategic Plan

In many instances, launching the strategic plan involves using the organization's many departments and resources in the form of projects. Note that projects are meant to support specific objectives, and usually a strategy involves multiple objectives. In this medium, it is appropriate to note that project management can—and must—play a critical role in planning and executing the projects associate with the new strategic direction. Consider Figure 3.2 that indicates when project management obtains the handoff from the strategic plan.

The discipline of project management is well suited to driving projects from the strategic plan.

The principles of project management can be utilized to ensure that strategic initiatives are planned and implemented within budget, schedule, and quality parameters. To a large degree, project plans are very similar to the key components of a strategic plan.

In Figure 3.2, a utility is sensitive to criticism about electricity generated from coal power plants. It has decided to diversify the sources of power and to close some of its older, less efficient coal plants. This diversification would represent a major, strategic shift for many utilities as the cost of replacing this power could represent billions of dollars in alternative electric generation (and in turn, higher electric rates). This change requires the kind of careful review, planning, strategy, and analysis that must

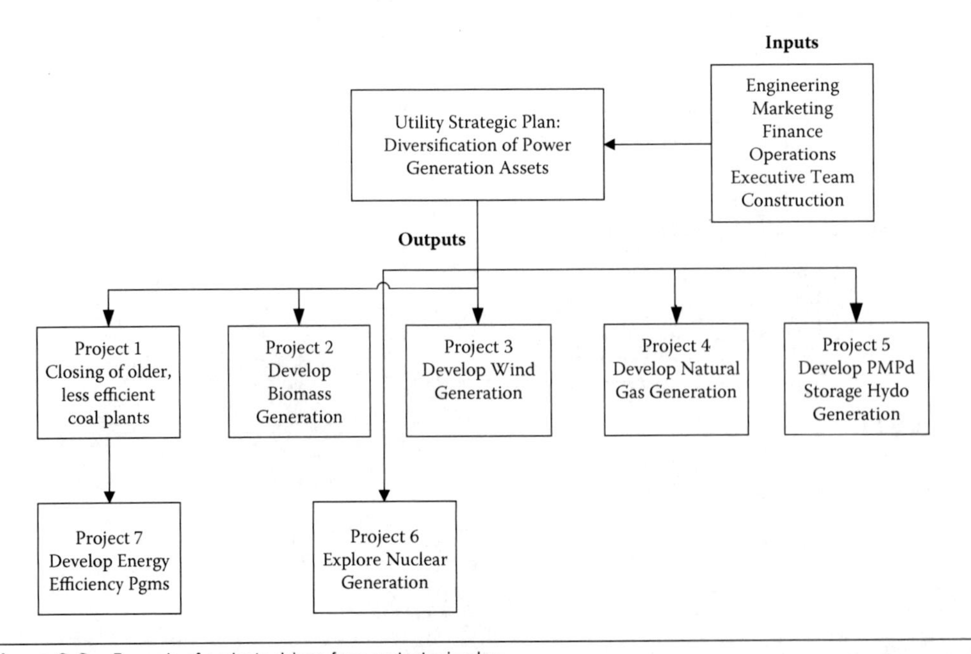

Figure 3.2 Example of projects driven from a strategic plan.

Table 3.1 Similarities of Strategic and Project Plan Elements

STRATEGIC PLANNING ELEMENTS (HIGH-LEVEL PERSPECTIVE)	PROJECT PLANNING ELEMENTS (DETAILED PROJECT-LEVEL PERSPECTIVE)
1. Goals and purpose	1. Project objectives
2. Timeline	2. Detailed schedule with milestones
3. High-level budget	3. Detailed budget with contingency
4. NPV and ROI[a] Analysis	4. Human and equipment resources
5. Resources needed	5. Communications management
6. Communicating the vision	6. Client acceptance criteria (Client: CEO)
7. Metrics	7. Procurement management
8. Risk Analysis[b]	8. Project risk reviews
9. Culture change	9. Quality management
10. Competitor analysis	10. Stakeholder management
11. Market trends	11. Project integration
12. Specific project assignments	12. Scope management

[a] NPV = Net Present Value (a form of cost-benefit analysis); ROI = Return on Investment

[b] Many companies do not perform risk reviews with their strategic planning; however, the author believes these reviews are an important component and should be included.

occur during strategic planning. Each source of alternative power must be considered both on its own individual merits but also combined as a whole with other power sources.

It is interesting to note that the project plans represent a "drill down" from the strategic plan. Further, if a project plan does not support the strategic plan, it should not be considered or funded.

To compare the components of a strategic plan with a project plan, consider Table 3.1.

There are many considerations in implementing the strategic plan. For example, will this require a culture change? Do we have the right people, internally, to implement the plan?

Implementing a Culture Change

It is important to note that fostering a company change in culture is one of the most demanding and challenging efforts for a firm to undertake. And, many times, companies either do not recognize they are moving toward a culture change or think that strong direction from the CEO will make this happen. For those strategies where a company culture change is needed, this will require significant diligence and *continuous* involvement of senior management. In the previous example, a utility is moving from traditional energy production to alternative energy production. This would require major changes in company operations, financing, cost–benefit analysis, regulatory compliance, and in the case of energy efficiency programs, a whole new way of thinking. For example, traditional "demand side energy efficiency programs" pay the customer to use less of your product—electricity. The implications of this strategy are huge from a financial perspective. There are a number of additional considerations that must be considered.

1. People do not like change. When a company's culture must evolve, people will resist; some even going so far as to attempt to sabotage the effort.
2. Communicating the plan is important. In order to achieve buy in, people need to understand why such drastic changes are occurring. It is incumbent on senior management (typically the CEO) to outline why the change is needed, when it will go into effect, the benefits that will accrue to the company and its employees, and how they will occur.

   It may be useful to put together a special team or task force to deal specifically with communicating the vision and to address employee concerns.
3. Key players may need to change. When company plans are developed to return to profitability, there are times when drastic action is needed (a turn-around scenario, for example). If key people on the management team cannot or will not buy into the changes, it may be necessary to make personnel changes. One of the highest issues for a CEO is the factor of trust. The employees must have faith and trust that the CEO is acting in the company's best interest. Conversely, the CEO must have subordinates that he or she can trust. If the CEO has doubts about his or her core team or key individuals, it may be necessary to replace them. This cannot be emphasized enough. Many companies have failed at implementing strategic changes because of weak follow through by disgruntled subordinates.
4. Seek out change agents. A change agent is one that knows change is required and is eager to assist. These are people, at all levels within the company, that can help bring about the changes with passion and in a timely fashion.
5. Bring the opponents over to your side. It is possible that you can bring your opponents onto your implementation team to bring about the change. If they can participate in the solution, it is hard for them to criticize the plan they have helped create. This is a valid strategy but does not always succeed. Proceed with caution.

The foregoing discussion highlights some of the known obstacles in pushing through a culture change. And, it is important to note that a risk review during the strategic planning process can assist with identifying and mitigating some of these types of issues. A "project" plan to implement the culture change may be appropriate.

This is an example of how a utility company strategy can generate projects. It is appropriate to note that each project will need to develop a project plan with an assigned project manager and supporting staff. While this strategy is deviating from the "way things are done" at this company, it is still in tune with its primary purpose, which is to produce reliable, cost-effective electrical power.

Beyond the above, it is still necessary to put together a project plan (or similarly, an implementation plan) in order to move the strategies forward.

Putting Strategy into Action

Putting strategic initiatives into motion is where the discipline of project management can assist greatly. Consider the following hypothetical scenario:

Because of the changing regulatory environment, it is becoming difficult and more costly for the Big Utility Company, Inc. to operate its coal-fired power plants. This is due to several factors:

- Unpopular political position of the plants because of potential global warming
- New emissions requirements are getting expensive to implement
- The uncertainty in the permitting and the regulatory arena puts these assets at risk
- A number of the plants are older and reaching the end of their operational life
- These older plants are less efficient

This puts the utility in somewhat of a quandary. As an investor-owned utility, it must generate a profit to stay in business and attract investors. At the same time, it is regulated by a state public service commission and is obligated to keep its electric rates reasonable, is mandated to serve customers living within its service territory, and must provide stable and reliable electric power.

During the recent strategic planning session, the senior management of the utility came up with the following objectives and strategies:

1. To continue to provide reliable electric power at as low a cost as possible
2. To be sensitive to the environment
3. To close the older, less efficient coal-fired power plants (replaced with?)
4. To develop alternative, cost-effective, cleaner sources of energy for electrical power
5. To investigate partnering with other utilities in building nuclear generation
6. To investigate the cost-benefits of subsidizing energy efficiency programs (energy audits, grants for homeowners to purchase efficient appliances, etc.)

It should be apparent that a strategic plan must be concise and clearly articulated.

Since developing new sources of electric generation takes years, this strategy has a 10-year time frame. While this strategic plan could generate up to 10 or 12 projects, the company may wish to focus initially on the top five projects as shown in the previous figure. For purposes of this chapter, consider the components of a project plan for the first project, "Closing of Older, Less Efficient Coal Power Plants."

While this may seem to be a simple act of "turning off the switch," it is much more complicated. Note that "Closing Plants" is a high-level statement. The project plan gets into the "devil of the details" as a result of this strategy. We can walk through how the project planning for this strategic initiative might develop. We can develop some initial talking points from the table of project plan elements.

Strategic Initiative No. 1: Close older, less efficient coal-fired power plants

(Note: A list of the plants considered for closing would have been generated in the strategic plan)

A project plan should now be developed that supports this strategic initiative as follows in Table 3.2.

Table 3.2 Project Plan Elements to Support Strategic Initiative Number 1

PLAN COMPONENT	ACTIVITIES
Project objectives:	1. Close three identified coal-fired power plants 2. Ensure compliance with all safety and environmental regulations during this process 3. Safely disconnect from the energy grid and dismantle the facilities 4. Follow all company policies and guidelines 5. Coordinate with other teams that will be replacing this lost power with alternative energy production
Schedule milestones:	Each plant will be staggered by a three-month timeline. That is, plant number 3 begins dismantling three months after plant number 2 starts; plant number 2 starts three months after plant number 1. The schedule template for each plant follows: September 1—engineering evaluation and recommendations for closure process October 1—develop list of qualified contractors October 10—develop list of any hazardous materials November 1—start work on detailed specification and the Statement of Work (SOW) December 1—issue specifications and Request for Proposals (RFP) to contractors February 1—award contract February 15—plant disconnected from grid and shut down March 1—contractors begin work under company supervision to dismantle plant March 1—plant employees relocated to other facilities or released from company September 1—plant dismantling complete The above represent high-level milestones; a more detailed activities list and schedule would also be developed.
Detailed budget:	The strategic plan would have identified a high-level budget to dismantle the plants, but the project team will need to determine, by detail, if the assigned budget is adequate. This will need to be communicated to the executive team.
Human resources and required equipment:	This consists of internal and external resources. Internally, the plant staff will be needed until disconnect from the grid, and a skeleton crew to supervise contractors then is required. It is also possible that resources from other departments will be needed: accounting, environmental compliance, safety, etc. The external resources consist of contractors hired to dismantle the plants. They will be responsible for their own staffing and equipment. The contractors should have a role in developing the detailed project plan.
Communications management:	How will we communicate internally and externally? Status reports—who gets them and how often? Who receives copies of the project plan? How will problems and scope changes be handled?
Client acceptance criteria:	The client, in this case the CEO and executive team, should indicate what constitutes a successful project conclusion. This might also be detailed by working together with the project team. Some initial thoughts: • The project meets the objectives on time and within budget • There are no accidents or injuries • There is minimal friction between the working parties during the project • The CEO signs off at the project's conclusion and congratulates the team

(continued)

Table 3.2 Project Plan Elements to Support Strategic Initiative Number 1 (continued)

PLAN COMPONENT	ACTIVITIES
Project risk reviews:	A formal project risk review process is in place and occurs on a regular basis. The impacts from the risk reviews are communicated to key stakeholders, and project scope, budget, and schedule are adjusted accordingly.
Procurement (also known as Supply Chain):	What resources (people and equipment) must be procured external to the company? When are they needed? Can we obtain volume discounts across three plants? Who will perform contract administration? Who has sign-off authority (time sheets, invoices, etc.)?
Quality management:	How will we manage the dismantling and disposal of hazardous materials? Do the contractors work to an acceptable quality plan? Who will inspect the work done by contractors? What materials are needed, and what is the level of quality required? Who will sign off on the standard of work performed? Which quality program will be followed (company or contractor)?
Stakeholder management:	This will consists of people both internal and external to the organization. Some of these stakeholders will have more influence than others. The initial list includes: External Stakeholders Regulatory officials (Local, State and Federal) Contractors Public utility commission Internal Stakeholders CEO and the executive team Procurement Deptartment Quality Department Human Resources Department Legal Engineering Department Operations Department Environmental Department Safety Department Accounting Department
Project integration:	How will we interface with the other strategic initiatives that are occurring? Will we be competing for the same resources at the same time? How will we communicate with other initiatives? Who is the point person coordinating strategy implementation across projects?
Scope management:	How will we maintain the project's scope? How will scope change requests be handled? Do we have a process in place? Who are the stakeholders with authority to change scope? What dollar authority levels are authorized to the project team? How will scope changes be incorporated into the project plan? Who can propose changes in scope?

It should be apparent that the project planning process is detailed but will go far in implementing and supporting the strategic plan. In the scenario presented with this company—the utility—it may be useful to develop an umbrella plan (a "program plan") to coordinate the seven strategic objectives with each other. Note that this is a major effort for this company and will entail significant dollars and resources to implement effectively.

Continuous Review of the Project versus the Plan

Once the project is off and running, how do we control the project activities? How do we ensure that we stay focused on the objectives and not wander off on tangents?

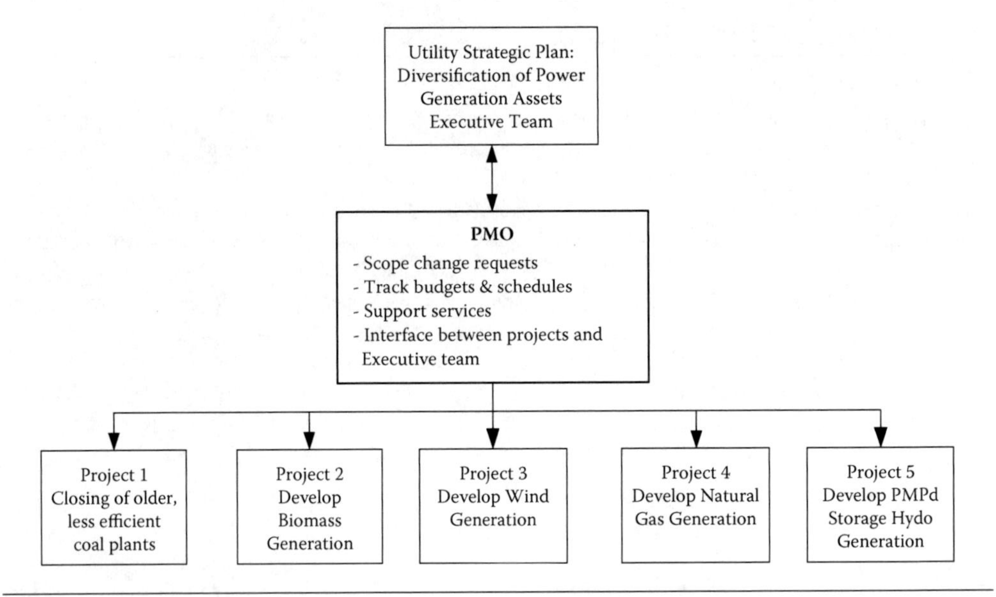

Figure 3.3 Role of the PMO.

A number of companies have implemented PMOs (Project Management Office) to assist in implementing or controlling company projects. A PMO may perform some or all of the following activities:

- Develop a methodology or process to plan, manage, and close projects
- Offer support in developing and tracking project budgets and schedules
- Provide general project management support services
- Provide training in project management
- In some cases, actually manage projects on behalf of management

For our utility company example, a PMO may play an advisory or managerial role. In Figure 3.3, the PMO plays an active role in managing the coordination of all projects affiliated with the strategic plan.

Depending on the company culture, the PMO may or may not be a welcome addition.

Changes to the Strategy

Staying focused on the objectives and strategy is important. However, there are times when the strategy may need to change and, in turn, the supporting projects. It is necessary to exhibit flexibility. In the world of power generation, for example, the regulations have changed dramatically over a short-time period. This includes environmental regulations, operating practices, financing options, etc. This means that the strategic plan should have some flexibility available to change course when needed. While it is desired to have some stability and predictability in implementing the strategic plan, it is not always possible. The use of risk reviews can help identify in advance potential problems, which then provide mitigating plans and flexibility to the strategic plan. Hopefully, this will minimize major disruptions to ongoing projects. This is where a PMO can offer optimum support by coordinating changes across all of the projects.

There are key tipping points when a company's strategy can be upset. These include:

- A merger or acquisition
- Transition from a publicly held to a privately held position (or vice-versa)
- Change in top leadership (different priorities)
- Changes in federal or state regulations
- Changes in market conditions

This list could certainly be expanded, but these represent the key drivers in strategy change.

As a project progresses, it is important that key milestone dates are met and budgets followed. There are a number of ways to track a project's progress, and they are common to the project management profession. They can include using earned value calculations, placing key milestones within a project schedule, and detailing project cash flows to some degree of certainty.

In order to minimize project costs, it is necessary for each project manager to monitor his or her projects and bring project changes (scope changes) to the attention of management. One key point: The author has many times seen project schedules changed to reflect project progress. This ensures the project and schedule always match, but it is not an accurate picture of project progress. The project should be compared to the planned schedule, not the other way around. Do not fall into this trap.

It is important to emphasize that project risk reviews, performed on a regular basis, can prevent problems from developing, or they can provide enough advance notice so that minimal disruption will occur. Project risk reviews are a strong management and problem prevention tool. Many times, project managers complain that they are constantly putting out fires. Risk reviews generally eliminate the need for constant fire fighting.

Summary

Project management provides an excellent vehicle for the implementation of strategic initiatives. A firm's strategic plan will generally focus on market conditions in the future and devise a strategy to adjust to these conditions. The senior management team performs strategic planning with a number of inputs from various departments—especially marketing.

The strategic plan components are similar to those of a project plan; hence the hand-off to project managers should be relatively easy. It is easy for company functional managers to get sidetracked with the day-to-day operations, but dedicated project managers can be utilized to stay focused. It should be noted that large organizations sometimes implement projects using a matrixed approach; that is, people from different areas are assigned to project teams on a temporary basis. This is a common and effective way to implement projects, but it can have its own set of issues—primarily, the team members report to their functional manager and not the project manager.

There must be some flexibility built into both the strategic and project plans. Because of changes in market trends, regulatory issues, or other actions, it may be necessary to change plans midstream. If flexibility is built into the planning stages, these changes can be dealt with easily.

Communicating concisely both the strategic and project plans is highly desired. It has been the author's experience that poor communications are one of the drivers associated with project failure. And, if a company culture change is required in order to implement the strategy, communications will play a larger role during implementation. It is the role of the CEO to clearly articulate the strategic plan's vision and purpose.

In summary, projects will be an output of a company's strategic plan. And well-planned and executed projects will effectively support the company's strategic direction in a timely, cost-effective manner, while reinforcing the purpose of the corporation.

4

Establishing a Governance Model for Strategic Portfolio Management

J. LeROY WARD, PMP, PgMP, CSM, PfMP

The essence of portfolio management is decision making. The organization, through its key executives, needs to decide what projects, programs, and other initiatives to invest in, delay, defer, terminate, change, or modify to meet its strategic objectives. It needs to make these decisions in accordance with a structured and systematic set of rules and criteria so that they are made rationally and logically, based on data and not exclusively on "gut feel." Most importantly, portfolio decisions must be made with due deliberate speed to take advantage of market-moving news and events. To do so requires a streamlined, customized approach, known as Portfolio Governance Management that works with, and not fights against, the culture and best interests of the organization. One definition of Portfolio Governance Management is "the structure and exercise of authority for the initiatives and the portfolios within the portfolio management domain, which defines and enables decision making; assesses metrics on initiatives value and alignment with business strategy; and is responsible for effective and legitimate oversight for the contributions to business success of these initiatives and portfolios" (Hanford, 2006, p. 10). Portfolio Governance Management is based on, and is a manifestation of, a Governance Model, the subject of this chapter.

The Project Management Institute's (PMI) *The Standard for Portfolio Management*, Third Edition, provides a foundational view of portfolio management in particular by identifying what can be interpreted as a process of portfolio management which, among other things, includes the following five key deliverables:

- Portfolio Strategic Plan
- Portfolio Charter
- Portfolio Management Plan (which includes a variety of subsidiary plans)
- Portfolio Performance Management Plan
- Portfolio (PMI, 2013)

Each of these deliverables is discussed in detail in *The Standard* and can be very helpful to an organization that is either just starting out to define and implement portfolio management, or one with a semblance of a portfolio management approach but wants to refine and improve it.

The key component of strategic portfolio management is the Governance Model (GM), which, according to PMI's standard, is included in the Portfolio Management Plan. PMI describes the GM as "the way the organizational assets and resources are planned to be managed within the portfolio according to the specific environment of the organization. It establishes and tailors the decision-making rights and authorities, responsibilities, rules, and protocols needed to manage progress based on portfolio risk towards the achievement of their organizational strategy and objectives" (PMI, 2013, p. 62). We can think of it in this way: the GM is the "control room" of the portfolio management process. Without it, portfolio management cannot happen in any structured, purposeful way. Moreover, "a rigorous governance model is critical to help enforce accountability, optimize cross-functional alignment, and escalate issues to the appropriate decision makers" (PWC, 2012, p. 3).

However, PMI places the development of the GM somewhat downstream in its process of portfolio management. In the author's view, this is an inefficient way to begin establishing a portfolio management process. Developing the GM is the absolute first activity an organization should do before anything else in portfolio management, and the reason is simple: the GM outlines and defines all the key elements and activities and scope and direction of the portfolio process. While it is a decision-making model, it is more than that; in short, the GM encompasses every aspect of *how* an organization is going to execute portfolio management. That is why it comes first.

The author acknowledges, and indeed has extensive experience in, managing portfolios where the GM was weak, underdeveloped, or only casually followed. In cases where governance is weak, portfolio management review meetings become events where the loudest voice in the room gets his or her way, regardless of whether it is directly related to portfolio decisions that commandeer the agenda, or the boss's projects are the ones that get funded regardless of their real business value. Needless to say, this approach is hardly a professional portfolio management process, and one that needs to restructure or re-charter itself by implementing a strong, agreed upon, and pragmatic GM.

Therefore, whether the reader is just starting out to develop a portfolio management process in his or her organization, or is in need of reconstituting or re-chartering portfolio management in a more professional manner, the GM is where we must begin.

Following are, at a minimum, the six key questions that the organization needs to answer in order to develop a pragmatic, sensible, and workable GM.

Question No. 1: What is our purpose and what portfolio(s) will we manage?

This is an important question as it forms the basis for the existence of the portfolio management group in the organization (more on the name of the group below). Will it be a decision-making or advisory body? What are the business outcomes we are trying to achieve? For example, are we looking at cost reduction, new market growth, risk reduction, improved quality, or some combination of these and other outcomes? In other words, why are we establishing the portfolio management process and what do we expect to gain from it?

As way of personal example, more than 15 years ago the author, a former key executive in a large training firm, was approached by one of his peers with a pressing problem. In the absence of any prioritization and rank ordering of its product development projects, the organization's Vice President of Product Development literally did not know which projects to work on first or next. Clearly, given her limited resources she could not work on all projects at once. Yet, each stakeholder held the very strong opinion that his or her project should come first, a common situation in almost every organization. This situation put enormous pressure on the Vice President and her staff. They realized that while they valiantly strived to perform at a very high level to accomplish what everyone wanted, they knew it was unsustainable. Something had to be done to help the Vice President prioritize the work so that the right type and number of resources could be assigned to execute the projects in the pipeline in a more orderly fashion.

To address the situation the author recognized that he needed to convene meetings on a regular basis with the firm's key executives in an attempt to prioritize the enormous amount of work in the product pipeline. That was the sole focus of the sessions initially. We did not have a charter, portfolio management plan, performance plan, or the like, nor did we feel we needed any of these governance documents. We had a serious situation that needed our specific attention and that is what we tackled first. Accordingly, that was our "purpose", and the portfolio to be managed was all of the company's customer-facing projects. It was a good first step in our nascent beginning of professional portfolio management. As time passed our "purpose" evolved and expanded to include a number of other business outcomes and objectives.

Once we have defined and agreed to our purpose, we now need to identify the portfolio(s) to be managed. In any organization, or subset thereof, such as a business unit or operating division, it is entirely normal and expected to have more than one portfolio. In fact, the author strongly discourages any organization from including *every* project and program in one portfolio. Greater efficiency, speed, and other benefits accrue from categorizing projects into multiple portfolios to be managed separately.

Consider a consolidated example a corporate training firm (example represents several training firms), which offers a wide variety of training programs and products to the global marketplace. This firm will have a portfolio of components that include such initiatives as new course development, assessment products, books, other publications, tools, games, and simulations. It may also include projects focused on the different types of training modalities such as on-line, virtual classroom, or video-based learning. The portfolio might also include projects associated with curriculum enhancements. We can see that a wide variety of projects all related to the "product set" of the company can constitute one portfolio.

However, this firm will also have a portfolio of other projects that focus on, for example, internal process improvements, to use a rather broad term. Projects in this portfolio may include the implementation of a new Human Resources personnel evaluation tool, use of a new Customer Relationship Management (CRM) system, delivering a new email system, and developing a new client-relationship management process to handle global accounts. These improvement projects and initiatives can be conveniently categorized in a second portfolio. The question is should both portfolios be managed as one? The answer is a resounding *no.*

This firm would be wise to have its products managed as one portfolio, and its internal improvement projects managed as a second portfolio. The reason is rather straightforward, and it has to do with focus. Making decisions about its customer-facing product set will, more than likely, include a different set of criteria, mix of resources, and risks than its internal improvement process projects. Also, combining too many disparate projects into one portfolio together can result in difficulties in bringing the right stakeholders together to make decisions, and once together can cause meetings to be become too lengthy with certain participants loosing interest when projects that they are not involved in or care about are discussed. Having the right group of people around the table is extremely important and will lead to better and more efficient decision making. One way to accomplish this objective is to only discuss common projects of mutual interest to all involved. This can be easily done by managing one specific portfolio at a time.

Also assume in the example above that it is a rather small company. As such, the same individuals who are selected to manage the product portfolio might very well be the same individuals who would manage the internal process improvement portfolio. This is perfectly acceptable, and in small companies this is often the case. Yet, the author still advises that the portfolios be managed separately for the reasons so stated.

The author acknowledges that at the highest level in the organization a particular person, or group, needs to provide oversight for all of its portfolios. At the highest levels, all the portfolios constitute *the* portfolio of the organization. This person or group will, among other important considerations, ensure that the entire collection of portfolio components aligns with the organizations corporate strategy. Figure 4.1 included here presents a structure of multiple portfolios in an organization, and how they may be managed at various levels. Portfolios One, Two and Three are managed separately; however, the Chief Executive, through the duly appointed Executive Committee provides executive-level oversight.

Thus, the first step in developing a GM is to define the purpose and vision of portfolio management and identify the portfolio(s) to be managed. We can then turn our attention to Question Number 2: Who will manage the portfolio?

Question No. 2: Who will manage the portfolio?

Most organizations form committees or task forces of one type or another to manage the target portfolio. They have a variety of names such as the *Portfolio Review*

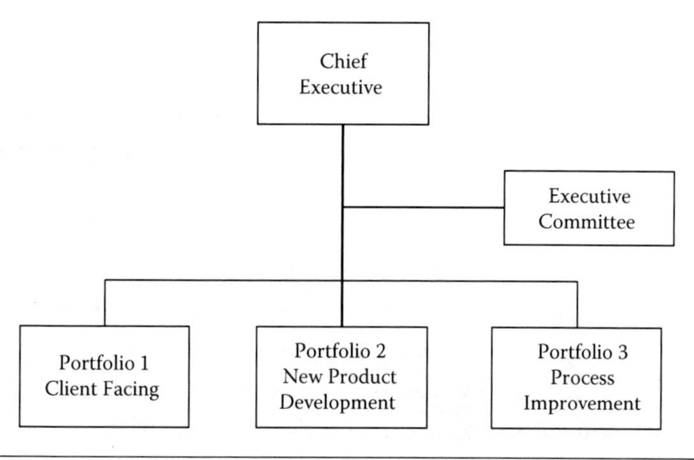

Figure 4.1 A suggested approach for managing multiple portfolios.

Committee, *Portfolio Board*, *Product Portfolio Committee*, *Portfolio Management Team*, *Innovation Board*, or a host of other monikers that generally describe the work of the collective team. Regardless of the name of the committee the key question is "who should be on the committee"? While it may seem like a simple question, the answer can be quite difficult for many reasons, not the least of which is the political ramifications that such a decision has. From this point forward, the author will use the term "Portfolio Committee" to identify said group or task force.

Best practice in this area suggests that we do not want too many people on the Portfolio Committee because that will slow down decision making and make the logistics of planning for, convening, and conducting meetings problematic. As anyone who has tried to plan and conduct meetings, we know it is much more difficult to convene a meeting (whether in person or virtual) with 20 people than it is with ten. The author contends it is not twice has hard; it is four times as hard!

At the same time, we do not want too few people on the Portfolio Committee because we need to make sure we have the right perspectives on the issues at hand and have a sufficient number of people who can represent the views of the organization at large. Also, if the team is too small, it will be criticized for serving the interests more of the members themselves than the organization as a whole. But this still begs the question "Who should be on the Portfolio Committee"?

The author's experience strongly suggests that the membership, or at least the "voting membership," (more on this idea below) be made up of *executives* in the organization, ones responsible for a large operational activity, such as a division or separate business unit, and especially those who manage a profit and loss statement (P&L). It is these executives whose performance is typically judged by the Chief Executive Officer (CEO), Corporate Board, or shareholders and who have a vested interest in the makeup and successful execution of the portfolio.

Second, key executives who support the P&L leaders, and whose organizational units will execute and support the work in the portfolio should also be included. Buy in to the portfolio components themselves is very important. These executives will have "skin in the game", and thus are more likely to provide whatever support is required to complete the projects in a timely fashion that are included in the portfolio.

Certain organizations include key technical personnel as part of the Portfolio Committee given their deep expertise in the subject matter of the individual projects that make up the portfolio or that are being evaluated for possible inclusion. While the author finds no particular fault with this approach, he strongly suggests that such technical personnel not have a "vote" as to the makeup of the portfolio. Portfolio management is a business-driven, decision-making process, not a technical one. Technical personnel can sometimes become enamored with technology and lose sight of the business rationale for investing in a certain project. If the reader's Portfolio Committee includes— or is contemplating to include— technical experts, the author suggests that they not be given a vote as to what the portfolio includes. Thus, a Portfolio Committee can consist of "voting" as well as "non-voting" members depending on the particular composition of the Committee itself (i.e., if technical personnel will be members).

The author once consulted with a client whose Portfolio Committee membership included key P&L executives, product development executives, and technical experts. The Portfolio Manager, who was the Chair of the Committee, complained to the author stating that the technical experts were making decisions based largely on the technical issues at hand, and their personal views on what was "good" for the company's future, rather than reviewing project proposals from a more business-focused perspective. The author recommended that the Portfolio Committee membership remain as constituted, but that an "investment subcommittee" be formed whose members should only include the key P&L executives. Additionally, the Portfolio Charter was changed to authorize this investment subcommittee, and only this subcommittee, to make investment decisions. The company enacted the change, and while the technical experts felt somewhat shunned at first, the change had the beneficial effect intended: implementation of a data-driven, not a technical-driven, decision process. We need to remind ourselves that portfolio management is business management, not technical management.

The thorniest issue regarding membership on the Portfolio Committee is whether to include members based on political considerations. In other words, should a particular person be invited to sit on the committee for "political" reasons? For example, there may be an individual in the organization who is very well regarded by staff, such as a senior technical expert, whose membership would be held in positive regard by the company given this person's technical expertise. There may be an individual in a business unit who has great influence over the decisions of a key business unit head, a key staffer perhaps. The key staffer's membership may help the Portfolio Committee as a whole because that person would see to it that his or her boss would make decisions quickly. Perhaps the key staffer could be persuaded by the other Committee members to "help" the business unit manager make the "right" decision. There are no "absolutes" in the business of portfolio management, and the author cannot say in any given case that these individuals should or should not be part of the portfolio committee other than cautioning the reader that too many members will slow down decision making.

The one advice the author can provide is that if an individual under consideration for membership has influence or direct authority over the budget, personnel, or physical assets affecting the ability of the organization to execute the portfolio that that person's involvement on the Portfolio Committee is generally warranted. The author can also say without hesitation that we should never include representatives on the Portfolio Committee out of professional courtesy, or to be nice, or to make sure an individual does not feel "left out." The author suggests meeting with the individual to explain the selection criteria, and why he or she was not asked to be involved. In most cases, once explained the individual will accept the rationale, and nothing will come of it. After all, the author does not know one person who wants to attend any more meetings than they have to these days!

Heather Colella, Research Vice President at Gartner. suggests answering the following four questions as it relates to "who will manage the portfolio," which can help the reader get started right away.

1. Who will Chair the group?
2. What are the positions of voting members?
3. What are the positions of the nonvoting members?
4. What are the roles of certain specific members (business process, industry, and technology, legal or financial expertise)? (Colella, 2014, p. 17)

The author also suggests appointing an "Executive Director" of the committee. This individual is responsible for coordinating all the activities of the group including, among other things, documenting the action items and decisions made in each meeting as well as distributing meeting minutes to the appropriate parties. The Executive Director will establish the meeting calendar for the group, remind the participants of each upcoming meeting, collect agenda items (which will be reviewed by the Chair), distribute any documentation for review to each member prior to the each meeting, secure appropriate meeting space, and all the other logistical activities of coordinating this important function. Having an Executive Director will allow the Chair, as well as all the Portfolio Committee members, to concentrate on the decisions that need to be made rather than on the logistics and mechanics of conducting said meetings.

Should the amount of work be too great for the Executive Director to handle comfortably, the author suggests assigning an Administrative Assistant to help. The Executive Director and Administrative Assistant are not full-time positions in most organizations. Such responsibilities can be accommodated by existing staff.

Figure 4.2 is an example (consolidated from multiple real-life implementations) of a Portfolio Committee organizational chart that addresses a number of the characteristics discussed above.

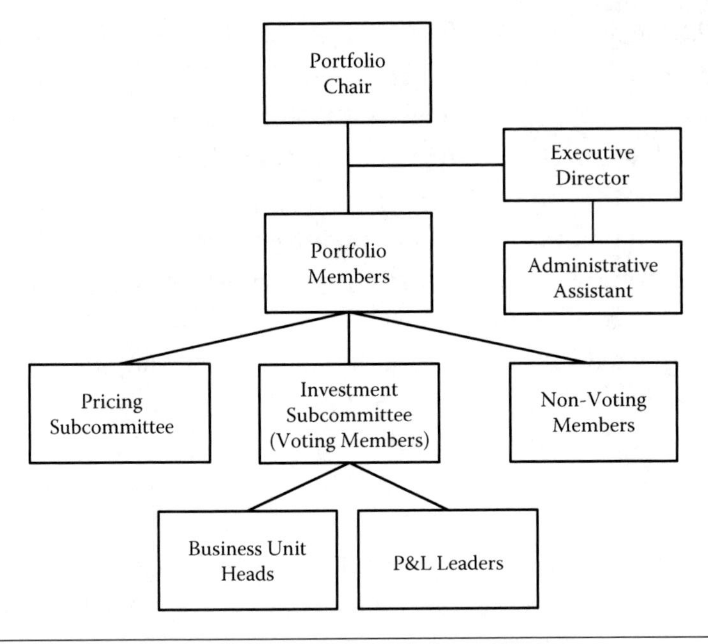

Figure 4.2 Example of a portfolio committee organizational structure.

Question No. 3: How will projects be selected for the portfolio?

Now that we have identified the portfolio(s) to be managed, and individuals who will be managing the portfolio, our next decision is to determine the criteria used to decide how any proposed project, program, or initiative (collectively referred to as the projects), will become part of the portfolio. This decision is the primary responsibility of the Portfolio Committee and ensuring that there is a transparent process, agreed to and followed by all, is of paramount importance in its work. It is a core element of the GM.

In fact, a wide variety of techniques and criteria are, and can be, used by organizations to select projects for their portfolio. They fall into two categories: quantitative and qualitative. The most common quantitative criteria are financial and include any one of a number of financial analysis models that help the Portfolio Committee decide where to invest the organization's limited financial resources. Such financial techniques include such approaches as net present value, break-even analysis, payback period, benefit/cost ratio, internal rate of return, economic value add, and others that may be unique to the organization. Certain organizations establish a "hurdle rate" that a project must exceed in order to be considered further for execution. Many organizations employ more than one financial model when evaluating a portfolio component.

In addition to such quantitative measures, qualitative measures can, and will be applied as well. Increase in customer satisfaction, employee morale, enhancement of good will, and reputation are but a few such measures. While best practice suggests that all benefits be quantified, there are occasions when that is not possible.

Therefore, selection of portfolio components almost always includes both quantitative and qualitative factors.

These quantitative and qualitative factors are an overall part of a business case that is developed and presented to the Portfolio Committee for review and approval. The business case is a valuable document in that it is also used throughout the project's execution to ensure that the benefits initially identified will be delivered based on current performance and in line with financial estimates. A deteriorating business case provides strong justification for terminating any project and removing it from the portfolio.

One other criterion for selecting certain projects has to do with regulatory compliance. For example, organizations in the financial services industry, such as investment banks, hedge-funds, commercial banks, insurance companies, mutual funds, and the like are governed by a broad array of state and Federal regulation in the U. S. Many countries, and political entities, such as the European Union, also have strict regulations governing compliance. As new laws and regulations governing the operations of financial institutions are written, new projects to satisfy compliance regulations will be added to the portfolio.

For example, in the U. S. the Dodd-Frank Act (officially known as the Dodd-Frank Wall Street Reform and Consumer Protection Act), requires financial institutions to operate in specific ways creating a host of regulatory compliance provisions that need to be incorporated into their operations. Such compliance requirements are satisfied through various projects, which are included in the portfolio simply because it is mandatory that they be done. If not done, the company would be subject to fines and penalties by the U.S. Government. The only decision the company has to make is *how* to do the project not whether it should be done. The project will therefore be included in the portfolio so that it can be carefully monitored by the Portfolio Committee.

The GM not only describes the measures and metrics the organization will use to select any individual project into the portfolio, it will also describe the various methods that will be used to rank the projects in order of importance. This second process is critical to developing the Portfolio Roadmap (PMI, 2013). Müller, Martinsuo, and Blomquist in their worldwide study of 133 organizations found, among other things, that successful organizations have a defined process to select *and prioritize* (emphasis added) projects supporting organizational strategy (Müller, et al., 2008).

There is one other point to be made regarding the specific projects that will make up a portfolio. The author has observed that even when an organization has a mature portfolio management process, not *every* project in the organization is formally included in the portfolio for professional management purposes. There may be projects of such low monetary value that specific selection rules do not apply, yet they are needed as part of a larger effort. To be sure, this is completely at the discretion of the organization. The author consulted with one organization where projects valued at less than $200,000 were not included in the portfolio for oversight and monitoring purposes at the Portfolio Committee level. These projects fall under a different and separate management regimen monitored through other management means. Of course, such projects are part of the organization's portfolio, but they do not necessarily have to be part of a formal portfolio process managed by a Portfolio Committee.

The reader can readily see that the rules of the road for defining the universe of projects that will be part of a portfolio can be as varied and unique as the organization itself. This is an important aspect of defining a GM, and the organization's leaders need to decide just which projects will become part of a portfolio to be managed in a structured, systematic way. These decisions are not easy, but they must be made at the outset to get the portfolio process under way. The organization must also recognize it can change the rules at any time as well if, for whatever reason, the selection and prioritization process is not efficient or is not achieving its intended results.

Question No. 4: How will we make decisions and resolve conflicts as a Portfolio Committee?

As stated earlier, the essence of portfolio management is decision making. What projects should be included or not? What projects should be delayed or deferred? What proposals for new portfolio components should an organization entertain? And the list goes on and on.

The author's experience in establishing and coordinating a Portfolio Committee for more than 15 years clearly shows there is no shortage of decisions, large and small, of critical importance or completely mundane, that needs to be made by such a Committee. In order to make decisions quickly, there has to be clear rules that everyone agrees to and follows. One thing we never want to do is to subjugate our decision making to a financial model or other analytical or financial decision-making technique. Models and techniques do not make decisions, people do. Models and techniques though are helpful for the Portfolio Committee as a way to start a conversation about what decision to make.

Even if the Portfolio Committee is presented with clear and compelling financial justification (e.g., easily clears hurdle rate) indicating that a particular project should be included in the portfolio, a "slam dunk" if you will, certain Committee members will disagree and will proceed ahead preferring that other projects be included instead. Many discussions at Portfolio Committee meetings are not necessarily about proceeding ahead with one project or terminating another. They are more about the broad strategy and direction of the organization and how the collective projects are meeting those goals. Regardless of the nature of the conversations or disputes, we have to have a way to resolve these conflicts. So, how do we do it?

As stated earlier, there are no "absolutes" in the business of portfolio management. The author can no more recommend a specific way to resolve such conflicts other than to suggest that an escalation process needs to exist to resolve disputes, otherwise a stalemate will occur, and time will be lost. The rules established must be palatable to all, and in fact, should be developed by everyone on the Portfolio Committee representing its broad consensus. Establishing rules and forcing them on the members will only create antagonistic feelings among the group. Members will either aggressively, or worse, passively, resist the process causing even greater problems and delays in the future. How an organization establishes such rules can oftentimes be more important than what those rules actually are.

Some Portfolio Committees operate on a very simple model: the majority rules. This is an easy and quick way to make decisions, but the organization needs to have the discipline (i.e., intestinal fortitude) to abide by the voting outcome. Too often, these votes are nothing more than "soft decisions", which are then reviewed at the next Portfolio Committee meeting and possibly overturned. Other Portfolio Committees will cast an informal vote that will then be reviewed by a more senior executive or a Portfolio Oversight Committee, one level above the Portfolio Committee. The final decision will rest with either of these two. As such, the Portfolio Committee is then acting in an advisory capacity to a more authoritative entity rather than a decision-making body (see Question Number 1).

The author has had experience with the latter. One client, a major pharmaceutical corporation, had a process where each member of its Innovation Management Board (IMB—its version of a Portfolio Committee) voted on which projects to initiate or kill, but in the end, all major product decisions, those that involved a substantial amount of investment, or would otherwise redirect the product strategy of the organization, needed to be reviewed and approved by the CEO (who did not sit on the IMB). Of course, this process caused delays, but it was an acceptable price to pay to make sure the CEO agreed with the proposed action.

The reader's GM must have a way to break logjams and stalemates around the table. The process for doing so must be discussed and then put into practice by consensus of the Portfolio Committee and other key executives. If no method exists to address these difficult situations, then individual members of the Portfolio Committee will lobby key stakeholders to advance their agenda. This lobbying effort can result in hard feelings by promoting a zero-sum, winner take all, approach to decision making. Moreover, while the result might be in the best interest of certain stakeholders, it may not be in the best interest of the organization as a whole. We should avoid the practice of "horse trading" because this practice promotes the interests of individual members of the Committee and certain stakeholders quite possibly at the expense of the needs of the organization.

The author believes that a portfolio management process should not necessarily be an entirely democratic one. At some point, an individual with a strong personality, and the will to exercise it, combined with a senior title, will need to step in and make the final decision. That person will be either a senior executive or an oversight committee that essentially serves as a proxy for a senior executive. However manner in which decisions are made, it is important that the approach be clear and transparent to all.

Question No. 5: How often will we meet and what rules will govern our meetings?

Portfolio management is an active and continuous management process. Accordingly, the Portfolio Committee should meet on a regular basis to discuss matters of importance including:

1. Reviewing projects proposed to be included in the portfolio
2. Reviewing and evaluating ongoing projects to decide if they should be continued, delayed, or terminated
3. Assessing the current order of priority to see if that should change
4. Discussing resource issues affecting the scheduled completion of the various portfolio components
5. Monitoring the identified portfolio risks to determine if they are likely to occur or to eliminate ones that no longer pose a threat
6. Scanning for opportunities for greater efficiency and effectiveness in portfolio performance
7. Assessing the quality of the portfolio management process used by the Portfolio Committee
8. And, any other matters of particular and unique interest to the Portfolio Committee as it relates to the portfolio and the portfolio management process

The key question is how frequent is frequent enough? The author suggests that the Portfolio Committee meet no less than once per quarter in any given calendar (or fiscal) year. To be sure, in an ever-changing business environment, much can happen in three months, and we need to keep our collective fingers on the pulse of the organization's portfolio activities. In certain business environments, meeting monthly might be advised. The frequency of the meetings will be dictated by the portfolio components themselves. If, for example, the organization's projects tend to run three months or less, then monthly meetings should be held. If they are longer, then less frequent meetings will suffice. In no case though, should the Portfolio Committee meet any less than once per quarter.

The author suggests convening meetings on the same day, time, and for the same time period each time a meeting is held. For example, the Portfolio Committee will meet on the second Thursday of every month for two hours from 9:00AM – 11:00AM. This approach allows the Executive Director to place this meeting on everyone's calendar, and it is firmly established for each Committee member for the year.

Meeting attendance should be made mandatory and only under extraordinary circumstances should a voting member miss a meeting. Having key executives unable to attend slows down the decision-making process. Time is of the essence in certain businesses, and decisions need to be made quickly to capture market share or beat a competitor to market. Oftentimes an executive will send a "key staffer" in his or her place. The author's experience shows that this does little to speed up decision making. At best, these individuals are "note takers" and "seat fillers" for the executive. This does little good to the Portfolio Committee, which is charged with making key, and difficult, decisions.

The author suggests that if a key executive is unable to attend a Portfolio Committee meeting, then he or she should inform the Portfolio Chair well in advance (especially if meetings are held only once per quarter) so that they can meet separately to discuss the issues that will be raised at the meeting and express his or her opinion on key decisions that will be made. Of course, should a rather significant number of voting members not be able to attend any specific meeting, it should be re-scheduled when a quorum will be in attendance. A "quorum" should be defined in the meeting rules.

The meeting agenda should be standardized for efficiency. The author suggests starting each meeting with a review of the minutes of the last meeting and the status of any action items assigned. There are two broad categories of agenda items that will follow the review of the meeting minutes: one is a review of the status of ongoing portfolio components, and the other is a review of proposed projects to be included in the portfolio. The order in which these are discussed is immaterial.

Regarding the review of ongoing project status, we must take care to discuss only projects that, for whatever reason, are of interest or concern to the Portfolio Committee. From a pragmatic perspective, we will probably not have sufficient time in the meeting to discuss all projects' status, and we do not need to do so. To be sure, we should review projects that are on a "watch" list, are in danger of not meeting their goals and objectives, whose priority might have changed because of certain market conditions, or whose variances to time, cost, and scope have exceeded acceptable tolerances.

Additionally, to accelerate decision making, the Portfolio Committee members should receive project dashboard information on all projects prior to the meeting, and, they should read the reports! If this important information is provided in advance, less time will be spent in the meeting discussing the information itself, allowing more time to discuss how to address the issues at hand.

With respect to the discussion on proposed projects, prior information provided to the Portfolio Committee is, again, helpful. Organizations would be wise to have specific rules on what data and information the sponsor needs to provide the Portfolio Committee in order to make its decision. The author suggests inviting the sponsors of the proposed projects, if different from the executives on the committee, to "pitch" their ideas to the Portfolio Committee. This will enable the participants to have a full discussion on the merits of the proposed project.

If a proposed project appears to satisfy the merits of the business case, the Portfolio Committee can decide at the meeting to include it in the portfolio. However, in many instances, the Committee will need time to think through the proposal, and how one project might affect the priorities of other portfolio projects. Even if a project appears on its face to be beneficial to the organization, its inclusion might require a re-prioritization of a number of portfolio components that might be very disruptive. Portfolio management is not as easy as some would make it out to be because it is not about the management of a collection of individual projects; its focus is on assembling that collection of projects whose total combined output provides the greatest benefit to the organization as a whole. The main guidance as to what to include or not include in any portfolio, above all other guidance, is the stated strategic goals of the organization. This is the first test to be applied to any project. How does this project support the strategic goals of the organization?

Question No. 6: What will be the role of the PMO in portfolio management?

Having worked in the project management profession for almost 40 years, the author has observed many different implementations of Project Management Offices (PMOs) globally and in all industry sectors. Clearly, there is no "standard" PMO, notwithstanding the fact that many PMOs provide the same types of services to the organizations they serve. There are certain PMOs, which are, in fact, actively involved in portfolio management. These tend to be PMOs that have been in existence for a number of years and ones that have direct management oversight of the project and program managers in the organization. In addition to being responsible for the execution of the projects and programs in the portfolio, their involvement in portfolio management tends to be with providing the Portfolio Committee with the important status information regarding the various portfolio components. Typically, a PMO head is not a voting member of the Portfolio Committee, but its Executive Director or advisor may be on the Committee. Additionally, the PMO may be called upon to provide assessment services helping the Portfolio Committee analyze a project proposal to determine if it passes muster.

Other PMOs have had little if any involvement with portfolio management in their respective organizations. These tend to be very small PMOs of only two or three people and have no project operational responsibility for project execution. They exist primarily to promote the standardization and application of project management practice, software, and other tools, and to coordinate training activities.

There also exists somewhat of a hybrid type of PMO as illustrated by a former client with whom the author consulted for more than ten years. The client, a global information technology organization, had an EPMO (Enterprise Program Management Office) headed by a vice president who reported to an executive vice president who was the P&L leader of one of the organization's major business units. The EPMO employed approximately six professionals focused on several key areas of project management: training, methodology, tools, and the "health of the portfolio." It was this last responsibility that brought it squarely into the activities of portfolio management of more than 2,000 active projects generating multiple billions of dollars in revenue worldwide.

As regards the "health of the portfolio" this EPMO was a participant in helping to select various projects for the portfolio and then providing data and information on the status of these ongoing projects to upper management through a sophisticated dash boarding process. The EPMO would continuously assess the health of the portfolio's various key projects by conducting audits and risk reviews and reporting its findings to the organization's key executives. For many years, the EPMO was seen as performing a critical role in the organization's portfolio management process. As can be seen, here we have a PMO acting in as a "staff" function operating at a very high level in the organization working side-by-side with key business unit executives to manage its vast portfolio of information technology service projects.

If the reader is developing or improving a portfolio management process certainly he or she would be advised to review the operations of any PMO within the organization to decide if its involvement would be beneficial to the overall implementation and application of portfolio management. In many cases, a decision regarding including a PMO or not in portfolio management is based on the professional experience and business acumen of the person or persons working in the PMO, rather than on the fact that there is a PMO.

Discipline and Encouragement: Essential Ingredients in Implementing a Governance Model

Portfolio management is one of the most important critical functions of any executive team. When done well, it results in the selection and execution of the optimum collection of projects, programs, and initiatives designed to meet the organization's strategic goals and objectives. And by best, the author means that collection of work that offers the maximum return that conforms to the organization's risk profile.

It is not easy to establish a portfolio management process for the simple reason that everyone seems to have their own ideas as to how it should work, what criteria should be used, who should be involved, what portfolios should we manage, how should we make decisions, and all the other issues that go along working with our peers to achieve a common vision. But simply because it is challenging does not mean we should not try to do it, and do it well.

Establishing rules is one thing, abiding by them is another. Organizations can tackle the hard part of the process by answering the six questions presented above. But the harder part of the process is the diligence and discipline required to make those rules come to life and work for the benefit of the organization. Executive "bad behavior" is the root cause of failure of any portfolio management process. Executives can sometimes place themselves above their very own rules because of their sense of self-importance, omnipotence, impatience, or just an "I know better" philosophy.

Time and time we see executives skirting the rules about providing business cases for the projects they want to do, initiating projects without due authorization, re-prioritizing work based on their own agendas, and only paying lip service to the portfolio management process. This is organizational self-defeating behavior because it affects everyone involved in project execution.

The only way for portfolio management to succeed in an organization is for those key individuals who will be responsible for portfolio management to be actively involved in developing what the process is going to be. And, they then need to "police" themselves to make sure they are all abiding by their collective agreement. One way to make the process much more palatable is to introduce "just enough" portfolio management to get started. Do not introduce a very heavy, documentation driven, portfolio process into an organization that has been used to making decisions on the fly with just a few people and no records! Everyone around the table needs to be convinced that professional portfolio management is in their best collective interest.

References

Colella, H. (2014). Top recommendations to design an effective governance process. Gartner Webinar. March 27.

Hanford, M. F. (2006). Establishing portfolio management governance: key components. IBM Corporation White Paper. Retrieved on March 3, 2014, from http://www.ibm.com/developerworks/rational/library/oct06/hanford/

Müller, R., Martinsuo, M., Blomquist, T. (2008). Project portfolio control and portfolio management performance in different contexts. *Project Management Journal,* 39(3), 28–42.

Project Management Institute. (2013). *The standard for portfolio management.* Third edition. Newtown Square, PA: Project Management Institute.

PricewaterhouseCoopers. (2012). How governance and financial discipline can improve portfolio performance. Retrieved on March 10, 2014, from http://www.pwc.com/us/en/increasing-it-effectiveness/publications/strategic-portfolio-management.jhtml

5

IT Governance as an Innovation Tool

DR. SUBBU MURTHY

Introduction

The goal of this chapter is to present Information Technology (IT) Governance as a valuable tool for organizations to leverage their processes, people, and technology to foster innovation. This chapter identifies the process of governing IT and the implications for the enterprise. The quality of information and its related systems revolves around the use of technology. IT is an integral part of an enterprise corporation's business and is fundamental to support, sustain, and grow the business. Typically, an IT group manages the transactions, information, and knowledge necessary to initiate and sustain the core activities of the enterprise.

Need for IT Governance

It is not very frequent to gain consensus in IT, but all agree that there is always more work to do than the resources available. Finance, operations, marketing, and sales all have IT requests that simply add up to create unending work for the IT department. On top of all that, IT managers must stay current with customer requirements, regulatory needs, as well as technology changes like Software as a Service (SaaS) and Cloud Computing. Is the solution to just pump in more money? How do we find the right balance? The challenge is to verify that IT dollars are spent on the right things, and just as importantly, at the right time. The answer to this important question depends on how IT is governed in the enterprise.

The need for IT governance is widely recognized as essential in delivering technology and technology-related services aligned to the enterprise. Without proper and effective IT governance, key IT decisions are left solely to the company's IT professionals. This lack of governance leads to a mismatch between IT priorities and the needs of the enterprise. IT governance helps facilitate decision making across all stakeholders. This prevents IT from independently making, and later being held solely responsible for, poor decisions.

Evolution of IT Governance

The primary goal of IT governance is to help manage and align IT activities to best meet the business requirements. For most enterprises, governance methodologies started with managing user requests using a ticketing/services support system. As IT evolved in organizations from the back-office enterprise resource planning (ERP), to measuring business performance, and finally to providing strategic value to the enterprise, the need for governing IT has also evolved from service management to integrated project/resource management, and finally to measuring IT alignment using benchmarks and scorecards. The role of the IT management has also evolved over these three stages. From being the head of operations to managing critical enterprise projects, the Chief Information Officer (CIO) is seen as the executive driving innovation and change within the enterprise.

The implementation of IT governance, however, is varied from informal manual processes to tool-based structured processes built on industry standards such as Information Technology Infrastructure Library (ITIL) or Control Objectives for Information and Related Technology (COBIT). The key challenge is to develop a set of processes that fit the particular enterprise's culture.

Issues with Implementing IT Governance

In a perfect (if imaginary) world, the CIO develops a perfect IT strategy, which is perfectly aligned to the business strategy, and implemented perfectly within the allocated budget and is on time.

In real life however, CIOs are faced with meeting demands from customers, managing a complex supply chain with both global and local providers, managing a robust and secure IT infrastructure, a complex set of applications, a constant flux of technology changes, and the organization's demand to deliver more for less. Integrating tools to provide a dashboard view are complex and expensive. In place of onerous processes and systems that can hide inefficiencies, what is needed is a flexible IT governance framework that adapts to the enterprise, and yet, provides complete visibility.

The key to practical governance is to "keep it simple," provide for a flexible approach, and ensure that the underlying processes and controls are not onerous and can readily be adapted to meet the enterprise size, culture, and technology environment.

The Innovation Conundrum

In the past, the scope of IT Governance included all technology components (IT infrastructure, telecommunications, and software applications) and resources (IT personnel and user-IT teams) necessary to effectively develop and manage technology initiatives and resources within the enterprise. The principal objective of the governance process was to manage the IT function efficiently. Though a decade old, Weill and Ross (2004) summed up the issue of IT governance as not being a static one size fits all, but one that needs to be tailored to meet the enterprise needs. While many of the ideas of IT governance Weill and Ross identified are as applicable today as they were in the past, one major shift in the role of the CIO from Chief Information Officer to the Chief Innovation Officer is driving a major change in the way the CIO is running the business of IT.

Enterprises seek more from technology than just simple efficient operations. They expect technology to provide value, and perhaps, a competitive edge. Aligning technology to business requires transparent IT planning and budgeting processes to ensure that all stakeholders contribute to the IT strategy.

In the *Harvard Business Review* blog, Daniel Burrus (2013)* in an informal blog identifies that technology is not just changing the business but is transforming it. Transformation requires the role shift of the CIO to focus on innovation. The challenge of course is that the CIO cannot shift attention to innovation if day-to-day and tactical projects are at the forefront!

This paradox leads to a very interesting conundrum – you need processes and structure to run IT, while also enabling the enterprise to be nimble and innovate. Andrew Horne and Brian Foster† identify why IT governance is onerous and killing innovation. Quoting them (2013), *"Most CIOs will tell you that they have no shortage of ideas to invest in— the hard part is whittling down to the right ones. Push that a bit further and what most CIOs say is that those ideas are in the form of project requests from business partners. The problem is that these "bottom-up" project requests often miss the big picture as too many are incremental or uninspiring. Yes, while most of these requests are vetted for alignment with corporate strategy, what's often missing is how these requests fit within a broader context of how the business overall generates value."*

The primary hypothesis in this chapter is that IT governance can be used as an innovation tool but to accomplish it requires innovation in IT governance itself. This chapter outlines the Guiding Principles, Typical Organizational Components, and the IT Governance Framework and tools that foster innovation.

Guiding Principles

The challenge is to implement IT governance to maximize the value from technology investments without creating process which can be onerous on the users, IT staff, and

* Daniel Burrus is an American futurist, business advisor, author, and frequent speaker about business strategy and innovation.

† Andrew Horne is a Managing Director at the Corporate Executive Board (CEB), who works with the CEB's membership network of Chief Information Officers and leads global research teams in producing best practice case studies, benchmarks, implementation tools, and executive education. Brian Foster, a Managing Director in CEB's IT Practice, oversees a global team that provides advice and consultation to a network of more than 2,500 IT leaders, including CIOs, enterprise architects, applications, infrastructure, security, and Project Management Office executives.

management. To achieve a nimble, flexible yet structured management process, six principles should guide the development of the governance process.

a) Facilitate the alignment of IT with the business;
b) Deliver optimum value to the business units by focusing on managing IT as a portfolio of investments;
c) Manage risks effectively;
d) Establish a governance process that is flexible, agile, and can be easily adapted to implement new innovative initiatives rapidly;
e) Implement integrated governance that involves the orchestration of policies (including delineation of roles, responsibilities, and authorities), processes, and tools for successfully meeting enterprise requirements; and
f) Create a governance structure, controls, and transparency at a level that appropriately fits the personality and readiness of the enterprise without injecting onerous processes that can stifle productivity or innovation.

Typical Organizational Components

Organizations can use one or more of the following structures to implement IT governance.

- Project Management Office (PMO)
- IT User Governance Committee (ITUG)
- IT Steering Committee (ITSC)
- Technology Innovation Center (TIC)

The responsibilities of the PMO, ITG, ITSC, and TIC across the various objectives of IT governance are shown in Figure 5.1.

Project Management Office (PMO)

The Project Management Office is responsible for establishing guidelines/standards and overseeing project implementations. The PMO is typically managed by the CIO for small to mid-sized organizations and by dedicated teams for larger organizations. One of the key charters of the PMO is to manage quality, costs, schedule, and risks and to develop standard procedures that improve quality, assessment of cost and schedule variance, and managing risk. As projects are undertaken, a certain level of risk is assumed. Active portfolio management gives the visibility needed to assess and communicate risk.

IT User Group (ITUG)

The ITUG is the organizational component that has responsibility for prioritizing and scheduling the user service and new project requests. It is typically made up of IT

Charter Area	Objectives/Responsibility	P– Primary, S – Support			
		PMO	IT User Governance Committee	IT Steering Committee	Technology Innovation Center
Alignment of IT to Business	Align to business objectives		S	P	S
	Manage project portfolio	P	S		S
	Manage service portfolio		P		S
	Enforce business guidelines	S	S	P	P
	Prioritize using Pareto Principle	S	P	S	S
Getting Value from IT Investments	Understand impact	P			
	Develop IT innovation frameworks	S	S		P
	Establish enterprise IT architecture	S	P		S
	Manage innovation portfolios	S	P		P
	Measure value	P	S	S	S
Using IT Resources Responsibly	Manage resource capacity, skills, and budgets	P	S	S	S
	Manage skill deficiency with right sourcing	P	S		S
	Track resource capacity and availability	S			
	Align resource utilization to priorities	P	S	S	S
Mitigating Risk	Perform SWOT analysis	P	S	S	S
	Develop risk mitigation plan across the IT function	P	S		S
	Manage risk	P	S		
	Implement a security strategy	P	S		
	Establish a control framework and IT processes	P	S	S	S
	Implement and manage change	P	S	S	S
	Manage major IT innovation effectively	S	S		P

Figure 5.1 IT governance responsibilities.

managers and users. If IT has a PMO, invariably selected members of the PMO will be part of the ITUG. Usually the CIO is also a member, except for large enterprises, where the CIO may identify a senior IT manager to be part of the ITUG.

The charter of ITUG is to assign priorities and resolve IT problems/issues and user change requests to the IT infrastructure and IT applications, prioritize, sequence, and oversee new project requests. The IT Governance Committee will have full transparency on exactly what projects are active, are on hold, and are pending approval. This will enable ITUG to make clear business decisions on what the priorities are and how effectively the resources are utilized.

IT Steering Committee (ITSC)

The ITSC is the organizational executive structure that provides the overall guidance on IT strategy with oversight over both the IT organization and the IT Governance Committee. Executives from different functional areas, the CIO, and perhaps even the Chief Operating Officer (COO)/CEO are part of ITSC.

The charter of the ITSC is to oversee the IT Governance Committee, break deadlocks between requests for resources if required, provide interface with other organizations to standardize the IT group, and oversee the IT budgets and value delivered by the IT group to business. The IT Steering Committee will provide increased visibility into IT spending. As a part of overall corporate governance, it is critical to have an understanding of where money is being spent throughout the organization. The committee will oversee the portfolio management function, including those projects that have been approved, those that are planned but not yet approved, and the remaining available budget. Effective IT portfolio management provides the enterprise with the information needed to make more effective decisions regarding their investments and to provide transparency into how those decisions were made.

Technology Innovation Center (TIC)

The TIC is responsible for creating exciting, high-impact solutions, which require existing and leading-edge technologies to be combined in ways that challenge the status quo. The resultant capabilities have the potential to transform traditional business practices and create tomorrow's high-growth opportunities.

The TIC is also responsible for establishing guidelines/standards and overseeing innovation. Innovation in the industry, such as shift to mobile platform, social media, analyzing the vast amounts of data, understanding on-demand services and on-demand SaaS solutions and how they will impact the enterprise's eco-system, is critical. Innovation is not limited to within IT itself; even "rouge-IT" (Holmes, 2012) can be leveraged to innovate.

The two key questions that TIC should address include: 1) What new revenue streams and business processes will be enabled?, and, 2) What new IT products and services are needed to support these next generation business processes? Answering these questions requires an integration of marketplace insight and technology analysis in a nimble, innovative, delivery-focused environment. Some of the areas that foster innovation include:

- Cloud and social networking for automating IT and business processes
- Integrated security that secures the emerging, global business and social ecosystem

- Mobile device business applications that introduce new ways for end users to interact with providers of goods and services
- Big data/analytics to mine the information to create the ability to channel products and services at the individual level

Common IT Governance Frameworks

There are many frameworks that can be deployed. While they address different needs, they do share the common goal of building and delivering IT efficiently, effectively, and securely to the enterprise. They should always be "adapted" to the enterprise to meet the specific business and cultural tenets cherished by the enterprise.

Some of the common ones are:

- The Balanced Scorecard
- Control Objectives for Information and Related Technology (CoBIT)
- The Information Technology Infrastructure Library (ITIL)
- Committee of Sponsoring Organizations (COSO)
- Capability Maturity Model Integration (CMMI)

The Balanced Scorecard[*]

This framework examines where IT makes a contribution in terms of achieving business goals, being a responsible user of resources, and developing people. It uses both qualitative and quantitative measures to get these answers. The authors of the balanced scorecard share their views on how it could be used as a strategic management system (Kaplan & Norton, 2007). They define four processes to leverage the balanced scorecard: (1) translating the vision, (2) communicating and linking, (3) business planning, and, (4) feedback and learning. Another example of the IT balanced scorecard[†] is shared by the Information Systems Audit and Control Association (ISACA) in building the IT strategy.

Control Objectives for Information and Related Technology (CoBIT)

The ISACA is a set of guidelines and supporting toolset for IT governance that is accepted worldwide. It is used by auditors and companies as a way to integrate technology to implement controls and meet specific business objectives.

[*] Refer to Balanced Scorecard Institute for getting the basics. The link below shares Robert Kaplan's and David Norton's framework. https://balancedscorecard.org/Resources/AbouttheBalancedScorecard/tabid/55/Default.aspx

[†] ISACA provides a framework for developing the IT Balanced Scorecard. A link to the IT Balanced Scorecard is included below. http://www.isaca.org/Journal/Past-Issues/2000/Volume-2/Pages/The-IT-Balanced-Scorecard-A-Roadmap-to-Effective-Governance-of-a-Shared-Services-IT-Organization.aspx

The Information Technology Infrastructure Library (ITIL)

The ITIL from the government of the United Kingdom offers eight sets of management procedures in eight books: service delivery, service support, service management, infrastructure management, software asset management, business perspective, security management, and application management. ITIL is a good fit for organizations concerned about operations.

Committee of Sponsoring Organizations (COSO)

This model for evaluating internal controls is from the Committee of Sponsoring Organizations of the Treadway Commission. It includes guidelines on many functions, including human resource management, inbound and outbound logistics, external resources, information technology, risk, legal affairs, the enterprise, marketing and sales, operations, all financial functions, procurement, and reporting. This is a more business-general framework that is less IT-specific than the others.

The Capability Maturity Model Integration (CMMI)

The Capability Maturity Model Integration method, created by a group from government, industry, and Carnegie-Mellon's Software Engineering Institute, is a process improvement approach that contains 22 process areas. It is divided into appraisal, evaluation, and structure. The CMMI is particularly well suited to organizations that need help with application development, life cycle issues and improving the delivery of products throughout the life cycle.

Analytics Based IT Governance—A New Framework for Managing IT

The unique thing about the next generation of IT governance is to use the very same technologies that are enabled by the CIOs for the organization.

As an analogy, one of the first steps in defining organizational Key Performance Indicators (KPIs) is to start with identifying the strategic business issues that can benefit from predictive analytics. Issues such as optimizing the supply chain, forecasting revenues, getting value from investments, acquiring new customers, retaining existing customers, increasing revenue per employee, etc. have all led CIOs to build the solution for the enterprise that help address these issues. The growth of analytics is strongest in financial management and budgeting followed by operations and production, strategy and business development. Sales and marketing and customer services are target areas of growth.

However, these techniques are seldom using in managing the business of IT. Significant is the lack of use of analytics in IT workforce planning and allocation. A paradigm shift may be needed to bring about a change in the way IT is managed. If we compare the CIO to the CEO of an IT firm, then the CIO is in essence running an IT services company, and thus the CIO can simply tailor and reuse the available scorecard measures.

There are many reasons for this discrepancy, not the least of which is being the perceived importance of IT in the organizations. IT is by and large treated as an expense and not an investment. The role of IT analytics can be better understood by making the distinction between micro-analytics (individual KPIs) and macro-analytics (combination of multiple metrics).

Micro versus Macro Analytics[*]

ITIL provides a set of metrics, which are very specific and cover a few specific measurements. These analytics can be termed microanalytics.

A balanced scorecard view provides macro analytics. Macro analytics is analogous to macroeconomics—dealing with the whole whereas microeconomics deals with a specific domain or area.

Macro IT analytics can be classified into four groups:

- Managing the Services Delivered (user perspective)
- Managing the Resources (staff perspective)
- Managing Value Delivered to the Enterprise (financial perspective)
- Managing External Factors

User-centered IT analytics include customer happiness, response time, ability to triage and prioritize requests, and defect management. Staff-centered IT analytics include staff morale, resource utilization, transparency, productivity, and turnover. Financial IT analytics includes the number of capital projects, return on investments (ROI), variance, and percentage of IT expenditures on new projects. External IT analytics includes outside threats, disaster recovery (DR) incidents, technology adoption velocity, and percentage of budget allocated to contingency. These four groups can be combined to provide a benchmark of IT itself.

Developing Micro IT Analytics

ITIL offers over 100 KPIs for assessing the value of IT delivered. They classify the KPIs into several areas:

1. Service Portfolio Management
2. Financial Management
3. Service Level Management
4. Capacity Management

---

[*] These are recreated from the original blogs written by the author and can be found on ugovernit.wordpress.com.

5. Availability Management
6. IT Service Continuity Management
7. IT Security Management
8. Supplier Management
9. Change Management
10. Project Management (Transition Planning and Support)
11. Release and Deployment Management
12. Service Validation and Testing
13. Service Asset and Configuration Management
14. Incident Management
15. Problem Management
16. Service Evaluation
17. Process Evaluation
18. Definition of CSI Initiatives

Examples of KPIs include the following:

Number of Releases: Number of releases rolled out into the productive environment, grouped into major and minor releases.

Adherence to Project Budget: Actual versus planned consumption of financial and personnel resources.

Gaps in Disaster Preparation: Number of identified gaps in the preparation for disaster events (major threats without any defined counter measures).

Number of Planned New Services: Percentage of new services, which are developed being triggered by Service Portfolio Management.

Duration of Service Interruptions: Average duration of service interruptions.

Developing Macro IT Analytics

These analytics must mirror the culture of the organization. In cases where IT is strategic and the IT changes are rapid, the analytics will focus on how agile the IT organization is to change/implement new systems at a high velocity. For example, one could define a macro analytic effectiveness of change by combining the number of changes per unit time (velocity of change), the quality of change (number of errors per change), and cost of change.

For stable organizations, where user satisfaction and availability of services are critical, analytics will focus on users including response times, quality of resolution, speed of resolution, etc. For example, the effectiveness of a change analytic could be modified to include the response time.

Macro analytics can typically drill down to the specific metrics that make up the macro analytic. The ability to drill down provides the opportunity to undertake root-cause analysis.

Macro analytics provide a measurement-based approach for benchmarking IT. A classic example of a macro analytic is the IT benchmark. This analytic combines the four-macro analytics identified earlier. Combining the four macro analytics can vary from simple weight based methods to more complex algorithmic based models.

Weight based methods typically score the IT with a formula:

IT Benchmark score equals:

$(W_1 * U)$ plus

$(W_2 * S)$ plus

$(W_3 * F)$ plus

$(W_4 * E)$

Where U is the macro analytic for user centered metrics, S for resource (IT staff or consultants) centered metrics, F is financial metrics, and E are the external factors. Weights will vary based on the importance the enterprise places on the specific analytic. For example, cost conscious enterprises may give more weightage to $W_{3,}$ whereas service centered enterprises may give importance to W_2.

Algorithmic methods are more complex and require the development of insight from the past through quantitative methods such as regression or qualitative methods such as case studies. Insight can also be gained from industry comparisons. These methods can help develop very robust benchmarks.

Challenges with Macro Analytics

There are four key challenges in implementing macro analytics. The first one is quite obvious. They are only as good as the metrics we have established to identify a benchmark score.

The second challenge is to come up with a meaning behind the scores. As an analogy, credit scores have meaning because institutions that use them can categorize scores into excellent, good, fair, etc. Quantitative IT scores are new, and such a rating scale does not exist.

The third challenge is that tools for developing analytics are very expensive. Enterprise tools require a large footprint of infrastructure. These are powerful but expensive and take a long time to make changes. The challenge is data migration across to the cloud. As more and more enterprises become more cloud savvy, the problem of data integration will be trivial.

Future of Macro Analytics

IT benchmarks provides insight to improve the effectiveness of technology and services delivered by the IT group.

Prior to using analytics IT balanced scorecards helped IT organizations think of IT in terms of four areas Financial, Customer (Users), Internal Business Process, and Learning and Growth. However, these were largely qualitative assessments. Macro analytics provide an objective measurement.

Research into the metrics that make up the macro analytic, and the research into scoring methods will help us build better macro analytics. Keynesian economics and Monetarists have fought long and hard over different macroeconomic models; CIOs will also fight over IT benchmarks for the foreseeable future.

Implementing Analytics Based IT Governance

Implementing analytics based IT governance is a difficult challenge for CIOs. Processes are still deemed to be onerous, and blind adoption of IT standards has failed to gain acceptance. Furthermore, tools to support IT governance have not made the leap to meet today's demands. Current solutions are fractured and costly. Typically, you have different tools for help desk/ticket management, project management, resource management, asset management, configuration management, and workflow coordination. For small- to mid-sized enterprises. tools for planning and managing IT budgets are still pre-dominantly Excel based.

The old adage "the Cobblers shoe is always torn" is very much true with respect to IT governance—a small percentage of IT organizations adopt technology such as dashboards and scorecards to assess how well they are delivering technology and value to the business. Using tools will bring efficiencies in managing IT departments for all levels from the CIO to the technical staff. Governance tools need to have at least three components:

1. A Service Management System that provides the basic ticketing system, user facing wizards to help users quickly and easily make user requests, a mobile module to help IT manage user requests quickly, and asset and change management modules to ensure the integrity of delivered services.
2. A PMO toolset that integrates project management and resource management powered by a project request and triaging system to help prioritize the project request and help management track the project's costs, schedule, risk, and assess the project performance.
3. A set of customizable IT dashboards based on business intelligence using IT analytics to help assess how well IT is performing, assess the ROI of IT investments, manage a portfolio of projects, conduct IT assessments, and maintain an IT balanced scorecard as a benchmark of the value created by technology investments.

Summary

CIOs are driven to help organizations grow. It is no different than running an IT services company. They are faced with meeting demands from customers, managing a complex supply chain with both global and local providers, managing a robust and secure IT infrastructure, a complex set of applications, a constant flux of technology changes, and an organization's demand to deliver more for less.

For enterprises, the guiding principle is to deliver value to the business without injecting onerous controls that stifle productivity. To achieve this business value, the IT governance framework should provide complete transparency on IT activities and make it simple for users to make, monitor, and prioritize IT requests. In order to achieve transparency, IT will need to establish controls and processes to deliver quality technology solutions on time and within allocated budgets. From a management perspective, it is critical to effectively allocate and track resources and costs.

Tool-based systems help provide complete transparency into the IT initiatives, resource allocation, project prioritization, issues and challenges faced, and the planning and budgeting process. Through use of dashboards and IT analytics, CIOs will gain integrated intelligence about IT to manage resources efficiently, manage project portfolios, manage risk appropriately, and facilitate alignment to business needs. This will help the CIO evolve from the old-school ("Chief Information Officer") to the new-school ("Chief Innovation Officer").

References

Burros, D. (2013). Today's CIO needs to be the chief innovation officer. 2013. *Harvard Business Review* Blog Network (http://blogs.hbr.org/2013/07/todays-cio-needs-to-be-the-chi/).

Holmes, R. (2012). Rogue IT is about to wreak havoc at work. August 9. *CNN Money* (http://tech.fortune.cnn.com/2012/08/09/rogue-it/).

Horne, A. & Foster, B. (2013). IT governance is killing innovation. August. *Harvard Business Review* Blog Network (http://blogs.hbr.org/2013/08/it-governance-is-killing-innov/).

Kaplan, R.S. & Norton, D.P. (2007). Using the balanced scorecard as a strategic management system. July. *Harvard Business Review. 85*(7–8), 150–161.

Weill, P. & Ross, J.W. (2004). *IT* governance on one page. Center for Information Systems Research Working Paper 349/Sloan School of Management. Cambridge, MA: Massachusetts Institute of Technology. Working Paper Number 4516-04.

6

Governance and Portfolio Management

LAURENCE LECOEUVRE, PhD

Introduction

More and more companies are using the project management discipline to manage multiple projects with limited resources in a competitive environment; this is certainly the reason why Association for Project Management (APM, UK) issued a booklet called: "Directing change: A guide to governance of project management"; this can help to fill the gap in the governing surveillance of project activities (Wideman, 2005). This chapter aims to question the relevance of governance to portfolio management. Mainly based on a literature review, the chapter gives a view point and proposes "Governance Project Portfolio Management" (GPPM) structure to optimally ensure a company's sustainable performance.

Project Portfolio Management

We can differentiate project management from the management of projects. Indeed, the management of projects includes project portfolio management that links corporate strategy to project management (Morris & Jamieson, 2008; Turner, 2009). Therefore a project portfolio is a collection of projects that are carried out in the same business unit sharing similar strategic objectives and the same resource pool (Rad & Levin 2006; Turner, 2009). Portfolio management is considered to be a dynamic decision-making process with a business's list of active projects that is constantly updated and revised. In a project-based firm, project portfolio management (PPM) focuses on the threats and opportunities that project initiatives and projects promise for future business success. PPM should therefore include the search for new customers as well as maintaining the enterprise environment and project capability. It should also include logistics support of past projects and the maintenance of relationships with customers (Lecoeuvre, Turner, & Patel, 2014).

Additionally, emphasis is given to managing a company's strategic portfolio of projects at an aggregate level, which includes project prioritization, project review, project re-alignment, and project re-prioritization; and knowing that continuous monitoring of each project's contribution and alignment to the enterprise goals are a key part of PPM (Rad & Levin, 2006).

Accordingly, PPM is a strategic approach concerned with an enterprise as a whole. Portfolio management is much more encompassing than project or program management since it focuses on identifying enterprise opportunities and then selecting the programs, projects, and other work to meet them. Once this identification is done, the next step is to plan and execute the work with a continual focus on benefit realization and sustainment to ensure organizational success (Rad & Levin, 2006). The primary objectives of PPM include maximizing the value of a portfolio, balancing it in many dimensions, and linking it to a strategy that reflects alignment between projects, strategic content, and resource allocation intended per the strategy of the business (Lecoeuvre, Turner, & Patel, 2012).

Moreover, successful organizations have a defined process to select and prioritize projects supporting organization strategy (Müller, Martinsuo, & Blomquist, 2008). Indeed prioritization is the critical link between where an organization wants to be and what it is currently doing. It helps determine required initiatives and how projects deliver organizational strategy. In this way, PPM forms a quantitative base for removing redundant projects and optimizing available project expenditures to align resources with strategic priorities (Crawford, Turner & Hobbs, 2006), because organizations are open systems looking to reach their corporate objectives (Checkland, 1999). The tools and techniques that are used to achieve these objectives are a part of the PPM process. Thus, the fundamental output is a prioritized list of projects (Rad & Levin, 2006).

Beside project prioritization, a key element in PPM is implementing strategy throughout an organization. However, strategy is not a stand alone management process. Each organization's mission dictates why it is important and how it must be achieved through a concrete strategy (Kaplan & Norton, 2001). Achieving the mission through strategy is where governance plays a crucial role. An organization's governance style can influence the use of portfolio control and development (Müller et al., 2008). In addition, governance is considered necessary (Levin, Artl, & Ward, 2010; Frey & Buxmann, 2011) to such areas in portfolio performance management, risk management, and component identification.

Governance

The following outlines the need for governance in portfolio management.

According to Müller (2009), and Turner (2009), governance is an important feature for project management. Notably governance helps define a project's objectives, as well as the means to achieve them and outlines the ways to monitor the project's progress. Thus, governance makes a project viable and steers it along its life cycle (Müller & Lecoeuvre, 2014, forthcoming).

Project governance is "an oversight function that is aligned with the organization's governance model and that encompasses the project life-cycle [by providing] a comprehensive, consistent method of controlling the project and ensuring its success by defining and documenting and communicating reliable, repeatable project practices" (Project Management Institute, 2013, p. 34). The collective governance of a program or portfolio, as well as the collective governance of all projects in an organization, from the corporate or board-level perspective, is referred to as governance of a project (GoP). GoP takes a broader view than the individual project (Müller & Lecoeuvre, 2014, forthcoming). It also takes into account governance's hard aspects that include structural and controlling elements and also its soft aspects, such as people interactions (Walker, Segon & Rowlinson, 2008) by Müller's (2009) model describes four governance paradigms of governing projects and notes the model supports corporate governance orientation and control orientation at the organizational unit level managing a project. The model is based on the intersection of the shareholder/stakeholder orientation with the behavior/outcome control of the project's organization.

Generally speaking, many claim PPM alone is not sufficient for success; PPM must be complemented by a governance board or a comparable group because it connects PPM to stakeholders and consequently to business outcomes (Hajela, 2013). According to Wideman (2005), APM's view is that "effective governance of project management ensures that an organization's project portfolio is aligned to its objectives, is delivered efficiently, and is sustainable. Governance of project management also supports the means by which the board and other major project stakeholders are provided with timely, relevant and reliable information" (Wideman, 2005, p. 2). The APM view stresses that GoP should not interfere with "most of the methodologies and activities involved with the day-to-day management of individual projects" (Wideman, 2005, p. 3).

Governance and Project Portfolio Management

PPM can be considered as projects grouped together according to their skills and resource needs. Thus, PPM governance is not developed on the same scale as the governance of one project. Accordingly, "research indicates that project management becomes increasingly difficult when there are multiple overlapping projects, resulting in a need for enhanced governance controls to increase success rates" (Misner, 2008, p. 4), and that without suitable and adequate governance of a portfolio, organizations fail and suffer negative impact at a high probability (Miller & Hobbs, 2005).

The collective governance of projects within a portfolio, as well as the collective governance of all projects in an organization, supports the idea that "a project is not an objective in itself but a means of achieving strategic change or future benefits" (Klakegg, 2010, p. 27). According to the Project Management Institute (PMI) (PMI, 2006), the purpose of a governance of programs or portfolios, is to achieve an objective (or benefit) through the summation of outputs, interactions, and synergies of multiple individual projects.

The governance of projects is inherently a part of the corporate governance framework, which steers all activity in an organization and not just that of projects. It "comprises the value system, responsibilities, processes and policies that allow projects to achieve organizational objectives and foster implementation that is in the best interest of all the stakeholders, internal and external, and the corporation itself." (Müller, 2009, p. 4; see also Müller & Lecoeuvre 2014, forthcoming).

According to APM's guide (2004; Wideman, 2005), there are four main components focused on effectiveness and efficiency of:

1. *Portfolio direction*, which is the alignment of the organization's project portfolio with its key business objectives, financial control, and risk assessment.
2. *Project sponsorship*, which includes competent and accountable sponsors, who devote time and energy to align of the interests of key project stakeholders with project success.
3. *Project management*, which includes clear critical success criteria used for decision making, developing opportunities for improving project outcomes, clear key governance of project management roles and responsibilities, and authority delegated to the right levels for balancing efficiency and control.
4. *Disclosure and reporting*, which involves relevant and reliable information on project progress and significant project-related risks and their management, establishing both key success drivers and key success indicators by the organization, and sharing project portfolio status with key stakeholders.

Along with these four components, Wideman (2005, p. 6) proposes "a more holistic understanding" of PPM titled "a Project Portfolio Life Span (PPLS)" that has a holistic view and consists of the following five-phased components:

1. Identification of needs and opportunities (corporate fiscal planning)
2. Selection of the best combinations of projects (PPM)
3. Planning and execution of projects (project management)
4. Product launch and deployment of project deliverables (marketing and sales)
5. Realization of benefits (corporate due diligence and accounting)

Wideman points out the organization's responsibility must be "complete" to "verify the success of its projects, validate its project portfolio management assumptions and generally become a real "learning organization" (p. 6); and to "extend to the evaluation" (p. 7) to the return of benefits to the organization, from short-term benefits to long-term benefits.

Wideman's approach is similar to the Results-based Monitoring and Evaluation System developed by the Organization for Economic Cooperation and Development (Binnendijk 2000) and that of Kusek & Rist (2001), later used by the World Bank, which describes the need to do the planned work to achieve results. This work requires resources, which leads often to new assets from projects. The point is that "the number of projects running in parallel and the timely implementation of these projects not only depend on the availability of financial resources, but also on the availability of the necessary project staff" (Frey & Buxmann, 2011, p. 9). Thus an operative "organizational assignment" of the resources of projects is decisive for success (Engwall & Jerbrant, 2003) and sustainable performance. It suggests that to "achieve results, it is necessary to do the planned work. The planned work consumes resources to undertake activities [. . .]. The activities deliver outputs; in the case of a project this is a new asset. The operation of the output enables the investor to do new things, called the outcome. The ability to do new things leads to the ultimately desired impact. This can be the desired benefit but may also be higher order strategic objectives" (Xue, Turner, Lecoeuvre, & Anbari, 2013, p. 7).

Frey & Buxmann (2011, p. 9) point out that "centralized and decentralized arrangements can be distinguished" in the management of projects' resources; decentralized control implies that "project resources belong to the individual business units;" and centralized control means that "resources can be assigned to projects in various sectors." The results of their research, which focused on information technology project portfolio management, underlines that "whether certain decision-making rights should be centralized or not, seems to depend on contingency factors as the general organizational structure and the strategic focus of the enterprise as a whole," but there are some kind of "freedoms of design especially in regard to budget allocation and project selection" (p. 11).

Consequently, it is all a question of project portfolio governance. Still, other contexts are equally important such as the financial context. According to Iovino's study (2005), the main expectations of portfolio governance are:

- Effective investment decision process and recovered accountability
- Better information and transparency
- More direct involvement of leaders

Governance should fit with the "benefits" expected from a portfolio management. It should be a rational approach to selecting and managing the portfolio that leads to a better understanding of the adjustments of value, risks, and cost; optimizes resource use; reduces costs; and increases value from the portfolio budget (Iovino, 2005). He further proposes a portfolio management model (2005, p. 8) based on:

1. governance "to orchestrate the building, implementation and management of the prioritization model"
2. process "to allow organizational adoption, data availability and environmental linkage"
3. tools "to enable data analysis, subjective criteria and value maximization."

Iovino contends that a successful interaction of these three key elements enables an organization to optimize portfolio management. Of course, communication and stakeholder involvement are also essential.

Müller (2009, p. 94) stresses the importance of stakeholder involvement and shows that the "responsibility of the board of directors and other executive managers", in particular is "in setting the stage for good project management in their organization". Governance of project and stakeholder management should have the same level of importance in order for a project to have success.

Conclusion: Toward a GPPM

A Project Management Office (PMO) is considered as "a centralized unit within an organization or department that oversees and improves the management of projects" (Grey and Larson, 2006, p. 561), Similarly, Misner (2008, p. 7) declares that "the project management office, or PMO, is a tool that addresses the need for selecting and managing multiple simultaneous projects in such a manner as to maximize the value obtained". A PMO answers the need for an "added governance structure" to increase control of "project efforts' (Misner (2008), p. 7).

Misner (2008, p. 15) also wonders if it is possible to coin an "acronym" for a Portfolio Management Office; if "all are variations of the same function", (of PMOs) the Portfolio Management Office acronym must emphasize the difference "in the scope of their responsibilities."

Similarly, we propose an acronym for Governance of Project Portfolio Management, "GPPM," which emphasizes the multiple environments that characterize portfolio management as it is defined as a multi-project environment. It would meet the need for organizations to be able to appraise complex systems from multiple views and perspectives (Miller & Hobbs, 2005). Misner (2008, p. 33) wrote that the "inability to recognize and react to the complexity can lead to compliance failures," and the GPPM would help a company to overcome this inability and assure a sustainable performance.

Misner (2008, p. 40) reviewed the literature and determined how "a PMO can facilitate the application of governance principals, in support of greater project success." He concluded that a PMO concept offers a chance to optimally use limited resources by advantageously "aligning project efforts with the greater organizational goals," and that such an organization allows close-fitting process controls for a higher success.

His literature review concludes the research on governance of projects, programs, and portfolios was immature. Thus there are many opportunities to study whether GPPM provides benefits; in particular, there is need to study the identification of the "overall strategic goals and weighing them against the perceived costs and benefits" of such an organization (Misner, p. 41).

In the results of their study on the impact of portfolio management, De Reyck & al. (2005, p. 524) presents "a strong correlation between (1) increasing adoption of PPM processes and a reduction in project related problems, and (2) between PPM adoption and project performance," especially when an organization performed governance at the portfolio level, which provided portfolio optimization. According to Hanford (2006), "in order to ensure that the proper results from an adoption of portfolio management are obtained, authority is delegated, entities or functions are created, and policies and practices are developed and implemented. Additionally, roles and responsibilities are defined and filled, and decision-making rules and powers are identified." This corresponds exactly to the concept of GPPM.

Hanford (2006) uses his experience and studies to conclude "portfolio management governance" should be established to ensure project. He lists these main components:

- "A set of governance principles
- A governance framework or structure
- Controls implementation that enables and sustains an oversight capability
- Definition of decision-making authority
- Ensuring the legitimate exercise of authority and decision-making
- Identification of needed enabling and sustaining functions, and governing roles and responsibilities"

Moreover the results achieved should provide rules, standards, structures, practices, expectations, and feedback for assessing effective practice and defining adjustments.

A process, based on a "new concept within the overall organizational governance effort" should be defined, approved and adopted (Hanford, 2006). For that reason an organization should have (1) a shared *vision* of governance, as well as of roles and responsibilities. Hanford adds that if this vision were documented, it could provide reference and guidance, for (2) "*visible leadership*" as "part of the organizational leadership group that governs the entire organization." Such buy in is not easy to gain! But it must be adopted by all parties. The effort undeniably demands "a larger work effort" requiring (3) a *structure* needed to determine and follow management processes to achieve organizational goals.

The GPPM is a continuing effort and investment for sustaining a company's performance. Performance is mainly contingent and contextual; thus to sustain performance, organizations must continuously improve. Performance in managing projects means not only improving practices but also training people in charge with implementation. People in charge of governance should also be trained and have a clear statement of roles and responsibilities.

The GPPM concept should be the subject of further research, mainly based on case studies in diverse sectors and industries, as GPPM is a significant topic for companies. Indeed governance then focuses on continuity, the ability to adapt and adjust, and to continue to evolve, all of which are needed to continue organizational processes (Hanford, 2006).

References

Association for Project Management (APM). (2004). *Directing change: A guide to governance of project management*. Association for Project Management, UK.

Binnendijk, A. (2000). *Results-based management in the development cooperation agencies: A review of experience*. OECD/DAC Working Party on Aid Evaluation. February.

Crawford, L., Turner, R., & Hobbs, B. (2005). Aligning capability with strategy: Categorizing projects to do the right projects and to do them right. *Project Management. Journal*, 37(2), 38-50.

Checkland, P. (1999). *Systems thinking, Systems Practice: Includes a 30-Year Retrospective.* New York: John Wiley & Sons Ltd.

De Reyck B., Grushka-Cockayne, Y., Lockett, M., Ricardo Calderini, S., Moura, M., & Sloper, A. (2005). The impact of project portfolio management on information technology projects. *International Journal of Project Management 23*, 524–537.

Engwall, M. & Jerbrant, A. (2003). The resource allocation syndrome: the prime challenge of multiproject management? *International Journal of Project Management*, *21*, 403-409.

Frey, T. & Buxmann, P. (2011). The importance of governance structures in IT project portfolio management. *ECIS 2011 Proceedings*. Paper 17. Retrieved from: http://aisel.aisnet.org/ecis2011/17

Gray, C. & Larson, E. (2006). *Project management: The managerial process.* Third Edition. New York: McGraw-Hill.

Hanford M.F. (2006). Establishing portfolio management governance: Key components, 16 October. Retrieved from: http://www.ibm.com/developerworks/rational/library/oct06/hanford/

Hajala, S. (2013). Project portfolio management governance. Retrieved from: http://www.cioindex.com/article/articleid/146214/project-portfolio-management-governance

Iovino P. (2005). Best practices in portfolio management—Portfolio prioritization & selection for business value maximisation. *UMT, Partner.* Retrieved from: nyspin.org/SpinPresentation20051011.pdf

Kaplan, R. S. & Norton, D. P. (2001). *The strategy-focused organization: How balanced scorecard companies thrive.* Boston, MA: Harvard Business Press.

Klakegg, O.J. (2010). *Governance of major public investment projects In pursuit of relevance and sustainability.* Doctoral theses. Norwegian University of Science and Technology.

Kusek, J.Z. &. Rist, R.C. (2001). Making M&E matter—Get the foundation right. *Evaluation Insights. 2(*2).

Lecoeuvre, L., Turner, J. R., & Patel, K. (2014). Is project marketing relevant to practitioners? 13th International Marketing Trends Conference, 23–25 January. Venice, Italy.

Lecoeuvre L., Turner J. R., & Patel K. (2012). Intergrating project marketing with project management. 12th *European Academy of Management Conference* (EURAM). 6–8 June, Rotterdam.

Levin, G., Artl, M., & Ward, J.L. (2010). Portfolio framework: a maturity model. Arlington, VA: ESI, International.

Miller, R., & Hobbs, B. (2005). Governance regimes for large complex projects. *Project Management Journal Research Quarterly, 36*(3), 8.

Misner J. (2008). The role of the project management office in a multi-project environment: Enhancing governance for increased project success rates. Retrieved from: https://scholarsbank.uoregon.edu/xmlui/bitstream/handle/1794/7663/2008-Misner.pdf?sequence=1

Morris, P.W.G. & Jamieson, A., (2008). Implementing strategy through programmes of projects. In Turner, J. R. (ed). *Gower Handbook of Project Management.* 4th edition. Aldershot, UK: Gower Publishing.

Müller, R. (2009). *Project governance*. Aldershot, UK: Gower Publishing.

Müller, R. & Lecoeuvre, L. (2014). Forthcoming. Operationalizing governance categories of projects. *International Journal of Project Management.*

Müller, R., Martinsuo, M., & Blomquist, T. (2008). Project portfolio control and portfolio management performance in different contexts. *Project Management Journal, 39*(3), 28–42.

Project Management Institute. (2006). *The standard for program management.* Newtown Square: Project Management Institute.

Rad, P. & Levin, G. (2006). *Project portfolio management tools & techniques.* New York: International Institute for Learning.

Turner, J. R. (2009). *The handbook of project-based management*. 3rd edition. New York: McGraw-Hill.

Walker, D. H. T., Segon, M., & Rowlinson, S. (2008). Business ethics and corporate citizenship. In Walker, D.H.T. & Rowlinson, S. (Eds.). *Procurement systems—A cross industry project management perspective*. Abdingdon, Oxon, UK: Taylor and Francis.

Wideman, M. (2005). Project portfolio governance guidelines. A review and commentary of a recent publication *Directing Change*, by the Association of Project Management, UK, 2004. *AEW Services, Vancouver, BC © 2005* Retrieved from: http://maxwideman.com/papers/governance/governance.pdf

Xue, Y, Turner, J. R., Lecoeuvre, L., & Anbari, F., (2013). Using results-based monitoring and evaluation to deliver results on key infrastructure projects in China", *Global Business Perspectives. 1*(2), 85–105.

7

Project Portfolio Governance

JENNIFER BAKER, PfMP MSPM,
PgMP, PMP, ITIL, MBB

Project portfolio governance involves the processes used to identify, select, prioritize and monitor projects within a company or line of business. The foundation of these processes should be the strategy and guiding principles of the company. With a firm foundation, ongoing governance and oversight are able to proceed down the path to arrive at the strategic destination. Project portfolio governance is a critical capability in the best of times. Project portfolio managers must navigate their projects toward achieving the "... overarching strategic objectives" (James & Ryan, p. 4).

A successful portfolio management strategy must comprise an end-to-end framework that methodically guides organizations from project selection through execution. Governance is a framework within which project/program decisions are made. Portfolio governance and organizational structures need to be thought through carefully to cater to the culture in which they are delivering. That culture should also demonstrate the governance expectations and provide input to the policies. While robust portfolio management is essential, "... a rigorous governance model is critical to help enforce accountability, optimize cross-functional alignment, and escalate issues to the appropriate decision makers" (PWC, p. 3). Strong governance also can help align communications, calendars, and strategies across business units (MacDougall, p. 2). A well-structured governance model provides:

- Accountability
  - Clear roles, responsibilities, and accountabilities
- Transparency
  - Clarity of stakeholders and financial authorities
  - Clear scope
  - Regular meeting schedules
- Integrity
  - Ethics

- Protection
  - Dispute and conflict resolution escalation channels
  - Empowerment of individuals to "do the right thing"
- Compliance
  - Clear procurement processes
  - Clear adherence to all regulatory, legal, and policy requirements
- Availability
  - Clear reporting and information flow
- Flexibility
  - Ability to adapt to changing business and organizational needs
  - Capability to accommodate shifts in organizational size and complexity
- Retention
  - Performance and benefits
  - Obvious delivery model overlays
- Disposition
  - Clear transition from project/program to operations

"The collective term 'governance, risk management and compliance' (GRC) has become a top business priority and an integrated approach to GRC is encouraged to reduce duplication of effort and cost." Figure 7.1 illustrates the holistic responsibility of corporate governance and how intertwined risk management and compliance truly are (Ho, p. 1). The governance framework must demonstrate this balance with

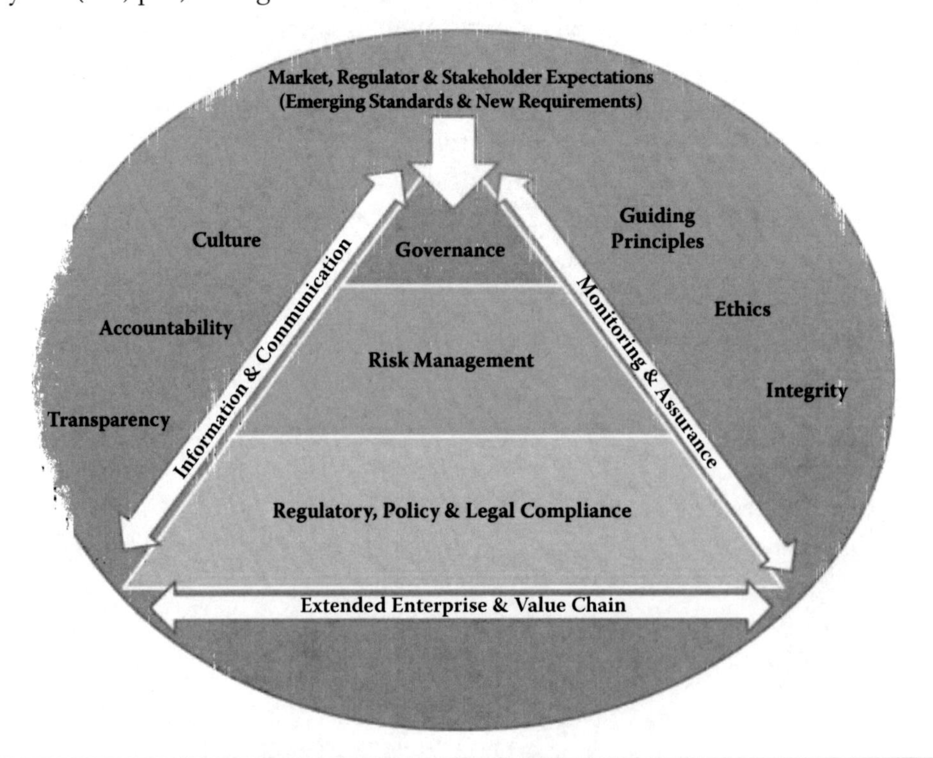

Figure 7.1 Organizational governance, risk management, and compliance (Significantly adapted. Ho, p. 1)

both accountability and transparency while providing risk and compliance intelligent governance.

Strategy and Guiding Principles

Project Management Offices (PMOs) that excel with the definition and execution of strategy have the ability to link the strategy of corporate entities to those of the larger corporate strategy. They articulate that strategy in a way that is not only operationally relevant but also is consistent with corporate culture. This allows portfolio management to make appropriate decisions regarding assets within their portfolio. The culture of the organization "... must reinforce all aspects of the project portfolio governance methodology" (James & Ryan, p. 10).

Successful PMOs start governance with project selection by ensuring that strategy alignment is in place before authorizing the project. Well-designed project/program acceptance criteria will impact, at a minimum, four disciplines: strategy, financial, risk, and technical (PWC, p. 7). Additional areas include time to implement as well as people and other resources. To ensure effective acceptance criteria for both projects and programs, portfolio leaders should consider the following:

- Alignment with corporate strategy or being an enabler to deliver corporate strategy (see Figure 7.3 below)
- Financial parameters and milestones (including appropriate KPIs to measure these)
- Risk parameters including the monetization of risk contingency and adequate management reserve (Jordan, p. 18)
- Structural architectural changes including the technology roadmap (Jordan, p. 302) (see Figure 7.2 below)
- Appropriate weighting for each of these considerations by stakeholder tolerances (see Figure 7.3 below)

Selecting projects with just a few obvious inputs or simply choosing projects because they are "... the squeakiest wheel..." is not the best technique to choose a portfolio component. These strategies may work sometimes, especially "... when tackling the low-hanging fruit..." but a more structured approach is required when the "... priorities are not so obvious" (Myers, p. 2). One of the biggest decisions that any

Architecture	Scope	Detail	Impact	Audience
Enterprise	Organization	Low	Strategic	Corporation
Segment	Division	Medium	Business	Business Owners
Component	Functional	High	Functional	Users

Figure 7.2 Architecture attributes including scope and impact within an organization.

organization would have to make is related to the project investments it undertakes. Once a proposal has been received, there are numerous factors that need to be considered before an organization decides to pursue a project. The most viable options need to be chosen, keeping in mind the strategic goals and requirements of the organization. Some additional variables need to be considered for each of the four disciplines: strategy, finance, risk, and technology. (James & Ryan, p. 6).

- Benefits and value: A measure of the positive outcomes of the project. These are often described as "the reasons why you are undertaking the project". This should include the enterprise and business value vs. spend ratio.
- Feasibility: A measure of the likelihood of the project being a success, i.e. achieving its objectives. Projects vary greatly in complexity and risk. By considering feasibility when selecting projects it means the easiest projects with the greatest benefits are given priority (Jordan, p. 53).
- Solution clarity: Is the solution already known? Does it support the current strategic goals? Does it align with the organizational culture and technology roadmap? (FEA-PMO, p. 8).
- Availability: Are there organizational resources (funding, staff, and equipment) available to complete the project objectives? How will this project compete with other higher priority projects?
- Stakeholders: Will the successful outcome of the project have a material impact on customers' (internal or external) perceptions of quality or performance?

These factors should be considered in a matrix that includes, at a minimum, strategy, finance, risk, and technology considerations that are unique to each business. Some of the most successful companies have developed robust project selection requirements with multiple reviews to ensure that the right projects are being done at the right time by the right resources for the right reasons (PM Solutions, p. 3). Strategy, finance, risk, and technology components as well as several other criteria are illustrated in Figure 7.3 below.

It is critical that organizations employ standardized financial targets as part of business-case reviews. Research shows that financial discipline at the beginning of the portfolio-acceptance process pays off in the long term. "Organizations with best-in-class project management review the potential for revenue or return 35% more than others before accepting or approving new projects. And when compared with other firms, companies that get portfolio management right tend to assess past performance as an indicator of future efficiency" (PWC, p. 8). The process selected should provide the organization with value. While the project itself may not have a high rate of return, it may be the enabler for other projects that will. There are a significant number of factors that should be considered and weighed in a manner that is most beneficial to the organization to achieve its strategy. With each project selection, the question needs to be asked, would this decision help me to increase organizational value?

Project Name	Business Value (Unweighted Score)	Ability to Execute (Unweighted Score)	Cumulative Score (40% Ability to Execute, 60% Business Value)	Investment Quadrant	Primary Financial Measures	BU Value	Enterprise Value	Readiness	Technical Characteristics	Net Present Value (NPV)	Alignment to BU Strategy	Improvement to BU Operations and Service	Operational Performance	Retain & Attract Talent	Sustainability	Regulatory & Legislative	Financial strength	Growth	Sponsorship	Business Resources & Skills	IT Resources & Skills	Change Enablement Impact	Project Rank	Endorsement Status 7/18	Endorsement Status 10/09	Same Status?	Critical?
Business objects replacement	5.2	3.8	4.6	Consider	1.0	9.1	7.0	4.5	3.0	1	10	7	7	1	1	10	3	1	3	7	7	1	3	Not endorsed	Endorsed	False	Y
Oracle upgrade	5.2	9.6	7.0	Go Do	1.0	9.1	7.0	9.3	10.0	1	10	7	7	1	1	10	3	1	7	10	10	10	10	Endorsed	Endorsed	True	Y
CorpTax system upgrade and rollover	5.8	9.1	7.1	Go Do	3.0	7.9	7.5	8.3	10.0	3	10	3	10	1	1	10	1	1	3	10	10	10	10	Endorsed	Endorsed	True	Y
Enterprise digital signage	3.9	8.1	5.6	Question	1.0	9.1	2.4	9.3	7.0	1	10	7	3	3	3	1	0	1	10	7	10	10	7	Endorsed	Endorsed	True	Y
Intranet enhancements	4.1	6.5	5.1	Question	3.0	5.8	3.9	6.0	7.0	3	7	3	3	1	1	7	1	1	3	7	7	7	7	Not Endorsed	Endorsed	False	Y
SAN replacement	4.6	7.6	5.8	Question	3.0	7.3	4.0	8.3	7.0	3	10	1	1	1	1	10	0	0	3	10	10	10	7	Endorsed	Endorsed	True	Y
Implement regulatory process enhancements	5.5	7.0	6.1	Go Do	1.0	10.0	7.0	7.0	7.0	1	10	10	7	1	1	7	7	1	7	7	7	7	7	Not Endorsed	Endorsed	False	N
Implement HR application reader software	7.1	8.9	7.8	Go Do	7.0	9.1	5.3	7.8	10.0	7	10	7	7	1	1	7	1	1	7	10	7	7	10	Not Endorsed	Endorsed	False	N
Internal email tracking & metrics	4.0	7.8	5.5	Question	3.0	7.0	2.4	8.5	7.0	3	7	7	3	3	3	1	1	1	10	7	7	10	7	Endorsed	Endorsed	True	N
Internal social media platform	3.1	7.4	4.8	Question	1.0	5.8	3.2	7.8	7.0	1	7	3	3	7	1	0	1	0	10	7	7	7	7	Not Endorsed	Endorsed	False	Y

Figure 7.3 Sample project selection criteria worksheet for an information technology (IT) PMO. This sample shows the business unit value, readiness, engagement and strategic alignment as well as other critical factors like resource skills and availability.

The rules around governance should lay out the framework for ownership of projects and portfolios, defining the steering committee structures and roles, and assigning responsibilities. Clarity about governance will be a critical first step to addressing decision rights and will provide a foundation for further improvements. "The policies to drive strategy should be concise and unambiguous." (James & Ryan, p. 6) They should set expectations at each stage of the project life cycle, effectively removing the element of surprise from project reviews for any project manager. The policies should contain a rule set, which defines where items are non-negotiable and where latitude is provided. These rules balance the need for local customization with guidelines that ensure adherence. Additionally, the portfolio manager should consider the portfolio roadmap in the processes and methodology surrounding the portfolio governance framework (PM Solutions, p. 3). See Figure 7.4 below showing the governance life cycle:

An example to demonstrate the concepts shown in Figure 7.4 would be the many regulated companies that complete projects in order to meet new or changing regulatory and/or legal requirements. By doing this, portfolio managers have selected the right projects to avoid regulatory fines and reputational risk. This is done most of the time as a project to ensure that the proper attention is paid to the work being completed. It also provides auditors the opportunity to review what was done and how it was done by reviewing project artifacts. Those artifacts should embody the standards for how the projects should be executed to provide value to the organization. This is executing projects the right way. The systems or "checks and balances" that are implemented to ensure compliance for the regulatory changes are included in the project selection would be the last arc in the circle. The core of this philosophy should be embedded within the culture of the organization.

The governance framework should consist of definitions, which guide strategic alignment, financial adherence, risk appetite, and resource requirements. Project

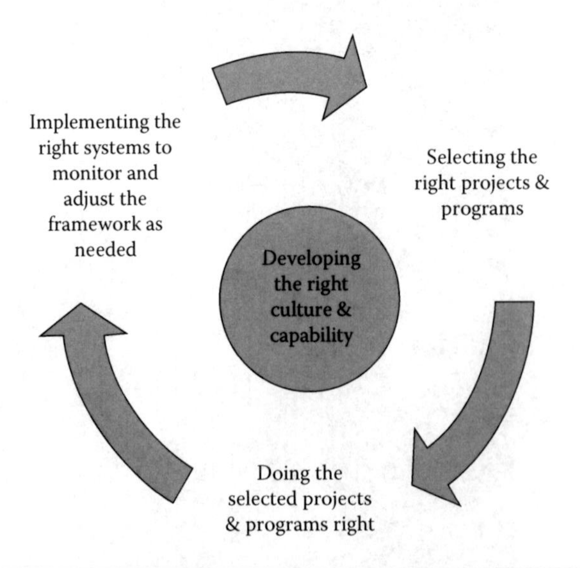

Figure 7.4 The "Right" Governance Life Cycle with Core Culture and Capability

scope should clearly articulate how investments should support the top business unit priorities. Financial direction should provide clear objectives in terms of investment performance (PM Solutions, p. 4). Stakeholder risk appetite should govern investment and management decisions. These principles should also provide clear expectations for maximizing capacity given scarce resources including financial, human capital, infrastructure, and partner resources.

Strategic shifts or changes need to be reflected within the governance model. The roadmap will be adjusted, which can have substantial downstream impact and trigger the need for reprioritization. According to the Project Management Institute (PMI) survey respondents, change in organizational priorities was the most frequently cited reason for project failure (PMI, p. 3). This research was echoed by PM Solutions showing that those PMOs that were most successful were able to consistently align organizational strategy (PM Solutions, p. 2). An example of shifting strategy could come from internal sources of change such as a merger announcement, a senior leadership change, or external sources of change such as a significant legal or regulatory shift or a change in a political party during elections. Portfolio leadership must be able to accommodate changes and shifting priorities in order to remain successful. "Guiding principles, combined with a comprehensive strategy, effectively shape all other elements of project portfolio governance. The decision-making frameworks, monitoring processes and, ultimately, culture all flow from these principles" (James & Ryan, p. 6).

Structured Processes and Methodology

A methodology is a set of rules associated within a given area of study or discipline. While a methodology does not provide solutions, it offers "best practices" for the work conducted—in this case a project or program. The methodology should provide a foundation with rules and policies but also provide flexibility to effectively manage that effort. Examples can be seen in multiple places and industries, and many times symbolize quality expectations such as the International Organization for Standardization (ISO) 9000 in quality management or board certifications within the medical profession. Another very relevant example is the rules, standards and principles set forth by the PMI for project, program and portfolio management practitioners where certification sets a standard and expectations. Having a systematic methodology to follow allows practitioners the opportunity to become more proficient, reliable, and effective as time passes. "According to Aberdeen Research, effective portfolio management can enable companies to achieve up to 25% more revenue from new products when compared with less successful competitors" (PWC, p. 3).

The rules around governance should lay out the framework for ownership of projects and portfolios, defining the steering committee structure, and determining roles and assigning responsibilities. They should define who is authorized to initiate, to continue, to amend and to stop projects as well as stipulate who sets standards and who controls the overall investment budget (PWC, p. 3). These rules need to be embodied in a set of governing forums, decision processes and tools. Process maturity has been found to yield "... more predictable project performance and lower direct project management costs" (MacDougall, p. 2).

Rules should also provide a basis for acceptable project performance with the rules that govern project milestones, deliverables, and reporting. These rules should define the way to monitor the progress of projects and introduce flexible options to the typical life cycle in order to accommodate different kinds of projects. Those rules should also define "... consequences for breaking rules or poor performance." (James & Ryan, p. 10) The portfolio manager must be able to influence the behavior of project managers. Rewards and penalties need to tie in with the governance framework and rules. For some companies, this risk and reward concept is implemented for the practitioners by inclusion in their annual performance appraisals. For example, one financial institution tied benefits attainment to PMO performance appraisals by tying the percentage of benefits attainment received to the percentage of targeted annual income adjustment. Another example is a utility that tied project team performance appraisals to safety incidents for its major projects.

Risk methodologies should also be well defined. The PMO should deploy decision tools to help evaluate the risk-adjusted financial performance of projects. PMOs utilizing best practices are starting to use simulation tools to quantify project uncertainty by generating a range of possible economic scenarios to estimate the distribution of project values (Jordan, p. 30). An example of one such tool is Monte Carlo analysis, which provides the probability of outcomes for the scenarios that are input into the tool. It is used in many industries and in government to provide the most and least likely path as well as all of the variations in between. While this particular type of model is used in many different types of portfolio analysis, it is particularly useful in the risk management area. These types of models evaluate the often-overlooked embedded value in projects by showing the probability of both risk and opportunity. One familiar example of this type of tool can be seen each summer on the weather forecast. Whenever a hurricane has formed, a simulation tool plots the most likely paths the storm will take based on the factors input into the tool. In this case that would be temperature, tides, etc. These simulation tools should have flexibility for usage with smaller projects but be robust enough to handle high risk and complexity. This is where benefits-realization success metrics prove their value. Enterprise Program Management Offices (EPMO) and PMOs that continuously track and manage the progress of benefits realization can more easily identify poorly performing projects early on to facilitate the process of managing risk and by doing so the EPMO/PMO can stop projects and programs before wasting additional resources (PWC, p. 9).

"Beyond governance, an unwavering financial discipline and regular reviews of portfolio performance throughout the entire process are necessary to guide informed decisions. This level of discipline will demand standardized key performance indicators (KPIs) and powerful analytics to deliver objective insight for proactive decisions" (PWC, p. 3). Also essential is a benefits-realization process that enables organizations to ensure projects yield the expected benefits and value so underperforming projects can be stopped early. The process should allow for both financial benefits and strategic value to be measured and monitored.

Portfolio performance itself is the key indicator for effective portfolio governance. With the utilization of portfolio performance tools, the portfolio manager can determine if governance and oversight is effective and identify opportunities to provide further guidance and support (James, p. 9). Some of these tools should include project/program reviews and portfolio performance dashboards. Project reviews are utilized to identify and help rescue important projects that may be in trouble. These types of reviews may also be scheduled for larger more complex efforts to ensure that they start out and remain on the projected course. A careful assessment can help correct or stop underperforming—and ineffective—projects early.

"The portfolio dashboard should be used to diagnose systematic problems that need to be addressed: e.g., problems with procurement practices, vendor performance, financial management and quality of testing. This approach to portfolio monitoring maximizes the chances of success for projects by providing the best possible development environment (e.g., for procurement, system build and testing)" (James & Ryan, p. 9). While the components of the portfolio dashboard vary from company to company, the standard KPIs typically include costs, benefits and value measured against milestones and expectations. See a simplistic version of a portfolio dashboard shown in Figure 7.5.

These KPIs should also be tracked over time for trends. A trend analysis chart is a report type that you can use to display the performance of one or more metrics over time. For example, you might use a trend analysis chart to display the historical performance of a key performance indicator (KPI) and to estimate future results of

Scope	Schedule	Cost	Quality	Risk	Overall
591 Green 6 Yellow 9 Red	539 Green 37 Yellow 31 Red	571 Green 19 Yellow 16 Red	80% Green 15% Yellow 5% Red	60% Green 30% Yellow 10% Red	531 Green 43 Yellow 43 Red
Average number of change controls per project = 1.03	Variance ranges High 1059% Low −57% Avg variance 3.68%	Variance ranges High 99% Low −158% Avg variance 8%	26 Projects entered testing 64 Projects left testing.	350 Escalated risks across 149 projects	617 Total Projects Total investment $1.3 Billion
All scope variances have pending change controls which also effects cost and schedule.	All projects that are red for schedule have pending change controls	Most of the projects that are over budget are due to changes in rates. The remainder are due to pending change controls.	All projects that are yellow and red for quality are pending retesting.	Escalated risks for 16 projects are for potential regulatory changes that impact project scope, schedule & cost.	23 of the escalated projects have no pending issues. All 23 have pending change controls.

Figure 7.5 Sample portfolio dashboard showing both actual count and averages for the selected KPI indicators.

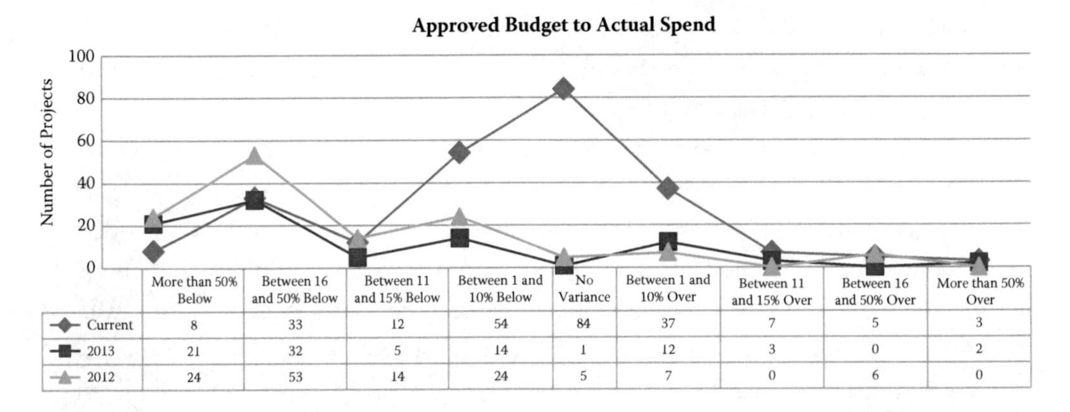

	More than 50% Below	Between 16 and 50% Below	Between 11 and 15% Below	Between 1 and 10% Below	No Variance	Between 1 and 10% Over	Between 11 and 15% Over	Between 16 and 50% Over	More than 50% Over
Current	8	33	12	54	84	37	7	5	3
2013	21	32	5	14	1	12	3	0	2
2012	24	53	14	24	5	7	0	6	0

Figure 7.6 Sample KPI element showing variance comparison year over year for actual spend to approved budget in banded categories.

that KPI. Because KPIs exist as elements in dashboards, a trend analysis chart must use dashboards as its data source. (Wise, p. 1) You can select more than one KPI for your report, but a trend analysis chart can display only one KPI—see Figure 7.6 and Figure 7.7. In Figure 7.6, the portfolio manager can see that there has been a historical trend for being significantly under budget. This would lead to some conclusions that the estimates provided for the project are using the maximum end of ranges for all items or that they are "padded" to ensure that the project comes in under budget. The increase in the percentage of projects that are at or within a 10% range in the current year illustrates that some method of corrective action has been taken to address the variance ranges. In Figure 7.7, the portfolio manager would see that projects generally start out the year well then to start to have problems in the second quarter, which peak in the beginning of the fourth quarter. By the end of the year, they appear to sort out their issues. This would indicate cyclical patterns in the organization that impact the project life cycle. The following are two suggestions to consider for completing KPI trend charts. First is to annotate the graph with any organizational events such as "Reorganization" or "Merger Announcement" so it is easier to explain any anomalies that are detected in the data. Second, do not aggregate the data if at all possible. In the aggregate, trends can hide insight and hence "dirty" the data. If you want to compare "clean" trends then your best option is to compare different segments within your data.

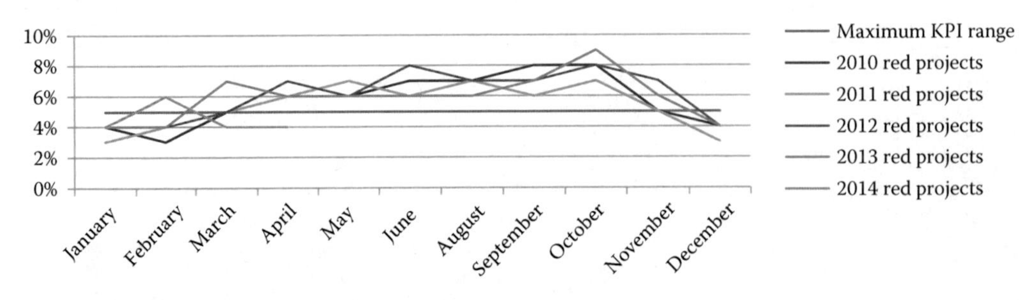

Figure 7.7 Sample comparison of KPI element showing the number of projects with red health status compared year over year as compared to the maximum range set by the PMO.

The benefits of standardized KPIs include improved consistency across projects, better managed expectations, and closer alignment of goals and accountabilities. Metrics can help ensure that appropriate levels of consistency and objectivity are applied when making decisions about an organization's portfolio. "KPIs tend to be fewer in number compared to their metric counterparts, but are much more targeted and related to the vision and goals of the company" (Wise, p. 1). Standardized, simplified KPIs also provide proactive insight into underperforming projects that enable management to take action during implementation (James & Ryan, p. 8). Because they are standardized, they can help PMO managers articulate to senior leadership the "big picture" rationale for portfolio decisions. Some examples here are shown in both Figures 7.6 and 7.7. For Figure 7.6, projects are expected to be completed within ±10% of their approved budget. For Figure 7.7, no more than 5% of projects should be in a red state at any given time. For both of these examples, it is easy to see where projects fall outside of expectations. In the portfolio dashboard, such as the one shown in Figure 7.5, those projects which fall outside of the standard can have explanations such as the one for project risk, which states there are escalated risks because of potential regulatory changes for 16 projects.

Many companies find that aligning metrics and standardize how they are calculated is a daunting task. Often, they do not agree on which KPIs to measure in each area of performance, and when they do concur, many organizations cannot reach consensus on what the thresholds should be for each one. "Research demonstrates that investing adequate time upfront to standardize KPIs will bring substantial benefits later. For instance, Aberdeen has found that 67% of best-in-class PMOs standardize performance metrics to assess portfolio health and value, compared with an industry average of 39%" (PWC, p. 8). One bit of guidance here is to look at the strategy and guiding principles of the company. That is a point of agreement from which the building can begin. Each group may have different interpretations of how to achieve those strategies but that basis of understanding should help determine which metrics need to be captured and how that transcends into a KPI that can be trended over time to show alignment and achievement of strategy. For example, a construction company or a utility wants to be a safe place to work and that is one of its guiding principles. These companies would capture metrics for different types of work accidents but would also capture information about "near misses" or potential accidents. They would also provide training based on the metrics captured to educate their staff in order to prevent incidents from occurring where there were an escalated number of "near misses". This behavior would support their guiding principle to be a safe place to work.

While this framework provides a good idea of what a comprehensive project management strategy might entail, every organization has different needs and objectives. Consequently, each will need a customized set of tools, processes, and methodologies to help realize the most effective portfolio. This framework must help control portfolio costs and ensure that strategy, execution, and results are properly aligned. In summary, utilizing structured processes and methodologies will enable portfolio governance capability. These processes should include and provide:

- Establishment of the process of project portfolio governance, the definition of roles and responsibilities associated with it, and its policies and procedures (James & Ryan, p. 6).
- Evaluation of project proposals to select those that provides the best investment of time and money for the resources available.
- Empowerment of individuals to increase decision efficiency and employee engagement (James & Ryan, p. 7).
- Definition of the desired outcome and value of proposed projects, the anticipated benefits, and implementation and ongoing costs.
- Control of the scope of projects as well as changes that are requested of projects.
- Monitor the progress of projects that have been approved and initiated ensuring they meet the defined goals and objectives.
- Measure actual effort and costs of project tasks, deliverables, and milestones against the planned values. This should include post-project reviews with a mechanism to implement lessons learned (James & Ryan, p. 11).
- Action to put projects "on track" as project measures demonstrate the need including the escalation of issues that cannot be mitigated by the project management team and the PMO (James & Ryan, p. 8).
- Improve the project delivery capability on a continual basis allowing it to complete projects in less time and for less cost while generating the maximum value.
- The culture of the organization must be embedded from the project managers through senior leadership demonstrating expected behaviors. Attention should be given to rewards and publicize examples of good behavior and not just penalizing poor behavior (James & Ryan, p. 10).
- Monitor project risks and contingency to know what risks are triggered and if that realization was as expected (Jordan, p. 18).
- "A portfolio-level performance dashboard that provides an accurate aggregated performance picture at different stages..." including measurement for strategic alignment and benefits attainment (James & Ryan, p. 8).

Stage Gates and Phase Gates

The intent of the gating process is to determine how the project is progressing. Gates provide various points during the process where an assessment of the quality of a project, program or initiative is undertaken. A gate meeting generally yields these decisions: go, kill, hold, recycle, or a conditional go. It includes three main components:

- Quality of execution: Are the deliverables being met as planned? Have risks been realized? What are the pending issues and risks? (Jordan, p. 18)
- Business rationale/justification: Are the KPIs being achieved as defined? Are benefits being obtained? Are there fewer or additional benefits? (PWC, p. 8)

- Action plan or next steps: Should this effort maintain the current prioritization? What is the approved action plan for the next gate, and a list of deliverables and date for the next gate? Are there any limitations?

The gating process should adapt to different project methodologies and life cycles. PMOs should also consider reducing the overall number of gates as much as prudently possible. Consider that the sphere of influence for the outcome of any effort is generally prior to execution. Figure 7.8 shows as the project or program progresses through its life cycle, the level of influence to modify the path of the effort exponentially declines. The life cycle pace for how this happens may alter greatly from one type of project to another. For example, consider if a web development project should utilize the same process and pace as a construction project or a project which is highly regulated such as in the healthcare, banking or energy fields. "... Projects using an Agile based project methodology might have fewer decision gates..." and documentation requirements than "...projects using the standard waterfall methodology..." whereas projects that have substantial regulatory requirements may need additional gating or checks between gates (James & Ryan, p. 7). Another consideration that should be discussed is for organizations that frequently work together, but have different methodologies and procedures, is the implementation of an interface or working agreement. An example is two separate PMOs within an organization or sister companies. The use of an interface agreement may allow projects meeting the parameters set forth in the agreement to go through only one set of stage gates rather than struggle with the duplicity.

Some questions that the portfolio manager should ask regarding the gating process:

- Do the gating rules include stipulations or limitations that are specific to when building or coding can begin?
- Do the gating rules include both guidance and procedures when a prototype or proof of concept is needed?

Think about a construction project where zoning/re-zoning or building permits may be required prior to construction. The "project build gate" or "building permit" often represents an onerous first milestone for projects. In order to obtain investment

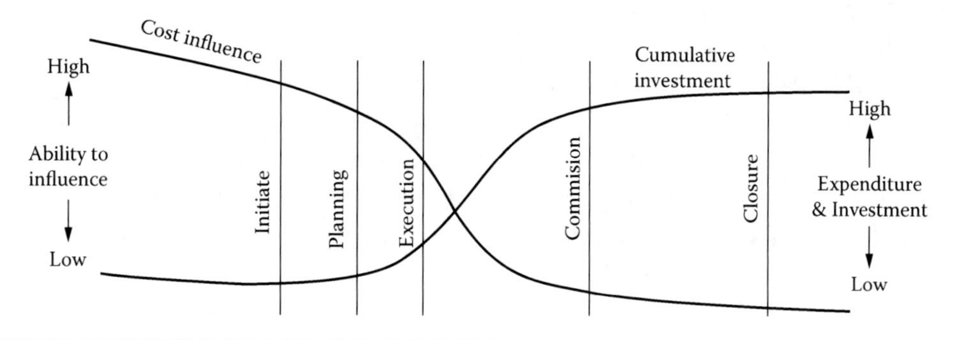

Figure 7.8 Influence curve. This figure demonstrates how the ability to influence a project's outcome diminishes as the cumulative investment increases throughout the project life cycle.

dollars, a project typically needs to show a detailed plan and refined business case. The opportunity cost for developing this material can be exceedingly high, so high, in fact, that it can discourage healthy risk-taking and innovation. An additional requirement may include a "zoning permit", which comes before the "building permit" to provide the project with the seed funding required to explore a proof of concept or prototype in order to warrant proceeding to the "building permit" stage (James & Ryan, p. 8).

Effective stage-gating authorizations require timely and concise communication between the sponsor, the finance group, the project team, and the often overlooked operations and maintenance partners to whom the asset will be turned over to at the end of the project. Part of governance procedures should include the concept of stakeholder and change management to ensure that these items have the proper visibility from the organization down through the projects.

The requirements for each stage gate should be clearly identified and explained. The project manager and project team should understand what is expected in order to be able to provide the information as defined with the appropriate level of detail for the life cycle of the project. Table 7.1 describes some of the concepts and questions that should be considered and understood as the premise for the stage gate as well as samples of documentation types that are typically required.

Governance Model Communication and Coordination

The portfolio needs to use tools and processes that are clearly understood and easily accessible. They should be easy to use—intuitive for both the teams providing the input and the sponsors obtaining the output. The lack of appropriate and effective tools to oversee and to manage project portfolios is a key contributing factor of project underperformance. For example, a report from Oliver Wyman found that "... most banks lack effective project governance processes, policies and monitoring mechanisms. Senior managers in charge of... hundreds of millions of investment dollars are often flying blind" (James & Ryan, p. 3). The reasons for this are varied—from lack of engagement to lack of understanding. Many times the leadership and management know what is needed to be tracked but are not sure how to measure it. They need tools to be designed to provide output that helps in the decision-making process. The tools should be conducive for the type of tools and process.

Leadership paired with governance and oversight can take many forms within an organization. For example, an EPMO aligns strategically with the organization and provides holistic management over multiple PMOs (both program and project management offices). In this capacity, an EPMO should have the ability to collect, analyze, and display data in a manner that enables executives to see at a glance how their projects, programs, and portfolios are running. A PMO is aligned with the organization with the purpose of providing management over multiple projects. (Hawald, p. 2) Similar to the EPMO, the PMO should have the ability to collect, analyze, and display data, but the data are project data and thus should be at a more specific and

Table 7.1 Stage Gate Considerations

This table shows questions that each stage gate should consider and answer with guidance for appropriate documentation requirements.

	INITIATE	PLANNING	EXECUTE	COMMISSION	CLOSING
Question for Gate Review	Does the idea merit moving forward?	Does the plan justify extensive investigation and execution investment?	Is the product, service, or asset ready to be commissioned?	Is the product, service, or asset ready to transition to normal operations?	How did the project perform based on projections?
Documentation Requirements	Documentation should include the initial project charter and high-level business case. Preliminary market or engineering assessments may also be required.	Documentation should include the detailed business case, quantifiable risks, milestone schedule, and value proposition. It may also include technical feasibility assessments and prototype requests.	Documentation may include a detailed implementation plan, change management plan, test plans, risk contingency, open issue log, etc.	Documentation may include validation steps and turnover and transition documentation.	Documentation may include signed accepted deliverables, lessons learned, and financial closure documentation.
Concept	Quick, inexpensive preliminary investigation and scoping of the project—largely desk research. The need for prototype development should have preliminary requirements.	Detailed investigation involving primary research—both market and technical—leading to a business case, including product and project definition, project justification, and the proposed plan for development.	The actual detailed design and development of the new product or service including the design and testing of the operations or production process required for eventual full-scale production.	Commercialization—beginning of full-scale operations or production, marketing, use, or selling of product or service.	Assessment of project or program execution performance including closure of finances and contracts.

granular level. A PMO provides insight in the schedule, budget, and risks of each project. Think of the EPMO as being the overarching umbrella that governs multiple PMOs, which in turn provide oversight over numerous projects. Another concept with increasing visibility is that of a Project Management Center of Excellence. These organizations typically are enterprise-level groups that implement and monitor enterprise project management standards. Their role frequently encompasses some level of training to share best practices and industry benchmarks throughout the organization. Typically, they are considered to be the governance and oversight leader of project management within the organization. In addition to the governance role, they sometimes introduce continuous improvement initiatives to bring strategic value as one of the company's organizational thought leaders.

The guiding principle of the communication policy should be to support the organizational strategy and the stakeholders; it ". . . is sometimes only as good as the principles that govern its application" (Needs, p. 1). It should be grounded in transparency, fairness, and reliability. One of the most important—and often most challenging—processes in PM governance is the escalation process. To escalate and resolve issues and mitigate risks in a timely and efficient manner, portfolio leaders must develop an easy-to-follow escalation process. They must also ensure that all stakeholders are aware of this process and agree to follow it. The tools and processes should provide a clear and understandable mechanism to escalate as needed. Organizations must clearly identify stakeholders who have responsibility for making decisions and define the escalation path for each type of issue and risk (James & Ryan, p. 10). An effective issue-escalation process also requires that the organization identify escalation dependencies such as tools, processes, and people who must be involved. Predictability and repeatability should be incorporated into the process so issues are resolved within a specific time frame with proper resolution communications. It should also include a feedback mechanism to sustain and adapt governance by providing the ability to identify and respond to new business, operational, competitive, and regulatory needs within the organization. An example here in its most simple form could include escalation check boxes within an issue log that feeds into a PMO or program issue log. Everything with a checkbox would "show up" on the escalated list for review by project leadership—whether it is the program management team or the portfolio manager.

This process has to include project, program, and portfolio sponsorship as applicable. An escalation process ensures that the next level of management is informed if an issue cannot be resolved at the lower level. Scope issues, major tradeoff decisions, and serious resource conflicts are all examples of situations that can require escalation to higher management (James & Ryan, p. 10). Regular stakeholder engagement with escalations will limit surprises. The escalation engagement level can be as simple as a regular review meeting for project needs such as escalated issues, risks, and change control approvals. A guideline with examples as to how a team can use "escalation processes" to raise project needs to higher authorities for timely resolution should be clearly outlined and readily available. Each PMO may define a separate process for escalation that works within its culture and structure. Examples vary from company to company and group to group. Groups with well-defined portfolio management tools may have ways within their toolset to "flag" risks or issues that need to be escalated. Another way to facilitate this process is to provide a decision matrix for project and program managers to use, such as the one shown in Table 7.2.

With escalation comes accountability. Has the organization's leadership created and encouraged a culture of accountability? Portfolio leadership also must create and nurture that culture of accountability (James & Ryan, p. 4). It is critical that objectives are aligned with the organization's performance goals and that portfolio success criteria are based on business value. Accountability must empower the governance steering committee with the authority to make decisions and hold people accountable. It is also important to ensure that position-based roles and responsibilities sufficiently encourage personal accountability. At the same time, incentive structures—return

Table 7.2 Decision Matrix

This decision matrix assists the project/program leadership with which decisions are appropriate to be made at their level and which decisions must be escalated and to whom.

	PROJECT TEAM	PROGRAM MANAGEMENT TEAM OR PMO	PROGRAM STEERING COMMITTEE OR PMO	SPONSOR	SENIOR/ EXECUTIVE LEADERSHIP
WHAT CAN PROJECT TEAMS DECIDE?					
Decisions that align with current, approved scope	Decide	Endorse	N/A	N/A	N/A
Decisions that align with corporate processes that do not impact items listed below	Decide	Endorse	N/A	N/A	N/A
WHAT MUST PROJECT TEAMS ESCALATE?					
Decisions that impact other projects or initiatives within the program or portfolio	Recommend	Recommend	Endorse or Decide	Decide	N/A
Decisions that impact schedule such as major milestones or critical path (positive or negative)	Recommend	Recommend	Recommend or Decide	Recommend or Decide	Decide
Decisions that impact scope other than what is approved	Recommend	Recommend	Recommend or Decide	Recommend or Decide	Decide
Decisions that impact regulatory, safety, or compliance	Recommend	Recommend	Recommend or Decide	Recommend or Decide	Decide
Decisions that impact the business beyond the original project/program management plan boundaries	Recommend	Recommend	Recommend or Decide	Recommend or Decide	Decide
Decisions that impact budget	Recommend	Recommend	Recommend or Decide	Recommend or Decide	Decide

on investment (ROI) and percentage of projects on budget, for example—must be properly aligned with overall portfolio goals.

The underlying need here is clear communication across the organization. Some questions to consider include:

- Are the communication channels well defined?
- Is there a strategy in place that is well understood throughout the organization?
- Are all of the stakeholder communication needs met?

To ensure that portfolio accountability is embedded in the corporate culture, portfolio leaders should develop an enterprise-wide communications plan that promotes the performance and business value of portfolio management. Communication is critical because the scope of most portfolios is distributed throughout the organization and is often global. Consequently, it is imperative that communications receive due consideration in portfolio governance planning. An effective communications strategy must consistently articulate portfolio management goals, mission, and vision. Communications plans, while centralized, should be customized for disparate stakeholder audiences and communicated via the appropriate channels. The communication strategy needs to include calendar alignment and process automation. Aligning calendars and timing across business units is a complex exercise that must be carefully planned. Portfolio leaders should identify and enforce the cross-functional principles that guide when, where, and how groups work together. EPMOs should create and practice calendar-alignment exercises across business units. At the same time, project leadership must also ensure that the business requirements are in synch with those of technical resources (IT or engineering). Successful execution will require cross-functional service-level agreements (SLAs) that take into consideration multiple hand-off points and dependencies (PWC, p. 7).

Effective governance requires automation of processes. Many times, significant efforts are made to develop, negotiate, and settle on a governance process, but the process only exists on paper and not in practice. “The policies must be conveyed automatically through standard PM communications tools such as a regularly updated FAQ policy portal” (PWC, p. 7). An investment in IT solutions may not be necessary because these processes can be automated using existing workflow platforms such as Microsoft SharePoint, which enables a simplified mechanism to providing and sharing information. Something to consider—Are you using these tools to ensure that there is both coordination and comprehension across portfolios and business leaders? Some minor adjustments may permit utilization of an existing functional tool with nominal additional investment. Cross-functional coordination is best achieved by establishing strong portfolio bylaws and a charter that governs portfolio management among disparate business functions and teams. The first step is to assess the maturity of the EPMO processes throughout the enterprise. Once a baseline is established, portfolio leaders should perform gap assessments and understand the desired future state. By doing this, they will have the critical information needed to begin standardizing the PM processes across all functions. Cross-functional coordination typically presents certain challenges. For instance, organizations often find it difficult to manage differing levels of maturity in portfolio management processes across the organization. When multiple programs from disparate groups are merged into a single EPMO, the challenges to coordinate resources can now be centrally coordinated and managed. These are critical issues because a lack of cross-functional coordination frequently results in duplicity for both management and maintenance of systems, people, and processes (PWC, p. 7).

Portfolio Performance and Diversification

Just as in any investment portfolio, there are some contributing factors for success. These typically include financial discipline and diversification. Some questions a portfolio manager should be asking about governance include asking how the governance and financial disciplines can improve portfolio performance. While there is no single solution, many organizations realize they can improve performance, curtail costs, reduce risk, and earn a greater ROI through better portfolio management (James & Ryan, p. 182). A successful portfolio management strategy must embrace an end-to-end framework that shepherds an organization from project selection to execution. For example, risk appetite is a key element that should be considered from when project selection occurs through the completion of the project. The portfolio needs to have risk included as a portfolio management governance component and provide tools and escalation within its framework to minimize risk impact across the portfolio. "The payoff can be significant... Aberdeen Research found that companies that excel at portfolio management typically complete projects on time and under budget while increasing ROI... In fact, an effective portfolio management program can enable companies to achieve up to 25% more revenue from new products when compared with less successful competitors, according to Aberdeen. These 'best in class' organizations typically improve project ROI by as much as 28%" (PWC, p.3).

Some questions to consider regarding the manner in which governance aids in project selection and prioritization:

- Is there adequate diversity within the portfolio?
- Is the taxonomy of project types appropriately defined?
- Do these categorizations accommodate risk profiles and ROI?

"There should be considerable portfolio balancing and optimization that is done on a regular basis which is determined by resource availability, cash flows of the organization, organizational risk and environmental changes" (MacDougall, p. 4). The governance process needs to accommodate and facilitate the portfolio prioritization process. Figure 7.9 shows how the diversification of risk allows the portfolio to reduce the inherent risk across the portfolio.

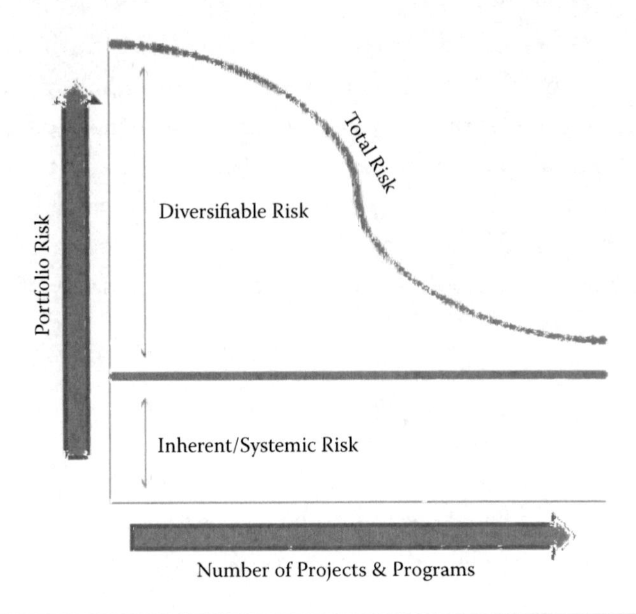

Figure 7.9 Depiction of risk diversity within a portfolio. While all portfolios have some element of risk, the selection of projects enables the portfolio manager to diversify the risk elements across the projects and programs within the portfolio.

References

Aberdeen Group. (2011). *Project portfolio management: Selecting the right projects for optimal investment opportunity.* March.

Ahles, E. (2011). *Simplifying IT using a disciplined portfolio governance approach.* Redwood Shores, CA: Oracle Corporation. April.

Federal Enterprise Architecture Program Management Office, OMB. (2007). *Value to the mission.* Retrieved from: http://www.whitehouse.gov/sites/default/files/omb/assets/fea_docs/FEA_Practice_Guidance_Nov_2007.pdf

Hawald, S. (2007). EPMO - *Enterprise program management office.* Retrieved from http://rg-epmo.blogspot.com/2007/05/epmo-vs-pmo.html May 14.

Ho, A. (2009). *Compliance management: A holistic approach.* ISACA Journal 5 May.

James, M, & Ryan, P. (2012). *Effective project portfolio governance.* Published by Oliver Wyman. Retrieved from: http://www.oliverwyman.com/insights/publications/2012/oct/effective-project-portfolio-governance.html

Jordan, A. (2013). *Risk management for project driven organizations: A strategic guide to portfolio, program and PMO success.* Plantation, FL: J. Ross Publishing, Inc.

MacDougal, C. & Cadilhac, C. (2011). *Project program and portfolio governance. Building 4 Business.* Retrieved from: http://www.building4business.com.au/publish.html

Needs, I. (2013). *Building an effective communication strategy for your PMO (program/project management office).* Retrieved from: http://www.business2community.com/communications/building-effective-communication-strategy-pmo-programproject-management-office-0718312#!FTflI

Center for Business Practices and PM Solutions Research. (2005). *Project portfolio management maturity: A benchmark of current best practices.* Retrieved from: http://www.pmsolutions.com/audio/Research-PPM-Maturity.pdf

Project Management Institute. (2012). *PMI's pulse of the profession.* Newtown Square, PA: Project Management Institute. March

PricewaterhouseCoopers, LLC. (PWC). (2012). *Strategic portfolio management.* Retrieved from: http://www.pwc.com/us/en/increasing-it-effectiveness/publications/strategic-portfolio-management.jhtml

Wise, L. (2010). *Using KPIs to identify trends in performance.* Retrieved from: http://www.dashboardinsight.com/articles/business-performance-management/using-kpis-to-identify-trends-in-performance.aspx

8

Business Intelligence Framework for Project Portfolios (BIPPf)

MUHAMMAD EHSAN KHAN, PHD, PgMP, PMP

Introduction

"If I had known the location of the hotel, I would have stayed at the airport hotel" was the first thought that came to my mind. I was traveling from Dubrovnik to Vienna after attending a conference. Based on the information available on one of the leading hotel booking sites, I booked a hotel near the airport for a night so that I could take the early morning flight to Dubai. The information available showed that the hotel was a two minute walking distance from the nearest airport shuttle station. It was not! I made my decision based on this incorrect information. Bad information, bad decisions. Needless to say that I had to book an early morning cab, had to wake up earlier, and ended up paying more money than I would have if I had stayed at the airport hotel. A decision based on incorrect information does more harm than good! Lesson learned.

Based on the information available, we take hundreds of decisions every day that shape our daily routine and in turn have an impact on our lives. Applying the same concept to business i.e., utilizing information to take informed decision can be termed as Business Intelligence (BI). BI, in a broader sense, is a set of concepts, methods, applications, and technologies, which are utilized to transform raw data into meaningful information that can be utilized by stakeholders to make informed decisions. Organizations use BI practices, tools, and techniques to understand their state of affairs, competitors, and market conditions. This enables organizations to address existing business concerns and develop organizational strategies that provide them with a competitive advantage in the market.

An organization's existing project portfolio depicts the organizational direction and strategy, thus there is a clear linkage between the project portfolio and BI. It is therefore important for organizations to apply BI practices and implement BI tools to the project portfolio, so that the leaders can understand the status of the portfolio, trends, history, and information about underlying components. Taking leverage of

BI concepts, methods, and tools, organizations can obtain important insight about the portfolio and its underlying components, empowering the portfolio management team with information that becomes the basis for better decision making.

The objective of this chapter is to propose a Business Intelligence Framework for Project Portfolios (BIPPf). The framework provides guidelines for implementing BI and related concepts on project portfolios. The framework will help the readers to:

1. Understand BI and the current trends.
2. Understand the difference between structured and unstructured data.
3. Understand the importance of Self Service BI.
4. Understand portfolio requirements for reporting and decision making.
5. Select portfolio key performance indicators (KPIs).
6. Understand the different user types and their business requirements from a BI implementation.
7. Understand the BI implementation process in project portfolios.

Before directing the discussion toward application of BI practices to project portfolios, we need to first understand the importance of BI.

Business Intelligence

What is Business Intelligence (BI)?

Business intelligence (BI), in the broader sense, is a set of concepts, methods, applications, and technologies, which are utilized to transform raw data into meaningful information that can be used by stakeholders to make informed decisions. Organizations utilize BI practices, tools, and techniques to understand their state of affairs, competitors, and market conditions. This enables them to address existing business concerns and develop organizational strategies that provide them with a competitive advantage in the marketplace. In order to make informed decisions, organizational data should be accessible to all relevant stakeholders. This accessibility ensures that the business users can perform their own analyses while freely sharing insight with team members.

BI systems manage organizational metrics and present them to decision makers through intuitive dashboards, reports, and self-service capabilities. The data management and analytics component of BI systems have the capability to consolidate complex internal and third-party data from multiple applications into a central framework for conversion into actionable information. This consolidated data can be analyzed in-depth and compared to other metrics or performance indicators.

BI has different meanings to different people. Some people equate BI with dashboards; others consider BI to be a software application. BI is not a tool and not a niche domain. BI is not a concept that is exclusive to information technology (IT) professionals. BI is, however, a simple notion that anyone can understand; it is about willing to base decisions on facts. While good information leads to smarter decisions and better results; important decisions of any nature in any business should be based on facts. BI supports the decisions that are made with facts and figures.

BI platforms are technology-based solutions that usually include the following components:

- Extract, transform, and load (ETL) tools,
- Master data management and governance tools,
- Data warehouse,
- Multidimensional database or cubes,
- Tools to design dimensions, attributes, facts, and/or measures,
- Reporting and dashboard tools or platforms,
- Analytics tools or platforms, and
- Self-service tools.

Reports are one part of the BI capability. Reports provide typical static information that is used by organizations to fulfill daily and weekly operational needs, whereas BI provides organizations with the flexibility or ability to look at the same information from different perspectives using granular capabilities without running a new report.

Dashboards are a mechanism used to present information in a more visually appealing, interactive manner in order to enhance the user experience and to present the information in concise manner. Organizations design dashboards in such a way that the most relevant information is provided to the relevant stakeholders in a presentable and understandable manner. Dashboards are one part of the BI equation, which are focused on providing information in a graphically appealing and understandable manner.

Predictive analysis capabilities available in BI tools can also be applied to utilize past information in order to derive models and analysis that help the portfolio and its underlying components and future outcomes. Applying BI provides visibility and assists in decisions that help in proactively managing projects, programs, and portfolios.

The world of BI and data has evolved in the last decade. The next section sheds light on this aspect of BI.

Transforming World of Data and BI

Several years ago, typically the only information source available to an organization was the data available in its systems. If there was data, it was available in silos. The world is now connecting rapidly. The data are connecting so rapidly that sometimes we feel an information overload. This information overload is influencing knowledge workers to spend too much time looking for information and not enough time analyzing it. We have all experienced this problem. Think back to your past few weeks, how much time did you spend searching for or asking others for data versus analyzing the information that was already available to you? Of all the people who could be using information more effectively to make better decisions, very few have any meaningful access to the required data.

On the other hand informed users are no longer satisfied with the static reports provided to them by the IT department. They are no longer information consumers. They are becoming data consumers who produce information. They want to look at data from different perspectives to gain new insight and take informed decisions. They understand their business, thus they appreciate the value of data that is within, as well as outside their organizations and tend to utilize generated information to take more effective decisions. Business requirements change too fast for the IT department to keep up with the demand. The dependency on IT to provide reports results in delayed decisions, which results in organizations losing a competitive edge.

Previously, BI was considered as a mechanism to:

1. Consolidate and connect to existing data sources
2. Stage that data for reporting and analysis
3. And then provision it from reports, dashboards, or cubes

However, the line between producers and consumers is being blurred, and end users in addition to the IT provisioned information have always wanted more flexibility and agility to do their own analysis and create their own reports. The two most common ways for end users to do this has been to use spreadsheets and a wide array of specialized tools for analysis and reporting.

In creating their own spreadsheets based on specialized tool-based BI solutions, end users have effectively found ways to get data directly from various data sources and bypass the formal IT solutions, making it challenging for IT to control, monitor, and manage these BI solutions being used within organizations. Here the concept of self-service BI comes into play.

Self-Service BI

BI, as discussed, is a capability for business. Therefore, it makes perfect sense that any tool implemented to support BI should have little to no dependency on IT or other business functions. Business users should be able to access the right information at the right time based on their role in the organization. Organizations need to empower the business users so that they can access, analyze, and collaborate on critical business information to make the best possible decisions.

For example, with self-service capabilities, a portfolio manager can directly use BI to review information to help identify the projects that are consistently performing well or those projects that are lagging behind for the last two months. The portfolio manager can then drill down to a project level to see which activities are causing these delays. This leads to informed discussion between the portfolio manager and the project manager allowing the underlying problems to be resolved more efficiently. Also with self-service access to real-time information, employees can quickly analyze the current business context and make informed decisions based on information that is current and relevant to their role.

Implementation Challenges

Although BI allows the consolidation and customization of information required for decision making, enabling the strategy process, and optimizing business operations, the information and figures needed for consolidation are spread out throughout the company. Moreover, finding the right scenario for every user is often more difficult than it sounds, because all users interacting with data do not have the same needs; top-level strategic insight and C-level day-to-day work require different information. To obtain BI, the challenge of producing reliable and relevant figures needs to be addressed.

In some cases, there may be multiple sources of information, which may result in conflicting reports. There could be scenarios where the information collection frequency and the content of the reports are not aligned with the stakeholder requirements. This generally results in a lack of trust of the business owners with the information obtained from the reports. Figure 8.1 depicts different issues that result in this confidence dearth.

In order to enhance the credibility of reports created by the portfolio team, it is important to ensure that the program team's reports meet the following requirements:

- Sources of information are identified and verified.
- The reporting mechanism is transparent, including a clear process through which the reports are created.
- Conflicting information between multiple sources is identified and resolved.
- The source which is used is accurate and relevant to the information required.
- The information is sufficiently specific and updated.

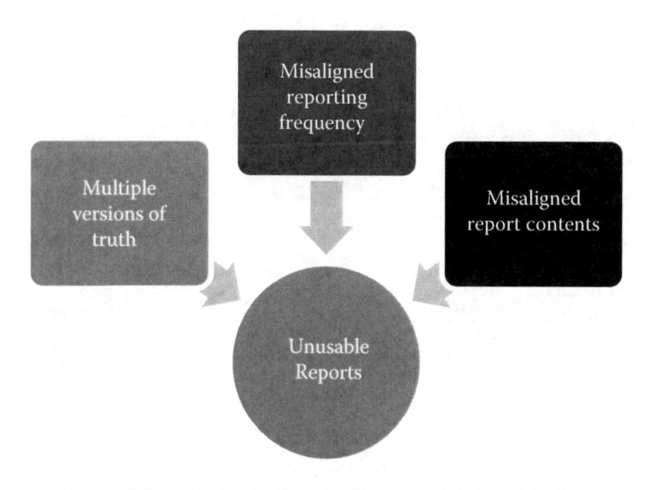

Figure 8.1 Unusable reports.

Another challenge is that organizations need to recognize that BI implementations are cross-organizational initiatives—not IT initiatives. Users of BI applications are business users and analysts who have a solid understanding of their particular business domain and who understand the data and enable decision making.

After having a thorough discussion on BI and related concepts, it is important to have a discussion related to project portfolios so a clear linkage of BI application on project portfolios can be established.

Project Portfolios

Projects and programs act as vehicles to drive and execute organizational strategy. Organizations group these initiatives under portfolios based on different criteria such as shared resources or aligned vertical markets. The Project Management Institute (PMI) defines portfolios as "a collection of projects, programs, subportfolios, and operations managed as a group to achieve strategic objectives" (PMI, 2013, p. 166). The objective of this grouping is to coordinate interfaces and prioritize resources between components in order to reduce uncertainty. The program and projects within a portfolio are temporary; however, portfolios are, generally, permanent within parent organizations, which are focused toward certain organizational strategic goals or objectives. Portfolio components may not necessarily have similar objectives, and the grouping can be based on vertical markets or organizational functions.

Organizations always have a limited availability of resources, and it is challenging for the management to execute every proposed initiative. Components within the same portfolio compete for scarce resources, which include finances, human resources, machineries, hardware, and other resources within the same organization. Thus it is important for the management team to clearly define the organizational strategy, so initiatives are selected for execution based on their alignment with strategic objectives.

An organization's project portfolio provides a clear snapshot of its current strategic direction; thus it is important that the portfolio manager, portfolio management team, organizational executives, and other stakeholders have accurate information about the portfolio's context and status. This information assists in making better, more-informed decisions.

Merging Portfolio Data Silos

Project portfolios generate substantial data on a daily basis. Some of the data is captured through software applications that are being utilized by organizations to manage portfolios and their underlying projects and programs. However, a considerable amount of data remains uncaptured and is never considered as part of the portfolio performance analysis. The uncaptured data are mostly unstructured, and capturing it will not add much value unless the data are organized and structured for portfolio analysis. Structured data can be analyzed; however, large amounts of unstructured data that can provide important insight of the portfolio are wasted.

This unstructured and structured data can be utilized in conjunction to understand portfolio performance from multiple dimensions. Consider a construction portfolio that has a project associated with the construction of a bridge in a populated area, performing within the defined thresholds of time and cost. However, analysis of tweets and Facebook feeds could reflect that the affected population is not pleased with the resulting noise and pollution. Such an analysis will allow the project team to devise a strategy to manage stakeholder expectations that will increase the chance of project success.

The evolution of technology has made data capturing and reporting a relatively simple endeavor. Organizations should select KPIs that have to be measured and reported so portfolio insight can be understood. These KPIs can act as a benchmark to capture data through the use of different tools and technologies.

Portfolio Governance and Reporting

Reporting of the component's progress and status to the portfolio management team is the responsibility of the component's management team. However, the portfolio management team needs to create high-level or executive reports that should be disseminated at the right time and with the right information to executive management and relevant stakeholders. This information delivery ensures that the executive management will be able to make accurate and timely decisions related to the organizational strategy and the portfolio in question.

At the portfolio level you need to ask two key questions:

1. What is important at the portfolio level? This information typically depends on functions executed by the portfolio and the KPIs that have to be reported at the portfolio level.
2. Which project/program data are important at the portfolio level? This information helps in deciding the specific features that can be utilized by interested stakeholders who can acquire the right level of information when required.

In addition to the answers to these questions, it is important that the communication frequency, method, and format, defined by the portfolio management team, meet the following requirements:

- The reporting should be aligned with the stakeholder needs.
- The reporting format is aligned with the reporting content.
- The reports should deliver information that is sufficiently specific and current.

Ensuring these measures increases the confidence of stakeholders in the reports and reporting mechanisms. It is also important to understand the timing and content of information because the validity of information is a key aspect.

The schedule and mode of communication from the portfolio management team depends on the communication management plan and is a management-level responsibility. The reports are reviewed and analyzed by the portfolio management team, and executive reports are generated. These reports are disseminated to the portfolio stakeholders based on the defined schedule. However, there is always a need for on-demand reporting, because different stakeholders, such as external regulatory authorities, may request additional communication and reports at different stages of the portfolio. It is the responsibility of the portfolio team to ensure that these reports are created and submitted to relevant entities. In cases where the portfolio management team receives continuous requests for additional reporting, it is imperative to review the communication plan and revise it accordingly.

Implementation of BI in project portfolios ensures that most of the reporting and communication is carried out utilizing the BI tools, and the information is pushed to the stakeholders instead of them pulling the information from different sources. A carefully implemented suite of BI tools provides a 360-degree view of portfolio status. The next section provides guidelines to implement BI in project portfolios.

Business Intelligence Framework for Project Portfolios (BIPPf)

Business Intelligence and Project Portfolios

Business intelligence is neither a product nor a system— it is a constantly evolving strategy, vision, and architecture that continuously seeks to align an organization's operations and direction with its strategic business goals. BI is a broad concept, which encompasses data capturing, cleansing, data warehousing, data mining, and performance management, along with reporting and dashboarding. This chapter, while focusing on the reporting aspect of business intelligence, proposes a mechanism to determine the reporting and communication needs of project portfolios and provides guidance on building a BI solution that can enable portfolio managers to make informed decisions. The application of BI to portfolio management helps portfolio managers, their teams, and other stakeholders to understand how the portfolio has been performing over time and what decisions should be taken to ensure that the portfolio balance is maintained and optimized.

Effective reporting and communication are considered to be key factors in improved decision making. It is important that information collection and dissemination frequency, the mechanism of communication, and the content of reports are aligned with the stakeholder requirements. The stakeholders should not search for the information; rather the information should be presented to them based on their needs. In addition, the reports should present information that will assist the stakeholders in making informed decisions.

Project portfolios create substantial information that is lost because of ineffective capturing and dissemination mechanism. Even the information captured is not circulated in an optimized manner. Portfolio BI capabilities and dashboards provide interactive, summarized information that consolidates, aggregates, and arranges project and portfolio measures that are important for the stakeholders. These capabilities can be utilized to display the right information to the right audience on a single screen.

The following sections explain specific steps to implement BI at the portfolio level.

Building the Business Case

The first step in any BI initiative is understanding the business needs of the end users. Based on those business needs the BI implementation team should identify the source systems from where the data will be extracted. From a business perspective, the following questions that need to be answered at this stage are:

1. The business problem and what you expect from the solution.
2. What would be the impact if the solution is not in place?
3. How is the business currently managing portfolios, programs, and projects, and what are the major pain points?
4. The value to the business the solution will provide, i.e., cost savings, redeployment of full-time equivalents (FTEs), increased productivity, improved customer service, etc.

Once the business need has been identified it becomes important to understand stakeholders' specific needs.

Stakeholders and KPIs

The second step is to determine what information is required at the portfolio, program, and project levels so the portfolio manager and portfolio team can manage, balance, and drive the portfolio more effectively. It is important to note that the information relevant at the project level could be insignificant at the portfolio level. One example is tasks that are delayed and do not impact the project's overall timelines; this information may be relevant to the project manager but will not be of much interest to the portfolio manager.

Traditional measures of time and cost, even though important, are not enough to measure the performance of the portfolio and its underlying projects. The importance of soft indicators, such as stakeholder satisfaction and quality of services or products, is of critical importance. Tracking whether the stakeholders are happy today is a key to project success and has an eventual impact on the portfolio level. It is important to do the following:

- Identify who needs what information, when, and in what format.
- Identify the core KPIs that are relevant to track component progress.
- Identify the core KPIs that are relevant to track portfolio progress.
- Categorize stakeholders in different groups based on the information needs.

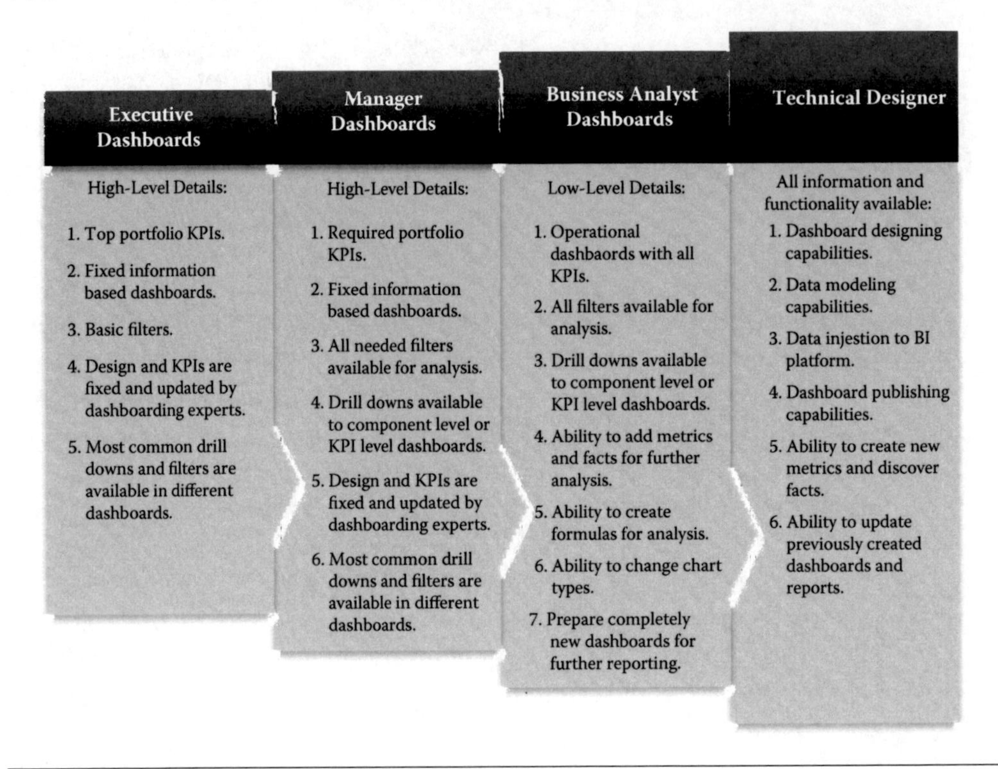

Figure 8.2 Information and functional requirements

BI users can typically be categorized into four groups, namely executives, managers, business analysts, and technical designers. These roles are defined based on different information and functional needs that these organizational users require. Figure 8.2 defines the roles and their information and functional needs:

From a KPIs perspective, based on experience and research, the following KPIs are of prime importance at the portfolio level:

- Deviation of planned hours of work for components.
- Resource utilization and availability.
- Average of the cost performance index (CPI), schedule performance index (SPI), schedule variance (SV), and cost variance (CV).
- Milestones missed and achieved on time.
- Average deviation of planned versus actual duration of components.
- Escalated risks and issues—their status and impact.
- Percentage of components with missed milestones.
- Stakeholder satisfaction (based on different categorizations).
- Percentage of projects that fit organizational strategy (all of them should but sometimes they do not).
- Portfolio balance indicators (high risk versus low risk, short term versus long term, and other attributes).
- Percentage of overdue component tasks.
- Number of new components in the pipeline with their attributes and information.

It is important to ensure that relevant stakeholders understand the KPIs and their meanings. Otherwise they will just see fancy numbers.

Capturing and Verifying Data

Once the information needs are identified, ensure that these data are captured through some means. As discussed previously, the data can be loosely divided into structured and unstructured data. Structured data are generated from applications already deployed in the organization. Unstructured data can be captured through analyzing memos, official letters, and publically available or organization-wide available data streams such as tweets, official e-mails, and other sources. Determine the source from which you want to obtain this information. When a system exists, integrate it, otherwise decide how you are going to capture this information.

It is also important at this point that the portfolio manager, portfolio team, and other stakeholders who will be making decisions based on these data understand it and have faith on the reliability and timeliness of the data. Otherwise, the captured data and resultant KPIs will be of no interest to these stakeholders. Problematic data can do more harm in the decision-making process than add value. Decisions based on faulty data can be detrimental to organizations. For example, when data are faulty, it is possible to terminate a component that was performing exceptionally well or possibly initiate a project that has low strategic value. This process takes time—start early and routinely check data health.

Designing the Dashboards and Analytics Layers

Once the stakeholder roles are identified, information needs to be finalized, KPIs agreed upon, and data made available. At this point, develop the portfolio dashboards and start performing analysis.

For a Tier 1 dashboard, focus on presenting charts and reports related to KPIs that are important at the portfolio level. Instead of making the dashboard cluttered with data, focus on the top five KPIs that are important to the current context. This will provide a snapshot of portfolio health—not details about each and every aspect of the portfolio. In addition, the focus is not on individual components, which is a project manager concern.

Figure 8.3 provides a sample dashboard that has high-level details of projects and their health, in addition to the high-level information related to the project portfolio.

One key aspect of this dashboard is the availability of actions that a stakeholder could decide to take based on the available information. Dashboards that simply display data do not add much value. The actions should include, but not be limited to, requesting a detailed status update, sending an e-mail, or requesting a meeting, etc. The KPIs shown in Figure 8.3 may be replaced with the KPIs discussed previously.

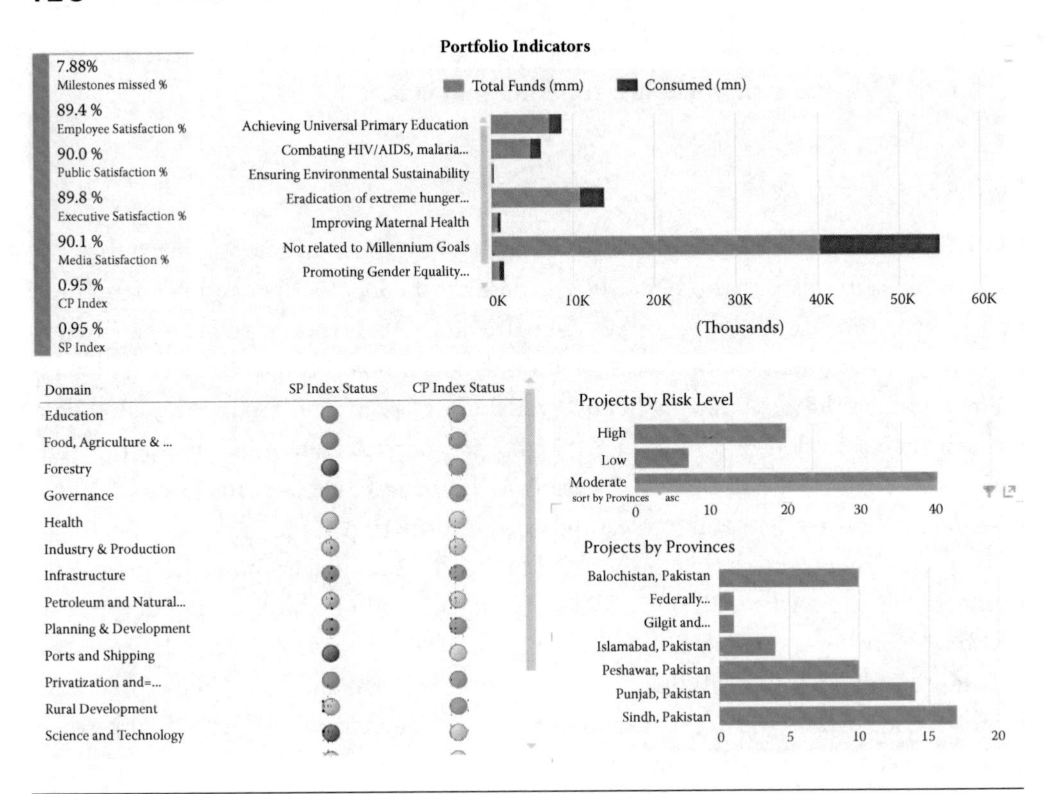

Figure 8.3 Sample portfolio dashboard.

Tier 2 dashboards should provide detailed information related to individual KPIs, where a user should be able to perform a detailed analysis related to a particular KPI. Another option to present at a Tier 2 dashboard is detailed information related to the component projects. This is useful when the stakeholder wants further details related to a specific aspect of a project, for example, resource utilization. The project dashboard provides information and KPIs that are important at the project level. Ensure that the navigation between different dashboard tiers is simple and seamless.

Figure 8.4 shows a sample dashboard for a project. In this case, the focus is on project-related tasks that were not important at the portfolio level.

When designing a dashboard, be sure to identify the relationships between the KPIs. KPIs that have dependencies on each other should be presented together so that the business user can make immediate sense of it. Examples of KPIs with dependencies are:

- Milestones missed and achieved on time, and
- Percentage of components with missed milestones.

These KPIs can be grouped together as both address the same business concern, that is, the *milestone* perspective of the portfolio.

A key aspect related to analytics is the self-service BI capabilities present in various BI tools. There is no longer a need to depend totally on IT—anyone can perform these analysis functions. These tools (after some backend work) can provide metrics

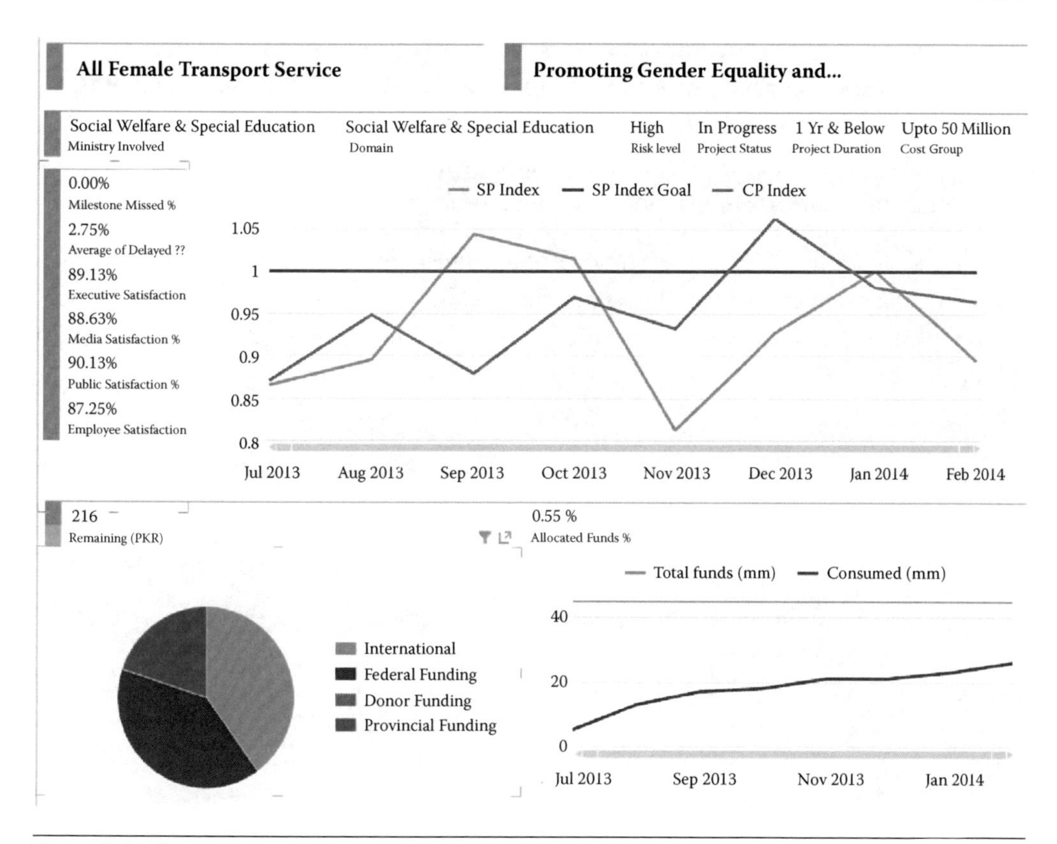

Figure 8.4 Sample project dashboard.

and dimensions for use in answering different questions through analytics. There is no longer a need to ask IT for another report. Create your own, learn insight, and share.

Conclusion

A portfolio's BI capability empowers stakeholders and improves portfolio management effectiveness. The portfolio's data (including external market data) are utilized effectively by converting the data into information and making this information available, discoverable, and transparent. Portfolio decisions are backed by insight into facts and figures resulting in a decision-making process that is effective and more reliable.

Context-driven dashboards ensure that the right information is presented to the right stakeholder in the right format and at the right time. With BI, portfolio success is realized through relevant, quick, and easy access to actionable information. A well-designed BI capability at the portfolio level ensures that portfolio managers and relevant stakeholders have the ability to:

- Utilize dashboards to help gain insight to make better, confident, and timely decisions.
- Drill-down and aggregate data on different dimensions utilizing self-service tools.

- Better manage the portfolio pipeline and effectively evaluate ideas by measuring the strategic contribution of competing requests to ensure strategic alignments.
- Effectively utilize resources by accurately measuring resource utilization and assignments based on organizational strategy.
- Maximize value from existing portfolio components.

In decision making, it is always important to value judgment and instincts. However, decision makers must eventually accept the results and consequences of their decisions. Complement the decision-making process with facts and figures. Implement BI and enable informed decisions.

References

Dickinson, M. W., Thornton, A. C., & Graves, S. (2001). Technology portfolio management: Optimizing interdependent project over multiple time periods. *IEEE Trans Engineering Management, 48*, 518–527.

Project Management Institute. (2013). *The standard for program management.* Third Edition. Newtown Square, PA.

9

Portfolio Management Success

CARL MARNEWICK PhD

Introduction

Kendall & Rollins (2003) state that there are four major reasons why portfolios are unsuccessful. These reasons are (i) too many projects in the portfolio, (ii) the wrong projects are in the portfolio, (iii) the projects are not linked to the strategy of the organisation, and (iv) the portfolio is unbalanced.

Portfolio success is measured in terms of the aggregate investment performance and benefit realisation of the portfolio (Project Management Institute, 2013b). This definition from the Project Management Institute (PMI) implies that the success of a portfolio is measured over an extended period and that the success is also linked to the strategic intent of the organisation itself. But to understand and analyse how the success of a portfolio is managed, it is important to understand what portfolio management entails. Portfolio management is focused on the achievement of the organisational strategies and objectives. The implementation of organisational strategies and objectives might take anything from months to years. In the case of large corporate organisations, the emphasis is more on years rather than on months. It also implies that to measure the success of a portfolio remains difficult as the strategies and objectives span across the entire organisation and all its divisions. The success of a portfolio is therefore complex and integrated.

The success of a portfolio is also measured against the benefits that need to be realised. Each and every project and/or programme that forms part of a portfolio will have an associated business case that promises some benefits. It is these benefits that need to be realised through the portfolio as these are only realised after the closure of a project and/or programme.

Jonas (2010) states that it remains difficult to capture the overall success or failure of a portfolio. That might be because portfolios are dynamic, multiple interdependent systems that constantly change and develop. There is a need for a comprehensive success framework that is capable to cover the portfolio in its entirety and additionally takes into consideration that changes made within a portfolio will take some time to have either a positive or negative effect. Portfolio success is therefore realized at different points during the lifespan of a portfolio (Jonas, 2010).

Literature also suggests that portfolio success should also be examined multi-dimensionally on the single project, portfolio, and organisational level (Blomquist & Müller, 2006; Müller, Martinsuo, & Blomquist, 2008). In contrast to the PMI definition, portfolio success is defined by (i) the average project success over all projects, (ii) the exploitation of synergies between projects within the portfolio that might additionally increase the overall portfolio value, (iii) the portfolio fit to the organisation's business strategy, and (iv) the portfolio balance in terms of risk, area of application and use of technology (Beringer, Jonas, & Kock, 2013).

Therefore, to assess a portfolio and its positive or negative effects on the organisation, the results have to be measureable and have to cover a wider perspective than the isolated project (Meskendahl, 2010). That raises the question of what constitutes the success criteria of a portfolio. The next section focuses on four major criteria, and it must be noted that they can be extended by organisations.

Portfolio Success Criteria

The discipline of portfolio management owes its origins to a seminal paper written in 1952, in which Harry Markowitz laid down the basis for the Modern Portfolio Theory (MPT). MPT allows determining the specific mix of investments generating the highest return for a given level of risk (De Reyck, Grushka-Cockayne, Lockett, Calderini, Moura, & Sloper, 2005).

The main portfolio success criteria according to Beringer et al. (2013) and Meskendahl (2010) are the:

1. maximization of the financial value of the portfolio,
2. linking the portfolio to the organisation's strategy,
3. balancing the projects within the portfolio taking into consideration the organisation's capacities, and
4. and the average single project success of the portfolio.

These criteria are mutually inclusive of each other. Figure 9.1 is a graphical presentation of these criteria and shows the interdependency between the four criteria.

Bannerman (2008) follows the same reasoning but provided no detail on how the portfolio should be assembled. He also states the business objectives that motivated the investment must be achieved. The achievement of these business objectives will lead to the success of the organisational strategy. This achievement of the business objectives is realized through (i) the various business cases which must be validated throughout the lifespan of the project, and (ii) the subsequent business benefits that must be realised.

Figure 9.1 Portfolio success criteria.

Financial Value of the Portfolio

The first major success criterion that a portfolio is measured against is that the portfolio must maximise the financial value of an organisation. The song by Kelly, "*Money makes the world go round*", summarises one of the reasons why organisations do exist.

Organisations are there to create profits for their owners and shareholders. The same applies to non-profit organisations. Although these organisations are not focusing on profits per se, they should still be sustainable, and this can only be achieved if the organisation makes a profit or is financially viable.

The aim of portfolio management is to diversify investments in such a way to reduce the total risk of a portfolio, but this must be done in an effective way and manner. In the realm of portfolio management, the aim is optimise the total financial value of all the projects within a portfolio, but at the same time, minimise risk exposure. There are various ways to measure the financial value of a portfolio, but it is not the purpose of the chapter to elaborate on these measures. The portfolio manager must make a clear distinction between 'financial portfolio' management and 'project portfolio' management.

The following measures can be used to determine the overall financial value of a portfolio (Reilly & Brown, 2012):

1. *Treynor Portfolio Performance Measurement* measures the returns earned in excess of that which could have been earned on an investment that has no diversifiable risk. This measurement does not take the diversification of a portfolio into consideration.
2. *Sharpe Portfolio Performance Measurement* is the same as the Treynor measure, but the focus is on the total risk of the entire portfolio. It examines the performance of an investment by adjusting for its risk. The ratio measures the

excess return per unit of deviation in an investment asset or a trading strategy, typically referred to as risk.

3. *Information Ratio Performance Measurement* measures the average return of a portfolio in excess of a benchmark portfolio divided by the standard deviation of this excess return.

The portfolio manager must examine all the financial parameters and determine which financial factors should be used. Financial factors include investment commitment, return on investment, and the investment period itself (Hill, 2007).

Strategic Success

Strategy maps are used to describe the vision and strategies of the organisation by means of processes and intangible assets. They can help align intangible assets such as information technology with the organisational strategies and ultimately the vision of the organisation. It is the duty of the portfolio manager to align all the portfolio components to the strategy of the organisation. One way of doing it is through the use of strategy maps and balanced scorecards (Kaplan & Norton, 2004).

A strategy map starts with a vision and follows a top-to-bottom and a bottom-to-top approach. This approach means that the vision dictates all the lower levels, but the bottom-up approach enables the organisation to link everything to the vision. Although the approach is a top-to-bottom approach, the lower levels must be linked to the upper levels to ensure that there is a consistent link between the upper and lower levels of the strategy map. The top-to-bottom flow enables an organisation to take the vision and break it down into its different components and ultimately into different projects. The purpose of the strategy map is to take the vision and break it down into measurable components. It must be made clear that not all the components need to be incorporated into the strategy.

The strategy map provides a framework for an organisation, but it is up to the organisation to determine which of the components it wants to use. The four perspectives of (i) financial, (ii) customer, (iii) internal, and (iv) learning and growth provide the different focal points of the strategy map. The strategy map makes use of the four perspectives. The financial and customer perspectives provide the strategies of the organisation, whereas the internal and learning and growth perspectives provide the business objectives associated with the strategies.

This criterion represents the highest level of benefit achieved by a project, despite the possibility of failures against lower level criteria, as recognized by external stakeholders, such as investors, industry peers, competitors, or the general public, dependent upon the nature of the project (Bannerman, 2008).

For each of the projects identified with a business objective, a business case must be drafted to ensure accountability.

Business Case The business case describes the justification for the project in terms of the value to be added to the business as a result of the deployed product or service (International Institute of Business Analysis, 2009). The purpose of the business case is to determine whether or not an organisation can justify its project investments to deliver a proposed business solution. Bradley (2010) states that the business case is a living document that needs to be constantly updated throughout the project life cycle. This update will happen at certain points within the lifespan of the portfolio and will determine the continuous inclusion of the project in the portfolio. The business case must drive the project activities and is used to determine whether the project is still desirable, viable, and achievable.

A business case exists to ensure that, whenever resources are consumed, this supports one or more business objectives. The implication is that a business case must be reviewed at the various stages during the project life cycle. Business cases are developed but are solely used to obtain funding approval for the huge up-front financial investment and not to actively manage the project (Eckartz, Daneva, Wieringa, & Hillegersberg, 2009).

It is common practice in organisations to approve projects based on a business case. Yet, it is indicated through research done by Eckartz et al. (2009) that many organisations are not satisfied with their business cases. Cooke-Davies (2005, p. 2) shows that many organisations find it difficult to state that projects are "approved on the basis of a well-founded business case linking the benefits of the project to explicit organization goals (whether financial or not)." Many organisations are also unable to state that they had a means of measuring and reporting on the extent to which benefits have been realised at any given point in time.

The Association of Project Management (APM) Body of Knowledge states that the business case "*provides justification for undertaking a project, in terms of evaluating the benefit, cost and risk of alternative options and rationale for the preferred solution.*" (APM, 2006, p. 129). The Office of Government Commerce (OGC) states that the business case must drive the project (OGC, 2003). Any project should not be started if there is not a satisfactory business case. In the case of PRojects IN Controlled Environments (PRINCE2), the business case is defined as the reasons for the project and the justification for the project based on the costs, risks, and expected benefits. PMI states that the business case determines whether the project is worth the investment from a business point of view (PMI, 2008).

Any decision taken about the inclusion of a project into the portfolio should take into consideration the promised benefits as stated in the business case.

Benefits Realisation Benefits should be identified and quantified before a project is initiated (Remenyi, Money, & Bannister, 2007). For a project to be judged a success, potential benefits need to be identified as early as possible and realised either during or after the lifespan of the portfolio (Remenyi & Sherwood-Smith, 1998). Benefit identification is the first step of the benefits management process and identifies and documents benefits that will be most relevant and convincing to decision makers (Bennington & Baccarini, 2004). In general, it can be stated that the proposed benefits from a project must link in some way to the objectives of the organisation itself (Dhillon, 2005).

Williams and Parr (2008) agree that the process of benefits management begins with the identification of benefits before a programme is initiated and continues with the measurement of the benefits even after the programme has been delivered. It can therefore be concluded that in both programme and project management, the benefits must be defined before the programme and project are initiated. The implication is that it is a function within both disciplines and is not mutually exclusive to either.

Benefits associated with a project are major determinants in its selection and funding. The selection of projects is difficult because there are various quantitative and qualitative factors to be considered such as organisational goals, benefits, project risks, and available resources (Chen & Cheng, 2009). The main reason that benefits are identified and quantified seems to be to gain project approval (Dhillon, 2005).

This implies that well-structured project selection criteria will help ensure that organisations select projects that will best support organisational needs. It further identifies and analyses risks and proposed benefits before funds and resources are allocated (Stewart, 2008). However, often benefits are primarily strategic or tactical in nature, and their financial rewards are difficult to forecast. The purpose of a formal project selection process is to weigh the risk and rewards of each identified project based on whether it is ideal and pragmatic to initiate (Williams & Parr, 2008).

The selection of projects based on its benefits is directly linked to the strategic objectives that guide the conceptualisation and selection of initiatives and/or projects. The ultimate benefit for any organisation is the realisation of its vision and strategies.

Selection and Balancing

The selection and continuous balancing of the projects that constitute the portfolio is the third success criteria. Projects should and must be selected based on the overall contribution to the achievement of the organisational strategies and objectives.

There are various ways and manners to select the components of a portfolio. Table 9.1 highlights the steps involved in the selection of portfolio components.

Table 9.1 Selection of Portfolio Components

	INTERNATIONAL ORGANIZATION FOR STANDARDIZATION (ISO) 21502 (GUIDANCE ON PROJECT AND PROGRAMME PORTFOLIO MANAGEMENT)	PMI'S STANDARD ON PORTFOLIO MANAGEMENT (2013)
1	Define the contribution of each component to the strategies and objectives	Identify the components
2	Rank the contribution of each component to the strategies and objectives	Categorise the components
3	Define the exposure to risk of each component	Score the components
4	Rank the exposure to risk of each component	Rank the components
5	Define the resource constraints	

The important aspect is that each component must be evaluated on a quantitative as well as a qualitative manner. The portfolio manager can use a variety of criteria to define the contribution of each component. Criteria that can be used include the following:

1. Alignment to the strategy
2. Alignment to the objectives
3. Benefits which can be defined as financial and/or non-financial
4. Market share
5. Risk exposure
6. Regulatory compliance

Based on the pre-defined criteria, a weighted scoring model can be used to determine each component's contribution to the portfolio as per Table 9.2.

Once each component's contribution has been determined, the next step is to select the appropriate components based on resource constraints. This is also where the balancing of the portfolio plays a major role. Organisations want to do all projects, but only certain projects can be done at a certain point in time based on the available resources. Resources that might have an impact on the amount of projects that form part of the portfolio include:

- The amount of human resources available to implement the various projects
- The financial status of the organisation
- Risk exposure

The ultimate goal is to have an optimised portfolio of projects that will be implemented within the constraints of the organisation. A master schedule of resource allocation is necessary to plan the consolidated demand for portfolio resources (PMI, 2013b).

Project Success

The fourth main criterion that contributes to portfolio success is the success of the individual projects itself that form part of the portfolio. The focus within portfolio management is that each project must be a success in its own right but at the end must and should contribute to the overall success of the portfolio.

Table 9.2 Weighted Scoring Model for Portfolio Component Selection

		COMPONENT 1		COMPONENT 2		COMPONENT 3	
	WEIGHTING	SCORE	TOTAL	SCORE	TOTAL	SCORE	TOTAL
Strategic alignment	20	80	4	40	8	70	14
Objective alignment	20	56	4	90	18	75	15
Benefits	20	70	4	90	18	50	10
Market share	10	80	1	65	6.5	80	8
Risk exposure	15	90	2.25	70	10.5	45	6.75
Regulatory compliance	15	30	2.25	30	4.5	60	9
TOTAL	100		17.5		**65.5**		62.75

For example, a project may have delivered a product that functionally exceeds the customer's expectations, but in its development it has exceeded the schedule and cost constraints of the accepted scope of the project. Should this project then be classified as a failure or a success? An example that comes to mind is the Sydney Opera House that was completed ten years late and over-budget by more than 14 times. But today it is an iconic site.

It is important that the concept of project success be defined in general as well as specifically for an organisation within a specific industry. Unfortunately literature remains vague regarding project success. Project success is defined by the (PMI, 2013a) as the quality of the product and project, timeliness, budget compliance, and the degree of customer satisfaction. According to the Projects and Program Management for Enterprise Innovation (P2M), "a project is successfully completed [when] it delivers novelty, differentiation and innovation on its product, either in a physical or service form" (Ohara, 2005, p.16). The APM Body of Knowledge defines project success in a similar fashion, stating that it is project stakeholders' needs that must be satisfied, and this is measured by the success criteria as identified and agreed upon at the start of the project (APM, 2006). PRINCE2, on the other hand, does not explicitly define project success but states that the objectives of the project need to be achieved (OGC, 2003).

Figure 9.2 is a summary of the success criteria of these major project management standards and methodologies.

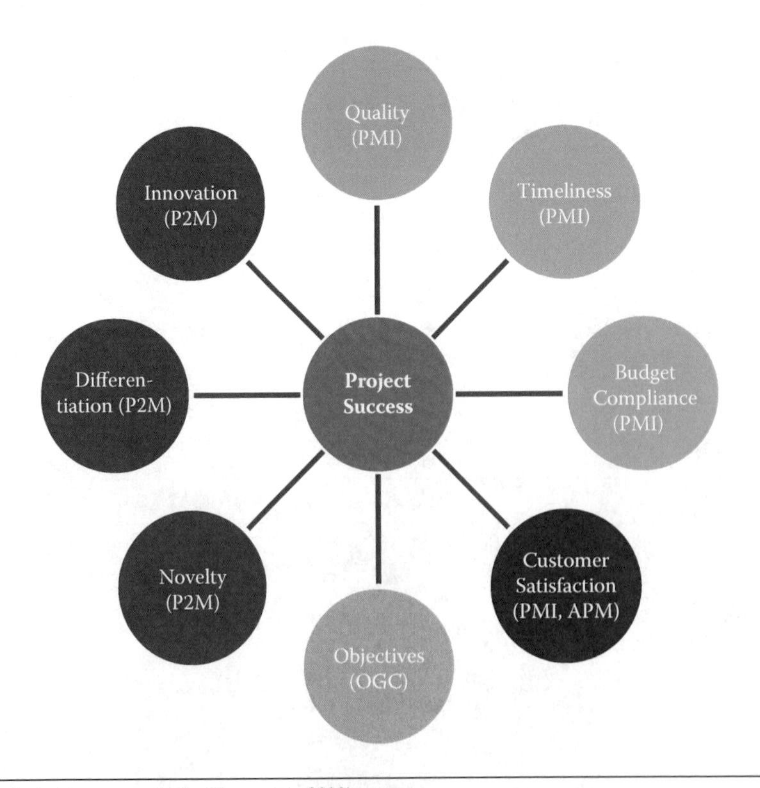

Figure 9.2 Project success criteria (Marnewick, 2012).

Figure 9.2 paints a picture of several criteria contributing to project success. In order to understand project success in totality, Hyväri (2006) suggests that the critical success factors (CSFs) must also be determined. If the CSFs are in place, then project success should follow as a natural outflow.

Hyväri (2006) suggests five CSFs, and each of these have sub-CSFs. These CSFs range from the project itself (size, end-user commitment) to the environment (competitors, nature, social environment). If companies and project managers focus on these CSFs, then project success should be assured. Dekkers & Forselius (2007) highlight the importance of scope management as a CSF where project managers can learn and embrace proven approaches that measure the size of software projects, streamline the requirements articulation and management, and impose solid change management controls to keep projects on time and on budget.

Projects and their subsequent products and/or services cannot be seen in isolation. According to Bannerman (2008), the success of the project can be measured on five levels: (i) process, (ii) project management, (iii) product, (iv) business, and (v) strategy. A project might deliver a service late and over budget, but it still delivers on the company's strategy. Is the project then a failure or a success? This multilevel view is supported by Thomas and Fernández (2008), who focus not on five, but three levels, i.e., project management, technical, and business.

It is all fine and well to understand how to measure the success of a portfolio, but at some point in time that alone is not enough to ensure the continuous success of a portfolio. Portfolio managers must also understand that there are organisational factors that either contribute to the success of the portfolio or do not contribute.

These factors need to be embraced by the portfolio manager, and where possible, they should be optimised or increased to ensure the continuous success of the portfolio.

Portfolio Success Factors

A portfolio success factor is a critical factor that is required to ensure the success of the portfolio. Although the absence or presence of a success criterion does not necessarily imply the automatic failure or success of a portfolio, it will certainly contribute to the success of the portfolio if all the success factors are addressed and followed.

Bolles & Hubbard (2007) identify various critical factors that might influence the successful implementation of a portfolio within an organisation. These factors include but are not limited to:

1. Total commitment and support from top executives. It is important that the executives of an organisation support portfolio management. If executives are not buying into the concept of portfolio management, how can we then expect that staff reporting to the executives will support the portfolio? Portfolio management spans across the entire organisation and touches the lives of every individual in the organisation. For this reason alone, executives should

support the portfolio management. The executives can make commitments around financial and human resources from the start of portfolio management. Executives will also be able to provide the business value of portfolio management to the organisation.

2. The selection of a portfolio sponsor is critical. Although executives will support portfolio management, one individual is needed to drive and ensure the continuous support of portfolio management. This individual, the portfolio sponsor, will be a strong proponent for portfolio management and will have some political cloud within the organisation. The sponsor will also serves as the liaison between the portfolio manager(s) and other primary decision makers within the organisation.
3. Full-time resources. A portfolio cannot be managed on a part-time basis by someone who does it whenever there is time for it. Dedicated resources are needed to manage the portfolio on a day-to-day basis. These individuals are employees with specific skills, knowledge, and principles that are needed to ensure the successful management of the portfolio.
4. Following the preceding argumentation, the impact of a strong project portfolio manager is highly dependent on the influence of associated management roles. The direct and indirect influences of the senior management and the line management on the project portfolio management system that goes beyond the execution of certain project portfolio management (PPM) tasks will in the following be considered under the term of management involvement (Jonas, 2010).
5. Adherence to processes and procedures (governance). Policies and procedures need to be adhered to especially when it comes to the selection and prioritisation of portfolio components. The portfolio sponsor and respective portfolio managers must ensure that the policies and processes are followed. By following this approach everyone is treated fairly based on the policies and processes. If the policies and processes are not adhered to, the credibility of the entire portfolio management process is under scrutiny.
6. Müller et al. (2008) show the positive relationship between strategic success and portfolio selection and project portfolio performance. A few other studies found project prioritization as part of the portfolio management process to be a key success factor (e.g., Cooper, Edgett & Kleinschmidt, 1999; Elonen & Artto, 2003; Fricke et al., 2000). As it is no end on itself, successful project portfolio management needs to contribute to the overall business objectives (Meskendahl, 2010). See Figure 9.3 for a graphical view of the portfolio success factors.

Maturity of Portfolio Management

Organisations need repeatable and successful performance of their portfolio(s). This should be accompanied by continuous improvements of all the portfolio management processes. The more uniform and consistently portfolio management processes are

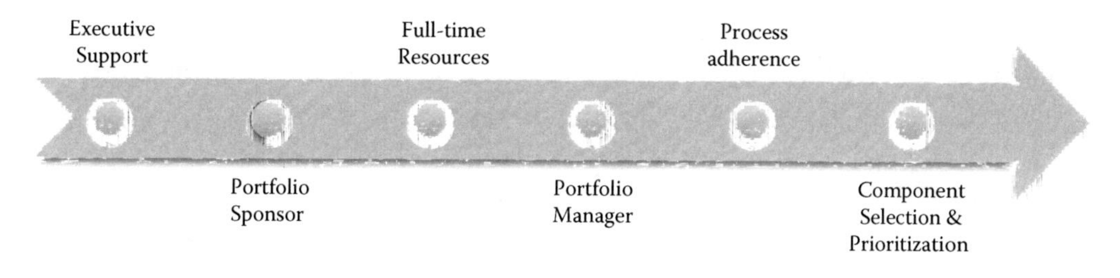

Figure 9.3 Portfolio success factors.

applied, and the more mature these processes are, the greater are the positive results and realised benefits (Bolles & Hubbard, 2007). The benefit of mature portfolio processes is that the consistent application of these processes results in the delivery of portfolio components within the allocated budget and time frame. This will then eventually result in the continuous success of the portfolio and ultimately the success of the organisation.

The development and sustainment of mature portfolio management processes requires the dedication, direction and involvement of the organisational executives and the portfolio sponsor.

Portfolio management should provide accurate, routine, consistent, and adequate information into the progress of each of the portfolio components in order for the executives to take effective action when there is a deviation from the approved success criteria. Executives however have difficulty to understand the benefits of mature portfolio management processes for the following reasons:

1. Executives cannot match the results of mature processes to the financial results.
2. Executives do not understand how the attainment of portfolio management capabilities at a predefined maturity level relates to the cost of effective operations.
3. Executives cannot determine when the processes defining a maturity level would be finally standardised and become part of the organisation.

A four-stage portfolio management maturity model is presented in Figure 9.4 for evaluating the progress made or not made with the implementation of portfolio management.

The first step is to determine the baseline of the portfolio management processes. This is used to measure the progress of the portfolio management maturity levels.

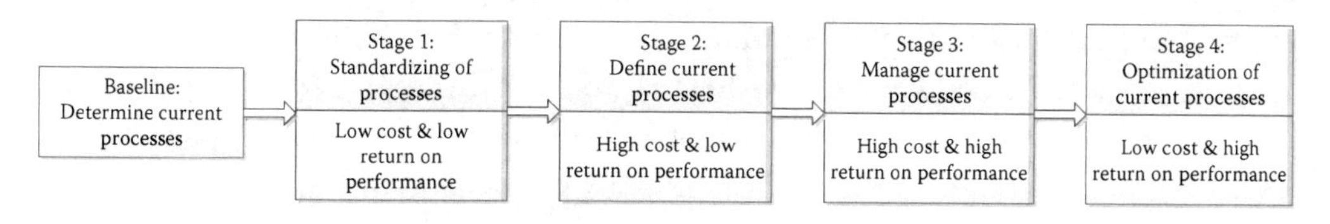

Figure 9.4 Four-stage portfolio management maturity model.

Stage 1 focuses on the allocation of additional resources and funds to develop the necessary processes to achieve stage 2. To move from stage 2 to stage 3, the newly developed processes must be documented and implemented to achieve the desired results and benefits. Stage 3 establishes the processes as part of the day-to-day operations of the portfolio. The level of required resources drops to normal operational costs as no additional resources are needed to maintain the processes. Stage 4 is the desired state where all the portfolio management processes are optimised and improved upon in a continuous way.

Auditing of the Portfolio

Müller et al. (2008) states that previous studies on portfolio management did not address the notion of portfolio control as a factor influencing portfolio management performance. However, they all include supporting evidence that points toward examining this phenomenon further. The auditing of the portfolio can be used as a control factor.

Project auditing and therefore portfolio auditing originate from financial auditing, which is defined as "an independent, objective assurance and consulting activity designed to add value and improve an organisation's operations" (Institute of Internal Auditors (IIA), 2013). In the case of portfolio management, the focus of the audit should be to determine whether the various portfolio components contribute to the overall portfolio success and ultimately the success of the organisation. It must also be noted that the portfolio audit must be done by independent auditors. The Project Management Office (PMO) can play a role in this regard.

A portfolio audit will help an organisation to accomplish its objectives by bringing a systematic, disciplined approach to evaluate and improve the effectiveness of risk management, control, and governance processes (The Institute of Internal Auditors, 2001). By nature, auditing is a reactive process, and the main objective of portfolio auditing is the early detection of errors and deviations (Kumāra & Sharma, 2011).

If the focus of attention moves to current portfolio and project management standards, then it is obvious that there is not really emphasis on the auditing of projects. Project audits are defined as the monitoring of compliance with project management standards, policies and procedures (PMI, 2013a). Another definition is that the purpose of a project audit is to provide an objective evaluation of the project (APM, 2006). Neither PRINCE2 nor the P2M make any reference to project auditing per se. *The Standard for Portfolio Management* does not make any reference to the auditing of a portfolio apart from the fact that the portfolio manager should have expertise and knowledge on the auditing of programmes and projects (PMI, 2013b).

If the advice of Gray & Larson (2008) as well as Hill (2007) are extrapolated to portfolio management, then the benefits of a portfolio audit should be:

- the continuous monitoring of portfolio components' contributions to the achievement of the business objectives;

- the identification and response to weak and troubled portfolio performance;
- the overseeing of quality management activities;
- the maintaining of professional and best practices within the portfolio and its various components; and
- the compliance with organisational policies, government regulations, and contractual obligations.

A distinction can be made between the various types of portfolio audits as discussed in the next section.

Types of Portfolio Audits

Figure 9.5 is a graphical presentation of the four major types of portfolio audits that an organisation can perform.

1. *Portfolio management audit:* this type of audit provides a comprehensive examination of the overall performance of portfolio management per se. The primary purpose is to ensure that the portfolio manager has put in place both business and technical processes that are likely to result in a successful portfolio.
2. *Portfolio performance audit:* In contrast to the portfolio management audit, the portfolio performance audit represents a detailed examination of the financial and business aspects of the portfolio. Typical elements that will be audited are the four portfolio success criteria as per Figure 9.1.
3. *Portfolio management methodology:* Another important audit is that of the portfolio management methodology that is used in the organisation. This audit validates the content and effectiveness of the adopted portfolio management methodology. The highest ranked benefit of this type of auditing has been found to be client confidence, followed by enhanced accountability, reduced project costs, and disputes in that order of significance (Sichombo, Muya, Shakantu, & Kaliba, 2009).

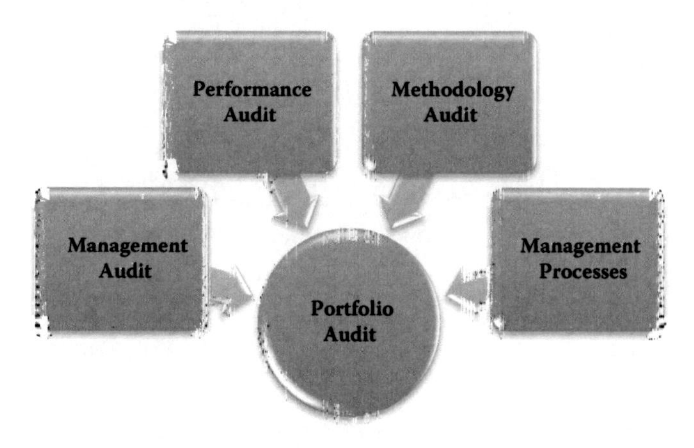

Figure 9.5 Types of portfolio audits.

4. *Portfolio management processes:* this type of audit audits the process of portfolio management itself and the focus should include the following processes:
   a. Strategic management: the audit will focus on the processes that are followed to develop the strategy and roadmap of the portfolio and the subsequent alignment of them to the organisational strategy and objectives.
   b. Governance: the audit should focus on the processes of selecting and balancing the portfolio as discussed in section 1.3. Emphasis should be on whether the processes are consistently applied, and where there are deviations, that they are properly documented.
   c. Communication: The audit should determine whether the processes are in place that are needed to develop the communication plan and the subsequent management of portfolio information. The communication processes must satisfy the information needs of all the stakeholders in order for effective and efficient decision making with regard to the performance of the portfolio.
   d. Risk: The audit should focus whether there is a structured process in place for the assessment and analysis of portfolio risks. The goal of portfolio risk management is to capitalise on potential opportunities and to mitigate those events or circumstances which can adversely impact the performance of the portfolio.

Portfolio audits, conducted routinely throughout the portfolio's lifespan, can help ensure organisational success through the identification of major risks that are likely to be faced by the portfolio manager, stimulating the portfolio manager and all the stakeholders to address the risks before it is too late to have a nagative impact (McDonald, 2002). A portfolio audit is of no value to the organisation if the audit findings are not addressed. The purpose of the portfolio audit findings is to identify lessons learned that can help improve the performance of a project or of future projects (Stanleigh, 2009). The lessons learned can be applied to all the portfolio components as well as the portfolio itself.

Intervals of Auditing

A survey done by Pricewaterhouse Coopers (PwC, 2012) stated that more than half of all respondents reported that they reviewed their portfolio on a monthly basis, and 20% reported more frequent reviews. Respondents with monthly review cycles reported significantly higher rates of portfolio performance especially on business benefits. The opposite was also true from the survey where quarterly or less frequent portfolio reviews are associated with a decline in the benefits of portfolio management.

The importance of frequent audits is highlighted by this report from PwC. But it also places a huge burden on the organisation itself to perform regular portfolio audits.

It is suggested that the following intervals are used for portfolio audits i.e., monthly and quarterly.

1. The monthly audit can evolve around the overall performance of the portfolio. The focus should be on auditing of the methodology and processes that are followed within the portfolio. It will also focus on the performance of the individual portfolio components as per section 1.4. The focus of the monthly audit is to ensure that everyone involved in the management of the portfolio are following the agreed-upon methodology and processes. This will in return improve the maturity of portfolio management as the audit will highlight deficiencies in the processes.
2. The quarterly audit can focus on the auditing of portfolio management itself and the overall performance of the portfolio. The audit will focus specifically on the success criteria of the portfolio as discussed in section 1. The reason for a quarterly audit is that changes to the financial value, strategic success, and portfolio construction are not seen and experienced immediately. Portfolio components are added on an irregular basis for instance every quarter or every six months.

It is important to highlight that these different types of audit activities require audit resources and should be part of the overall management of the portfolios. In fulfilling their duties, portfolio auditors must adhere to auditing standards, principles, and the code of ethics defined by different standard-setting organisations and bodies to which they belong. It is also important that the auditors are independent of the portfolio team. "An independent [portfolio] audit committee fulfils a vital role in corporate governance. The [portfolio] audit committee is vital to ensure the integrity of integrated reporting and internal financial controls and identify and manage financial risks" (Institute of Directors Southern Africa, 2009, p. 56). This statement implies that project audits fulfill a vital role in portfolio governance. Portfolio governance provides the structure through which the objectives of a portfolio are set, and the means of attaining those objectives and monitoring performance are determined (adapted from Turner, 2006). Portfolio audits, on the other hand, determine whether portfolio governance is in place, i.e., determining and attaining the portfolio objectives as well as monitoring the results of the objectives.

Governance

Governance can be defined as a system or method of management (Bolles & Hubbard, 2007). Portfolio governance establishes the roles, responsibilities and authorities of each individual, the rules of conduct, and management protocols. The establishment of portfolio structures within the organisation is needed as it will institute the planning and management processes of portfolio management. This view is supported by (Mosavi, 2014), who states that governance deals with roles and responsibilities, the decision-making frameworks, accountability, transparency, risk management, ethics, performance, and implementation of strategy.

Portfolio governance is the governance of the portfolio itself and focuses on the interrelationship between individuals, bodies, roles and responsibilities, decision-making processes, and other governance elements at the portfolio level (Mosavi, 2014).

Portfolio Management Roles and Responsibilities

Organisations must create the functional structures that own and provide portfolio management. These structures are needed for the adequate and effective planning, authorisation, and chartering of portfolios. The benefit of having these functional structures in place is:

1. The assurance that all the portfolio components are aligned with the organisational strategy and objectives.
2. Integration is provided for the monitoring and controlling of portfolio components.
3. There is ownership and maintenance of portfolio processes, procedures, and templates.
4. Identification and categorisation of each business objective into a specific portfolio occurs.

The following portfolio roles are necessary for the effective management of the portfolio:

Portfolio Steering Committee One structure that needs to be in place is that of the portfolio steering committee. Portfolio steering committees have the delegated authority to govern the portfolio but need to focus on the following aspects:

1. The portfolio steering committee must communicate and confirm the decisions that are made. These decisions might include the selection and optimisation of portfolio components, the general status of the portfolio, and the release of the various audit reports.
2. The portfolio steering committee also performs the role of negotiators. The negotiation is to determine whether a new project should be part of the portfolio and secondly when will it be become part of the portfolio.
3. The portfolio steering committee must also make timely decisions. It is up to the portfolio steering committee to make decisions that affect the overall performance of the portfolio, and these decisions must be made in a quick and decisive manner.

There are also other factors that play a role such as the frequency and duration of the portfolio steering committee meetings. These meetings must take place in accordance with the mandate. Depending on the complexity of the portfolio but also the maturity levels, meetings can be held monthly, quarterly, or every six months. The portfolio steering committee must take into consideration the various audit reports that need to be discussed as well as the selection and optimisation of the portfolio components when it decides the frequency of the meetings. Research has established the most portfolio steering committees are spending half a day to a full day on portfolio-related issues.

The portfolio steering committee normally consists of the portfolio manager, project owners, and representatives from the top management. One important aspect is that there must be trust and open communication between the various members of the portfolio steering committee.

Project owners Project owners are typically middle managers who are responsible for the implementation of projects and programmes. According to Blomquist & Müller (2006), these project owners are not formally part of portfolio management but are co-opted based on the state of the project. Project owners engage with the portfolio steering committee before and during the execution of the project. The role of the project owner is to identify non-performing projects, participate in the portfolio steering group, and perform administrative tasks relate to the management of a project. The aim is to improve the efficiency and success rates of a project, which is a criterion for a successful portfolio.

Top managers Top managers or someone representing them should also form part of the portfolio steering committee. Their role is to ensure that everyone understands the organisational vision and strategies. The purpose of portfolio management is to ensure that the strategies are successfully implemented. This is not possible if there is not a clear understanding of the organisation's vision and strategies. Top managers will ensure that the vision and strategies are foremost at any meeting. They will also inform the portfolio steering committee about changes to the vision and strategies of the organisation. Such changes will have an impact on the portfolio itself. Portfolio components will have to be re-evaluated, and new components might be selected in favour of current portfolio components.

Portfolio manager This role is assigned to one individual only. The portfolio manager analyses all future and current portfolio components and recommends a viable mix. The portfolio manager will monitor the performance of the portfolio on a daily basis and report to the steering committee if there are any deviations. The portfolio manager also plays an important role in supporting the overall strategy of the organisation and prepares information for the steering committee.

Organisations must also realise that the governance structures can and must be adapted to suit the needs of the organisation. There is no silver bullet telling organisations what the governance structures are and how they should operate.

Portfolio Management Office (PfMO)

The role of the PfMO within the realm of portfolio management is to facilitate the involvement of the executives and senior managers in the oversight of portfolio components (Hill, 2007). The PfMO can develop and implement processes and procedures for each of the portfolio management activities, and this should be consistent with the maturity levels. Hill (2007) and the PMI (2013b) state that the following activities should be performed by the PfMO:

1. The process and criteria for project selection should be developed.
2. Data are collected on portfolio component performance for comparison purposes.
3. There is alignment of portfolio components with the organisational strategy.
4. Gate reviews are established that are in line with auditing of the portfolio.
5. A process is implemented to review the availability of resources.
6. The real-time data for decision making are available.

In spite of all these activities that a PfMO must fulfil, Unger, Gemünden, & Aubry (2012) state that that the PfMO must fit into the overall organisational management framework. They continue to state that the PfMO must have a strong governance focus that ensures the implementation of the organisational goals. This governance aspect of the PfMO focuses on three tiers, i.e.:

1. Coordinating, which includes the facilitation of cross-portfolio components and the allocation of resources
2. Controlling, which focuses on establishing and updating information for accurate decision making. Control should also include advice on corrective measures to the portfolio steering committee.
3. Supporting, which provides services as per Hill (2007) to the various managers of the portfolio components and ultimately the cultivation of portfolio management standards and procedures.

The PfMO must first of all perform the three governance roles and then align the respective responsibilities to these roles. By doing this, it is assured that the PfMO performs its duties in accordance with the strategies of the organisation.

Portfolio Management Information System

A portfolio manager needs to understand what progress is made within the portfolio with regard to the success criteria, and this needs to be done fairly frequently to minimise risk exposure and to maximise opportunities (Kendall & Rollins, 2003).

Access to real-time portfolio information is also a key factor in the successful management of a portfolio (Rajegopal, McGuin, & Waller, 2007). Real-time information is needed to make decisions on unforeseen business changes as well as unexpected opportunities within the environment. The purpose of a portfolio management information system (MIS) is to get relevant information into the hands of decision makers as soon as possible.

Information availability for decision makers is shown as the most significant project-level factor contributing to portfolio management efficiency both directly and through project management efficiency (Müller et al., 2008). The following are the basic functionality that a portfolio MIS should provide to the organisation:

- Define and establish a common communication and reporting platform for all projects in the portfolio. Measure and compare projects along similar metrics.
- Take portfolio decisions in teams, after evaluation of the pros and cons of different mixes of priorities and go/no-go decisions, at the organizational and the portfolio level.
- Establish organizational routines to ensure project selection based on the organization's strategy, not the personal preferences of individual managers.
- Review portfolios periodically using comparable metrics.

There are various organizations that provide portfolio MIS solutions. Some of the organisations include Primavera (http://www.oracle.com/us/products/applications/primavera/p6-enterprise-project-portfolio-management/overview/index.html), Sciforma (http://www.sciforma.com/en-uk/), and Microsoft (http://office.microsoft.com/en-za/project/project-portfolio-management-for-the-enterprise-project-server-FX103802061.aspx). Figure 9.6 is a presentation from a Primavera screenshot illustrating how a portfolio is presented.

All portfolio MIS solutions should provide the following functionality:

- Assist in the planning, scheduling, and controlling of large-scale programs and individual projects
- Assist in the selection of the right strategic mix of projects
- Balance resource capacity
- Assist in the allocation of the best resources and track progress
- Monitor and visualize project performance versus planned performance
- Foster team collaboration
- Integrate with financial management and human capital management systems
- Simulate the impacts of a project on the entire portfolio
- Draw up a long-term project plan, using "what if" scenarios
- Breakdown the relevant sections of the long-term plan into operational projects
- Update the portfolio according to actuals
- Achieve a balanced scorecard by meeting the organisation's vision and strategies while managing the portfolios and the portfolio components

Organisations should use both the control mechanisms and the success factors and assign operational values to them. That allows tracking, setting, and measurement of goals and their achievement, which is the aim of portfolio management. The MIS systems can highly successful be used to achieve this objective.

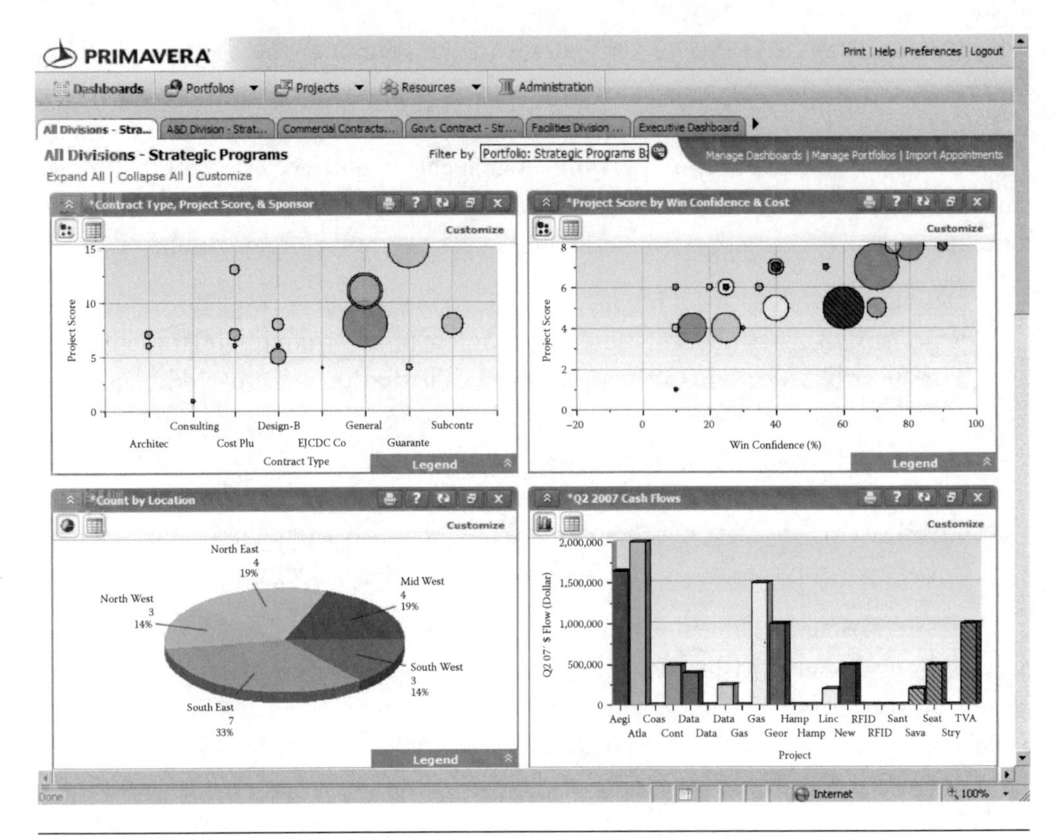

Figure 9.6 Primavera portfolio dashboard.

Sustainability

Sustainability is to manage the portfolio in such a way that it should be sustained indefinitely or in other words to implement the vision and strategies of the organisation without the depletion or destruction of resources.

Silvius, Schipper, Planko, Van den Brink, & Köhler (2012) list three dimensions of sustainability that need to be incorporated within a portfolio. The dimensions are social, economic, and environmental, see Figure 9.7. Harmony should be created between these dimensions.

The challenge a portfolio manager faces is how to balance the three dimensions within a portfolio. There are principles that a portfolio manager can adopt to ensure the sustainability of the portfolio itself and ultimately that of the organisation.

The first principle is to think about all three dimensions. The organisation's focus is to make profits and create shareholder value. It is the duty of the portfolio manager to incorporate the social and environmental dimensions as well. Two of the four success criteria are affected by this thinking. The first is the success of the individual projects. Project success should also be measured based on sustainability focusing on the environmental and sociable dimensions. The second criterion is that of strategic success. Organisations cannot be successful in the long run if they are not sustainable.

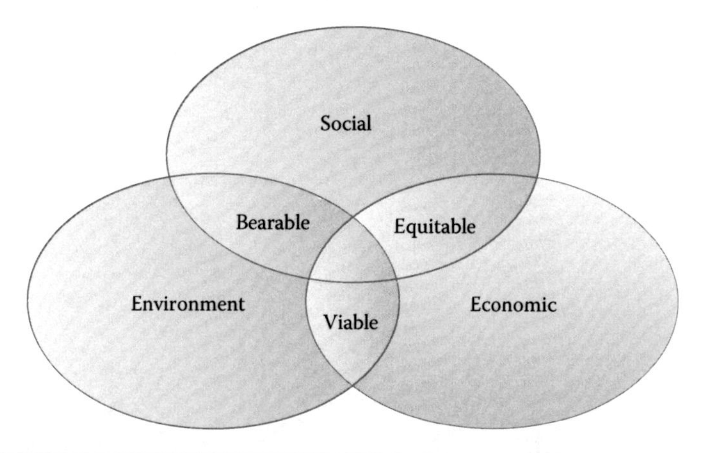

Figure 9.7 The three dimensions of sustainability.

The portfolio manager must think long term and focus on the benefits that each portfolio component will deliver. The benefits are normally reaped after the closure of a project. The benefits as stipulated in the business case should include economic, social, and environmental benefits. These benefits go beyond the quick wins of a successful project implementation.

Organisations are operating globally even if they are locally situated in a country or state. A portfolio might consist of portfolio components, which are implemented across the globe. The advent of technology can assist a portfolio manager to have virtual meetings and therefore reduce the emission of gas as travelling is reduced. Care should be taken in the selection and balancing of a portfolio where the impact of portfolio components must be measured across all three dimensions.

One of the benefits of a portfolio MIS is that resources are scheduled across all the portfolio components. A portfolio manager must remember that one of the principles of sustainability is that resources should not be depleted. This is especially the case when it comes to human resources. Resources should be managed in a sustainable manner where their performance on a current portfolio component should not have a negative impact on future performance.

Silvius et al. (2012) provide a basic checklist that can be used to incorporate sustainability into project management and ultimately into portfolio management. The checklist focuses on the three dimensions and allocates some aspects to these dimensions; e.g., under environmental sustainability one would get the notion of materials focusing on reusability and sourcing of said materials.

Conclusion

Portfolio management is complex and to sustain a portfolio that continuously contributes to the success of the organisation is just as complex. There is a magnitude of factors that need to be considered during the management of a portfolio. This complexity is illustrated in Figure 9.8.

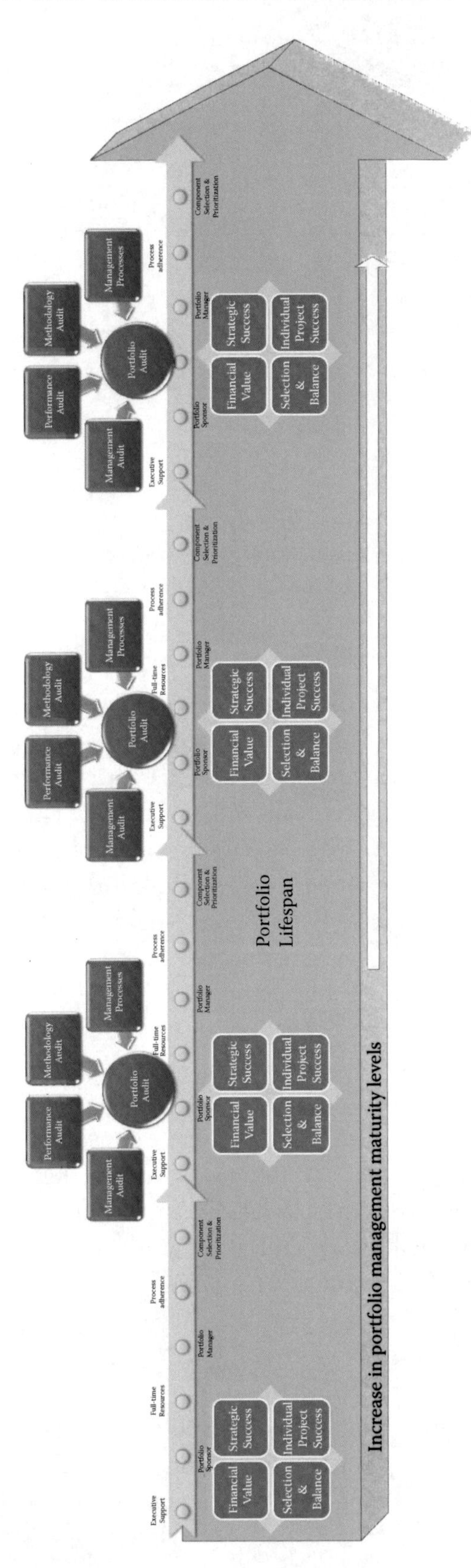

Figure 9.8 Portfolio success delivery model.

The Portfolio Success Delivery Model summarises the aspects that a portfolio manager must master in order to continuously deliver a successful portfolio. During the lifespan of a portfolio, the success of the portfolio must be measured at certain intervals. These intervals can be pre-determined for example once a quarter or once every six months. The entire portfolio must be measured against the four success criteria, under-performing portfolio components can be removed, and new portfolio components can be added based on the vision and strategies of the organisation.

The portfolio manager must also at these intervals institute the various audits. The audits can assist in the overall performance of the portfolio. The audits will highlight any deviations to the overall vision and strategies. The results of the audit will also highlight areas of concern with regard to the portfolio management processes that are adhered to or not.

Another aspect that the portfolio manager must take into consideration is the maturity level of the organisation with regard to portfolio management. This assessment will be done with the assistance of the PfMO. The ultimate goal is to improve the maturity levels throughout the lifespan of the portfolio. This must be done without interrupting the overall performance of the portfolio. In fact, the maturity levels must be improved in such a way that they contribute to the overall performance of the portfolio.

The success of the portfolio cannot be guaranteed with certain factors in place. A few factors have been discussed that can have a positive impact on the success of the portfolio. The portfolio manager must continuously engage with the various stakeholders and sponsors to ensure that the political will is there to support the portfolio even when some decisions taken by the portfolio manager are not popular. The portfolio manager must also observe the environment to determine whether there are new or different factors that contribute to the overall success of the portfolio. It is also of the essence to look at factors that do not contribute to the overall success of the portfolio and minimise the impact of these factors.

All this balancing and cross-checking must be done within the constraints of sustainability and governance. It does not make sense to drive the success agenda of the portfolio if it is not sustainable in the long run. Any decision made by the portfolio manager must be weighed against the three dimension of sustainability. None of the three dimensions are more important that any of the other dimensions. In that lies the difficulty for the portfolio manager as the economic dimension will always get preference.

The role of a portfolio manager is complex, and managing a portfolio in a successful way and manner is daunting. The Portfolio Success Delivery Model simplifies the complexity and highlights the aspects that the portfolio manager must focus on continually. The challenge for the portfolio manager is to ensure that all the aspects are taken into consideration during the lifespan of the portfolio.

References

Association for Project Management. (2006). *APM body of knowledge* (5 ed.). Buckinghamshire, UK: Association for Project Management.

Bannerman, P. L. (2008). *Defining project success: a multilevel framework.* Paper presented at the PMI Research Conference: Defining the future of project management, Warsaw, Poland.

Bennington, P. & Baccarini, D. (2004). Project benefits management in IT projects—An Australian perspective. *Project Management Journal, 35*(2), 20–31.

Beringer, C., Jonas, D., & Kock, A. (2013). Behavior of internal stakeholders in project portfolio management and its impact on success. *International Journal of Project Management, 31*(6), 830–846. doi: http://dx.doi.org/10.1016/j.ijproman.2012.11.006

Blomquist, T. & Müller, R. (2006). Practices, roles, and responsibilities of middle managers in program and portfolio management. *Project Management Journal, 37*(1), 52–66.

Bolles, D. L., & Hubbard, D. G. (2007). *The power of enterprise-wide project management.* New York: American Management Association.

Bradley, B. (2010). *Benefit realisation management: A practical guide to achieving benefits through change* (2 ed.). Surrey, England: Gower Publishing Limited.

Chen, C.-T. & Cheng, H.-L. (2009). A comprehensive model for selecting information system projects under fuzzy environment. *International Journal of Project Management, 27*(4), 389–399.

Cooke-Davies, T. J. (2005). *The executive sponsor – The hinge upon which organisational project management maturity turns.* Paper presented at the PMI Global Congress, Edinburgh, Scotland. Project Management Institute: Newton Square, PA.

Cooper, R., Edgett, S., & Kleinschmidt E. (1999). New product portfolio management: Practices and performance. *Journal of Product Innovation Management, 16*, 333–351.

De Reyck, B., Grushka-Cockayne, Y., Lockett, M., Calderini, S. R., Moura, M. & Sloper, A. (2005). The impact of project portfolio management on information technology projects. *International Journal of Project Management, 23*(7), 524–537. doi: http://dx.doi.org/10.1016/j.ijproman.2005.02.003

Dekkers, C. & Forselius, P. (2007). *Increase ICT project success with concrete scope management.* Paper presented at the 33rd EUROMICRO Conference on Software Engineering and Advanced Applications, 2007.

Dhillon, G. (2005). Gaining benefits from IS/IT implementation: Interpretations from case studies. *International Journal of Information Management, 25*(6), 502–515.

Eckartz, S., Daneva, M., Wieringa, R., & Hillegersberg, J. V. (2009). *Cross-organizational ERP management: how to create a successful business case?* Paper presented at the Proceedings of the 2009 ACM symposium on Applied Computing, Honolulu, Hawaii.

Hill, G. M. (2007). *The complete project management office handbook* (2 ed.). Boca Raton, FL: Auerbach Publications.

Hyväri, I. (2006). Success of projects in different organizational conditions. *Project Management Journal, 37*(4), 31–41.

Institute of Directors Southern Africa. (2009). *King Code of Governance for South Africa 2009.* Johannesburg: Institute of Directors Southern Africa.

Institute of Internal Auditors. (2013). *Definition of internal auditing.* Available from http://www.theiia.org/guidance/standards-and-guidance/ippf/definition-of-internal-auditing/?search%C2%BCdefinition. Accessed 31 March 2014.

International Institute of Business Analysis. (2009). *A guide to the business analysis body of knowledge (BABOK Guide)* (2 ed.). Toronto, Canada: International Institute of Business Analysis.

Jonas, D. (2010). Empowering project portfolio managers: How management involvement impacts project portfolio management performance. *International Journal of Project Management, 28*(8), 818–831. doi: DOI: 10.1016/j.ijproman.2010.07.002

Kaplan, R. S. & Norton, D. P. (2004). *Strategy maps: Converting intangible assets into tangible outcomes*. Boston, MA: Harvard Business School Press.

Kendall, G. & Rollins, S. (2003). *Advanced project portfolio management and the PMO*. Boca Raton, FL: J. Ross Publishing.

Marnewick, C. (2012). *A longitudinal analysis of ICT project success*. Paper presented at the Proceedings of the South African Institute for Computer Scientists and Information Technologists Conference, Pretoria, South Africa.

McDonald, J. (2002). Software project management audits—Update and experience report. *Journal of Systems and Software, 64*(3), 247:255.

Meskendahl, S. (2010). The influence of business strategy on project portfolio management and its success—A conceptual framework. *International Journal of Project Management, 28*(8), 807–817. doi: DOI: 10.1016/j.ijproman.2010.06.007

Mosavi, A. (2014). Exploring the roles of portfolio steering committees in project portfolio governance. *International Journal of Project Management, 32*(3), 388–399. doi: http://dx.doi.org/10.1016/j.ijproman.2013.07.004

Müller, R., Martinsuo, M., & Blomquist, T. (2008). Project portfolio control and portfolio management performance in different contexts. *Project Management Journal, 39*(3), 28–42. doi: 10.1002/pmj.20053

Office of Government Commerce. (2003). *Managing Successful Projects with PRINCE2*. London: Office of Goverment Commerce.

Ohara, S. (2005). *P2M: A Guidebook of Project & Program Management for Enterprise Innovation* (3 ed.): Project Management Association of Japan. Tokyo, Japan.

Project Management Institute. (2008). *A guide to the project management body of knowledge (PMBOK® Guide)* Fourth edition. Newtown Square, PA: Project Management Institute.

Project Management Institute. (2013a). *A guide to the project management body of knowledge (PMBOK® Guide)* Fifth edition. Newtown Square, PA: Project Management Institute.

Project Management Institute. (2013b). *The standard for portfolio management.* Third edition. Newtown Square, PA: Project Management Institute.

Pricewaterhouse Coopers. (2012). Insights and trends: Current portfolio, programme, and project management practices. PwC. Available from: http://www.pwc.com/us/en/increasing-it-effectiveness/publications/strategic-portfolio-management.jhtml

Rajegopal, S., McGuin, P., & Waller, J. (2007). *Project portfolio management: Leading the corporate vision*. New York: Palgrave Macmillan.

Reilly, F. K, & Brown, K. C. (2012). *Analysis of investments and management of portfolios*. Mason, OH: South-Western,Cengage Learning.

Remenyi, D., Money, A., & Bannister, F. (2007). *The effective measurement and management of ICT costs and benefits* (3 ed.). Oxford, United Kingdom: CIMA Publishing.

Remenyi, D. & Sherwood-Smith, M. (1998). Business benefits from information systems through an active benefits realisation programme. *International Journal of Project Management, 16*(2), 81–98.

Sichombo, B., Muya, M., Shakantu, W., & Kaliba C. (2009). The need for technical auditing in the Zambian construction industry. *International Journal of Project Management, 27*(8), 821–832.

Silvius, G., Ron Schipper, R., Planko, J., Van den Brink, J., & Köhler, A. (2012). *Sustainability in project management*. Surrey, England: Gower Publishing.

Stanleigh, M. (2009). Undertaking a successful project audit: Business Improvement Architects. Available from http://www.projectsmart.co.uk/undertaking-a-successful-project-audit.html. Accessed 31 March 2014.

Stewart, R. A. (2008). A framework for the life cycle management of information technology projects: ProjectIT. *International Journal of Project Management, 26*(2), 203–212.

Thomas, G., & Fernández, W. (2008). Success in IT projects: A matter of definition? *International Journal of Project Management, 26*(7), 733–742.

Turner, J. R. (2006). Towards a theory of project management: The nature of the project governance and project management. *International Journal of Project Management, 24*(2), 93:95.

Unger, B. N., Gemünden, H. G., & Aubry, M. (2012). The three roles of a project portfolio management office: Their impact on portfolio management execution and success. *International Journal of Project Management, 30*(5), 608–620. doi: 10.1016/j.ijproman.2012.01.015

Williams, D. & Parr, T. (2008) *Enterprise programme management—Delivering Value.* New York: Palgrave MacMillan.

10

Building the Bridge between Organizational Strategy and Portfolio Management

Alignment and Prioritization

MANUEL VARA, PMP

Introduction

Successful portfolio management requires programs and projects directly interconnected to the organization's strategy. As per *The Standard for Portfolio Management* – Third Edition (PMI, 2013) strategic alignment and prioritization analysis are two fundamental techniques to achieve this objective.

In most of the cases, the alignment of an organization's strategic portfolios becomes difficult mainly because of the poor execution organizations do when communicating their strategy. Hence the implementation of proven communication tools may help to address these issues.

On the other hand, another relevant issue is to have a detailed look as to how prioritization is done. What tools or methods are used by corporations to do so, and what type of governance is implemented to make sure plans are properly executed can be critical in order to have an effective prioritization process in place.

Simply rating competing projects with a rank number may be considered too subjective; more detail and some tools and techniques are required in order to back up relevant prioritization decisions. So these organizations may need more detailed prioritization models that employ categorization rules in place, the use of weighted rankings, or even different types of scoring techniques.

This chapter introduces a brief description of the importance of using appropriate tools and techniques in order to align strategy and project portfolio management and to prioritize portfolio components, in our case, programs and projects. While the usage of strategy maps and balanced scorecards are introduced in general management terms, they have also been successfully applied in other areas of business management. In this chapter, they are specifically applied to strategic alignment analysis of portfolios.

Considerations and general guidelines are explained to create prioritization plans including establishing domains, prioritization criteria, weighted values, and scoring anchors. Finally, we introduce a possible procedure to create a prioritization scoring model as a way to classify and prioritize portfolio components.

The management tools and techniques presented in this chapter have the aim of giving new ways to help organizations to coordinate plans, execute strategies, and/or help managers to better prioritize their analysis.

The Importance of Strategic Alignment and Strategy Prioritization to Portfolio Management

In general terms strategic alignment can be defined as "a mechanism by which an organization can visualize the relationship between its business processes and strategies. It enables organizational decision makers to collect meaningful insights based on their current processes" (Morrison, Ghose, Dam, Hinge & Hoesch-Klohe, 2011, p. 3).

Andolsen (2007, p. 35) defines strategic alignment as "the link between organization's overall goals and the goals of each of the units that contributes to the success of those overall goals."

These definitions can be translated to a portfolio management view as a mechanism to establish the relationship between the portfolio components and the organization's strategy. Portfolio strategic alignment focuses work; contributes to project and program coordination, such as outcome redundancy elimination; assists in eliminating conflict; and promotes quality assurance.

Portfolio prioritization is the process of classifying portfolio components based on a certain prioritization criteria previously defined. One of the most important tasks for portfolio managers is setting up priorities for portfolio components with the objective of ensuring the best selection among all the potential alternatives.

Dye and Pennypacker (2000) state project portfolio managers must continually ask themselves if they are assigning resources to the highest priority portfolio components. Hence, portfolio components must be prioritized based on their relative importance and contribution to the overall strategy, but unfortunately, priorities may change which may be due, for example, to technical and business environment changes.

Lindblom and Eberhardt (2011, p. 24) note "there is a tendency to assign all projects in the portfolio with the same priority, which means that all projects come to be considered high priority, or at least equally important when in execution." They also indicate that the prioritizing process should not only be executed in the project selection phase but continuously during the portfolio life cycle by comparing the portfolio components relevance to prioritization criteria.

Therefore, in order to avoid this tendency a "priority ranking" only based on classifying portfolio components and assigning a number from a scale is far too subjective. A much more relevant model may be needed in order to back up the portfolio's prioritization decisions.

Strategic Alignment Analysis

As stated in the Project Management Institute's (PMI) *Standard for Portfolio Management*, strategic alignment analysis "focuses on the new or changing organizational strategy and objectives. This analysis should indicate where there are gaps in focus, investment or alignment within the portfolio." (PMI, 2013, p.44)

There are two dimensions within this definition:

1. Alignment of the overall portfolio objectives and goals, and
2. Alignment of the portfolio components

In order to define and control a proper portfolio alignment process we propose to use strategy maps and the balanced scorecard as relevant tools for the portfolio's strategic alignment analysis.

Strategy Map

Strategy maps emerged as part of Robert Kaplan's and David Norton's seminal work with the balanced scorecard. A well-known tool in business strategy management, they are also considered to be one of the most relevant ways to improve strategy communication by using a visual representation of key elements of components of the organization's strategy.

At their simplest, strategy maps normally describe how the organization creates value—addressing any missing link between strategy definition and strategy execution, mainly ones that are due to the fact that strategy maps show cause-and-effect linkages between the next four perspectives:

- Financial situation—in order to succeed financially what shall be done? How should the information be presented to shareholders?
- Customer relationships—in order to satisfy customers, how should interaction and reaction be presented in front of them?
- Internal business process definition—in order to properly satisfy our shareholders and customers, what business processes are needed to excel?
- Learning and growth capabilities—in order to achieve our vision, how should we sustain our ability to introduce change in order to improve?

As per Kaplan and Norton (2000), a strategy map is a generic "architecture" for describing strategy that helps organizations develop an integrated and systematic way of viewing their strategy. Strategy maps force organizations not only to think about their strategy in each of the four perspectives previously described (Financial, Customer, Internal, and Growth), but also to develop all critical outcomes and drivers needed to implement the strategy previously defined and to draw the linkages between them.

Developing a strategy map forces an organization to first reach a consensus on its strategy and related objectives and then to develop expected targets and initiatives.

Once we have finished the cause-and-effect linking between the perspective's strategies, the resulting strategy map should facilitate the definition of the portfolio management plan.

Portfolio Strategy Map

Applying the same principles of strategy mapping to the portfolio level we propose the use of a portfolio strategy map tool that can help align the different portfolio components with the organization's strategy.

The strategy map for a portfolio can be similar to the generic business strategy map, and an organization achieves portfolio strategic alignment when the portfolio strategy map links to the organizational strategy map. For example, if the organizational strategy is primarily a productivity strategy emphasizing driving as much possible cost out of the business to provide customers with the lowest price, then a portfolio strategy must reflect this goal. Therefore, this portfolio strategy should also be a productivity strategy, focused on programs and projects to reduce costs. The portfolio strategy map follows the corresponding four portfolio perspectives, including:

- Portfolio value. Similar to a business, the portfolio's strategic goal is to create value via programs and projects. This value can come through projects enabling the business to develop innovative products and/or services, expanding markets (growth strategy), or helping the business become more efficient and cost effective (productivity strategy).
- Customer. At the same time, portfolio components have customers they must satisfy. Satisfaction can stem from the development side by delivering the right program objectives and delivering projects on time and on budget.
- Process excellence. Creating portfolio value and maintaining high levels of customer satisfaction requires excellence of the portfolio management process and project execution, which can be achieved following the standards, such as those of the PMI.
- Future orientation capabilities. Process excellence in portfolio management and project execution require a skilled and motivated workforce with clear goals and objectives that are aligned with the portfolio goals and ultimately with the organization's strategy. Furthermore, team members must be led by capable program and project managers and have the knowledge and access to the tools and information required to perform their jobs.

The Balanced Scorecard

The balanced scorecard (BSC) has emerged as a decision support tool at the strategic management level to comply with operational strategies. During their study on the evaluation of organizational performance, Kaplan and Norton (1996) introduced the concept of "Balanced Scorecard" (BSC) as a methodology aimed at revealing problem areas within organizations and pointing out areas for improvement. It was also promoted as a tool to align an organization with its strategy, by deriving objectives and measures for specific organizational units from a top–down process driven by the mission and strategy of the entire organization.

The fundamental premises that underlie the use of a scorecard are:

- It should help us in our process of defining strategies, objectives, measures, targets, and actions.
- It must provide communication of strategic management and help convey what each member of the organization should do in order that their individual actions contribute to the fulfillment of these objectives.
- It can allow us to compare the evolution of goals and compliance (real versus planned).
- It should be simple to understand, easy to use for the end user, and easy to maintain for administrators.

The BSC is a collection of measures, which is arranged in cards. The measures are related to four major managerial perspectives, and they provide a comprehensive view of the business. In general, the BSC merges business strategies into a comprehensive management system (see Figure 10.1).

The BSC introduces four management processes to join short-term activities with long-term strategic goals (Kuang-Hua, 2005). These processes are:

1. Clarify visions and reach consensus to ensure the duty of vision transformation.
2. Clarify objectives and join compensation with performance through communications and connections.
3. Recognize operation planning has several roles: setting goals, unifying strategic motives, distributing resources, and setting benchmarks.
4. Create and modify visions through feedback. It should be a base to facilitate strategic learning and auditing.

Strategy maps are most effective when used in conjunction with the BSC. The strategy map helps clarify the strategy and related strategic objectives. The BSC is then used to establish measures, targets, and initiatives, such as programs and projects, to align and manage the performance of the portfolio against those strategic objectives.

Balanced Portfolio Scorecard

Since portfolio managers must continually monitor the organizational strategy and changes to control the alignment of the portfolio components, we propose to adapt the BSC concept to assist those managing a portfolio of programs and projects to control its strategic alignment. We develop a balanced scorecard for portfolio management that

Strategy Balanced Scorecard

	Strategy Theme	Objectives	Measures	Targets	Initiatives
Financial	Improve profitability; Increase Revenue; Reduce costs	Profitable Business Growth	Operating Income	20% increase	
		Reduced Operational Costs	Operating Costs	5% decrease	
Customer	Improve shopping experience; Increase product quality	Quality product certified from a well-known association	Return Rate	Reduced by 50% each yr	Quality Management Program
		Improve shopping experience	Customer Satisfaction Shopping Index	5% change	Customer Loyalty Program
Internal	Increase Innovation; Increase operational Efficiency	Increase Innovation	% New Product Revenue	30%	NPD Program
		Increase Operational Efficiency	Waste Reduction	60% Reduction	Green & Cost Reduction Manufacturing Program
Organizational	Merchandise Buying/Product Planning skills; Improve Information Technology	Train & equip workforce	% Strategic skills acquired	85% target employees	Strategic skills improvement project
		Improve Information Technology	IT Maturity score	7.5	

Figure 10.1 A complete strategy-based balanced scorecard.

measures and evaluates portfolio activities from the following perspectives: portfolio value, customer, internal process, and future readiness.

This Balanced Portfolio Scorecard (BPS) can be the foundation for a strategic portfolio management alignment system provided that the portfolio strategy map themes are followed, objectives and appropriate metrics are identified for each theme, and initiatives are linked to the defined strategies.

To evaluate the attractiveness of program and project proposals, or the success of ongoing or completed projects, appropriate criteria should be determined. At the minimum, criteria should include items that managers feel are most important and for which they can provide hard data or firm opinions. It is also important that the criteria be complete but not redundant, and that they are linked to the short- and long-term objectives of the organization. To determine the criteria set for a portfolio component evaluation, we use a model based on the BSC approach.

Portfolios, for the purpose of BPS, can be considered "internal organizations" requiring the same strategy control as the whole organizational strategy. In fact, because programs and projects are typically more structured than organizations, they are even more suitable for evaluation.

PMI's (2013) framework includes a series of processes that are described in terms of their inputs, outputs, tools and techniques used to transform the inputs into outputs.

The objective of the BPS is to support the evaluation process during the different portfolio management process groups.

At the defining process group, where plans are defined and developed, and the program and project proposals are aligned and prioritized, the BPS can be useful to clarify and translate the vision and the organizational strategy of and to set the appropriate criteria and targets for a initiative's (programs and projects) attractiveness. Measures in this case would usually be forward looking, representing what is expected from these portfolio components.

At the aligning process group, the scorecard might be used to control and optimize the targets and the alignment of the portfolio components with organizational strategy and allocate resources within and among portfolio components. The BPS could be instrumental in providing a relative measure of performance, evaluating the value of the portfolio components in the face of changing circumstances and priorities, and communicating the results throughout the organization. The measures in this case would be a mix of forward-looking measures, as mentioned above and backward-looking measures that represent what has already been accomplished.

Finally, at the authorizing and controlling process group, the BPS can be used as a method to track and review the portfolio as a whole to achieve the metrics defined by the organization.

A key element to any BSC-based tool, such as in project management tracking, is to serve as the baseline or benchmark against which performance is measured. Without a baseline, evaluation is impossible. Once a baseline for evaluation is determined, the evaluation and control are done against the benchmark and the targeted plans.

At the highest level, portfolios can be managed as operational work being undertaken by the organization. This aspect means BSC perspectives can be used to control the alignment of the organization's portfolio.

The BPS Financial Perspective For each portfolio component the financial perspective examines the bottom-line in monetary terms. It reflects the profitability, cash flow, cost versus budget, etc. The financial objectives serve as the focus for the objectives and measures in all the other scorecard perspectives. Every measure in the scorecard should be part of a cause-and-effect relationship to improve financial performance.

We should be aware of not focusing only in short-financial terms such as profit and cost because there are other benefits such as intellectual property. To overcome the tangible and intangible benefits issue, the BPS presents three other perspectives that ensure a more balanced evaluation of the portfolio:

- Customer Perspective
- Internal Business Processes Perspective
- Learning and Growth Perspective

The BPS Customer Perspective The customer perspective of our BPS looks at the market value of the portfolio components deliverables, as well as stakeholder satisfaction with the final outcomes. Customers may be interested in how the portfolio components achieve their requirements (business and functional) for example timeliness, service, or quality that the portfolio component provides. This perspective can include measures taken from customer meetings, surveys, complaints, delivery statistics, etc. One could consider whether or not the programs and projects are considered a success by the customer. Time to market, quality, and performance, as well as the way the customer is treated and how his or her expectations are satisfied, are all relevant to evaluate the portfolio component.

The BPS Internal-Business Process Perspective This perspective measures the contribution of the portfolio component to the core competencies of the organization. It addresses the degree to which the proposed program or project supports the organization's mission and strategic objectives.

Obviously, it is assumed that top management has determined the strategic direction of the organization's internal processes beforehand. Considering the organization's internal desired goals, such to develop a new competitive advantage, it must establish specific measures to reflect this goal.

When the portfolio component fit is poor, the project or program proposed must be rejected, or the strategy must be rethought. Otherwise, the fit level, be it strong, good, moderate, or only peripheral, should affect the measure of the portfolio component attractiveness for prioritization purposes.

The BPS Learning and Growth Perspective The objective in the learning and growth perspective is to provide the infrastructure to enable the objectives of the above three perspectives. When the evaluation is solely based on the short-term financial perspective, it is often difficult to sustain investments to enhance the capability of the human resources, systems, and organizational processes.

Hence, this perspective looks at the long-range growth impact of the portfolio component. This includes checking whether the portfolio component is a platform for growth or not and looking at the durability of its effects.

Using a BPS Checklist For every portfolio component we can create a checklist using the BPS from the four perspectives, improving the strategic view of the project or program analyzed. It can be customized according to the organizational strategy so that it will define the business strategy depending on the situation of the portfolio component and its environment.

Amarasuriya and Jayawardane (2011) suggest the use of a checklist for every portfolio component to control its strategic alignment.

As shown in Figure 10.2 a portfolio component's BPS card contains the four BSC perspectives with several specific measures of the program or project analyzed.

Portfolio component	Objectives			Est.value
NPD Improvement Facility Project	Increase Revenue in NPD product line			$18 million
Perspective	Measure	Weight	Status measure	Avg. perspective
Financial	Has a minimum 80% project funding reserved for the Project?	60%		
	Will the economic situation affect the expected completion?	20%		
	Is there any other source of funding available?	20%		
Customer	Will the customer see this as the solution for their need?	100%		
Internal	Will the organization's existing capacity be able to complete this project?	50%		
	Will this project add competitive advantage to the organization?	50%		
Learning & Growth	Will this project bring new technologies to the organization?	40%		
	Will this project be an innovative solution for the existing deficiency?	60%		

Figure 10.2 Sample checklist portfolio component BPS evaluation card.

The color 'Green' means that portfolio component meets a certain measure aligned with strategy and stakeholder's expectations. 'Orange' means that portfolio component could have potential problems to meet a certain measure with strategy and stakeholder's expectations. 'Red' means that the portfolio component does not meet a certain measure, and therefore, it is not aligned with strategy and stakeholder's expectations. Every measure has a weight to get the average status of each perspective.

The introduction of the BSC perspectives enables improving the strategic vision of the portfolio management in order to analyze the strategic alignment of every project or program within the portfolio.

Prioritization Analysis

PMI (2013, p. 180) defines prioritization analysis as "a technique to compare and rank selected portfolio components, based on their evaluation scores and other management considerations, to ensure alignment with organizational strategy and objectives."

PMI also indicates that a Portfolio Strategic Plan should contain a prioritization model that guides the ongoing decisions about the portfolio components that should be added, terminated, or changed, as well as prioritizes and balances the component mix over time.

Managers directly or indirectly set the priority of each portfolio component through strategy definition. The portfolio manager uses this prioritization throughout the portfolio management process to properly plan the portfolio, evaluate the impact of strategic changes, and take corrective action.

Before designing a prioritization model for portfolio management we need to consider:

- Methods to define different types of project work
- Continually aligning prioritization decisions with business strategies
- A comprehensive set of programs and projects
- Accurate and complete program and project information
- Decisions should be fact-based and supported by a consistently applied process
- Difficult decisions will need to be made. In some cases lack of resources will lead to eliminating some portfolio components.

Portfolio Prioritization Plan

In order to ensure the proper prioritization we need to define a plan that specifies the domains, the prioritization criteria, the proposal requirements for each candidate project or program, the weighted values, and the scales for the scoring anchors. The following steps may be followed to create the portfolio prioritization plan.

1. Establish domains: A domain is the grouping of projects to which a standard set of criteria is applied for prioritization (Geoghan & Snow, 2001). Domains can be defined in many ways, some examples are:
   - Corporate focus
     - Target markets
     - Product/services types
     - Technical expertise
     - Industry focus
     - Investment goals
   - Business unit focus
     - Functional area
     - Customer type

  - Expense category
  - Internal/external customers
  - Geographic location
- Product focus
  - Customer base
  - Market segment
  - Competition

2. Define the project domain parameters: As Geoghan & Snow (2001) noted, the domain parameters are the set of unique characteristics that define which projects belong to a specific domain. To ensure the correctness of a domain we need to be sure of the following:
   - All project work fits within a specific domain
   - Each domain may emphasize certain business strategies differently
   - Project prioritization criteria and scoring anchors are domain-specific
   - A clear definition of the domain parameters is essential
   - Project characteristics can be used to define domain parameters:
     - Project type
     - Primary source of resources
     - Business contribution
     - Investment required
     - Size/level of effort
     - Duration of effort
     - Phase of the project life cycle
     - Degree of definition

   Once the project characteristics involved are determined, we also use the *Is/Is Not* technique to define domain parameters

IS	IS NOT
Characteristics are included within the domain	Characteristics are not included within the domain

3. Identify organizational strategies: If the organization has already done its portfolio strategic alignment analysis, using for example a portfolio strategy map and a BSC as tools, it will be easier to identify the strategies implied in the portfolio management. If not then we have to do it now. The current strategies could be discovered in annual reports, vision and mission statements, capability statements, corporate history legend, and corporate behavior.
4. Define the prioritization criteria: The organizational strategies must drive the prioritization criteria definition. We need meaningful criteria. Some characteristics to create these important criteria are:
   i. Few in number
   ii. No overlap

iii. Understandable
iv. Clearly measurable
v. Consistently applied
vi. Appropriate for each portfolio domain

The portfolio component criteria range between tangible and intangible, as shown in Figure 10.3. As Geoghan & Snow (2001) note, this is an important step because the prioritization criteria drive the portfolio management process, providing important information for the authorization of resource assignments, the project priority ranking, and the optimization and adaptive action activities.

5. Define the proposal requirements: To select the portfolio components we need to know more about them, so we should define the pre-selected program and project requirements in terms of the identified strategies.

   How does the project or program fit into or support the company's strategic plan? To get this information we can create a strategic business case, a project or program high-level plan, and a portfolio component definition.

   The high-level plan should contain key milestones, a schedule, resources estimates, key assumptions, major issues, and risk assessments.

   In the project or program definition we can consider the main objectives, goals, and expected outcomes such as the target market, market penetration, profitability, customer satisfaction, and regulatory compliance.

6. Establish the weighted values: The use of weighting and scoring techniques allow ranking portfolio components according to prioritization criteria (PMI, 2103). Some important considerations are:
   - Not all the strategies are created equal
   - Cross-functional considerations are critical
   - The relationship between the criteria is expressed using weights
   - Criteria common that are used in more than one domain within the same portfolio may have different weights
   - Criteria weighting should be simple; the basic purpose is to provide separation between projects as ranked

   For each prioritization criteria we can establish a weight from a scale, for example choose one value from 0.5, 1, or 1.5.

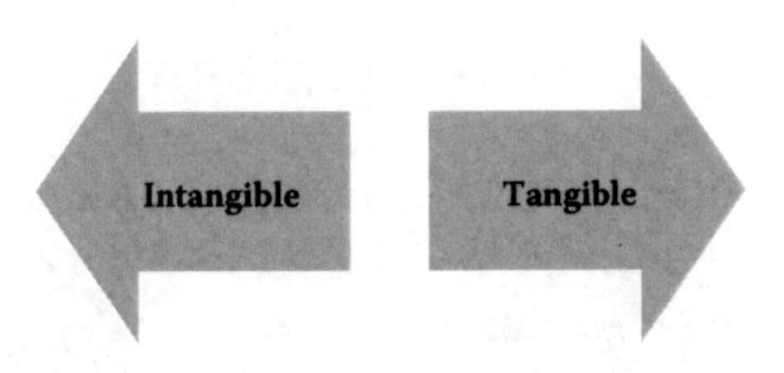

Figure 10.3 Portfolio component selection criteria.

7. Define the scoring anchors: We should rate portfolio components on a defined scale with an explanation of the meaning of each of the values. For example a three-value scale for each of the prioritization criteria:
   - Market penetration
     - 1 = No new markets
     - 3 = Growth in existing markets
     - 5 = Introduction to target market(s)
   - Return on Investment
     - 1 = Negative ROI
     - 3 = Break even
     - 5 = Positive ROI
   - Uses existing technology
     - 1 = Difficult to acquire
     - 3 = Easily acquired
     - 5 = No new technology

Create the Prioritization Scoring Model

Once we have the portfolio prioritization plan with chosen domains, prioritization criteria, organizational strategies implied, proposal requirements, weighting factors, and scoring anchors, we can proceed to create the prioritization scoring model with the following guidelines:

- Define and weight criteria for each domain
  - Criteria weighting could be a three-value scale: 0.5, 1, 1.5
- Determine scoring anchors for each criterion
  - Decide scoring scale for example from 1 to 5
- Rate each portfolio component
  - Applying selected criteria weighting and scoring anchors, rate each portfolio component completely before moving on to the next component.
- Compute the total scores
  - Finally, compute the total scores of each portfolio component to obtain a prioritized list

The application of this scoring model has several steps to complete taking information provided by the portfolio prioritization plan defined earlier.

1. Determine portfolio components by domain. We have to classify the proposed portfolio work for each domain:
   - Determine portfolio work: how many possible portfolio components do we really have?
   - Categorize work by domain: which portfolio components go in which domains?

- Require proposal submittal: ensure that every possible program or project has submitted a proposal as defined in portfolio prioritization plan.

2. Proposal quality review. Submitted proposals must have the mandatory content defined in the proposal requirements of the portfolio prioritization plan considering the following steps:
   - Ensure necessary information exists
   - Validate the proposal data for completeness
   - Use a standard proposal checklist
   - Perform an initial evaluation of proposals
3. Establish a portfolio component registry. Create a summary card for every portfolio component including:
   - Program/project sponsor name
   - Portfolio component priority ranking
   - Target milestone schedule
   - Resource requirements
   - Identified domain within the project portfolio
4. Score the portfolio components. Using a prioritization scoring model, such as the one defined above, every portfolio component is rated.

   Once we have the entire portfolio component scored, an opportunity for cross-functional consideration arises to optimize duplicate efforts identified in different programs or projects.
5. Prioritize the portfolio. A prioritized list of portfolio components emerges based on an objective review against prioritization criteria. This list will be a key component of the selection portfolio management process and of the programs and projects finally included in the portfolio.

Summary

The use of strategic alignment analysis and prioritization analysis are the two fundamental techniques to develop the Portfolio Strategic Plan (PMI, 2103).

Selecting the right programs and projects based on the organization's strategy is critical to ensure the business value. Tools, such as the portfolio strategy map, can help to improve strategic alignment and used in conjunction with the BPS, we can measure and control the dynamic changes in the strategy.

Prioritization analysis is a critical element for the selection process of the final components of the portfolio. PMI (2013) states the importance of using a prioritization model aligned with the organization's strategy, and we introduced the use a portfolio prioritization plan to ensure the contents of this model are well defined. As Lindblom & Eberhardt (2011) note, a prioritization process would not only highlight the most important projects in the portfolio but also would make sure that they also receive highest support.

References

Amarasuriya, U. and Jayawardane, A. (2011) Balance ScoreCard - A strategic project management tool for infrastructure development projects. *International Symposium on Social Management Systems,* 14-16 September 2011, Hotel Galadari, Sri Lanka.

Andolsen, A.A. (2007). Does your RIM program need a strategic alignment? *The Information Management Journal.* ARMA International, 35-40.

Dye, L. D. & Pennypacker, J.S. (2000). Project portfolio management and managing multiple projects: Two sides of the same coin? *Proceedings of the Project Management Institute Annual Seminars & Symposium, Houston, TX.* Newtown Square, PA: Project Management Institute.

Geoghan, T. & Snow, B. (2001). *Aligning projects with business strategies.* Book used in class "Mastering the Project Portfolio" taken at Stanford Advanced Project Management Program (2007).

Janhangiri, H. & Dashti, F. (2012) Model for performance evaluation of strategic information technology world. *Applied Programming. 2*(3), 175-183.

Kaplan, R.S. & Norton, D.P. (2000). Having trouble with your strategy? Then map it. *Harvard Business Review,* September-October, 1-10.

Kaplan, R.S. & Norton, D.P. (1996). *The balanced scorecard: translating strategy into action.* Boston, MA: Harvard Business School Press.

Kaplan, R.S. & Norton, D.P. (2007). Using the balanced scorecard as a strategic management system. July. *Harvard Business Review. 85*(7-8), 150-161.

Kuang-Hua, H. (2005). Using balanced scorecard and fuzzy data envelopment analysis for multinational R & D project performance assessment. *Journal of American Academy of Business.* 7, 189-196.

Lindblom, D. & Eberhardt, H. (May 2011) *A perspective on prioritization in project portfolio environment.* Stockholm, Sweden. Master Thesis. KTH Electrical Engineering.

Morrison, E.D., Ghose, A.K., Dam, H.K., Hinge, K.G., & Hoesch-Klohe, K. (2011). *Strategic alignment of business processes.* Australia, University of Wollongong. 7th International Workshop on Engineering Service-Oriented Applications, Paphos, Cyprus, 5 December.

Niknazar, P. (2011). *Evaluating the use of Bsc-Dea method in measuring organization's efficiency.* Borås, Sweden. Master Thesis. University of Borås.

Project Management Institute. (2013). *The standard for portfolio management.* Third Edition. Newtown Square, PA: Project Management Institute.

11

The Accuracy of Portfolio Risk Estimation

DAVID A. MAYNARD, BSEE, MBA, PMP

Overview

These portfolios, such as most business enterprises, are created to accept a calculated degree of risk for an expected reward. This reward can be received in short- or long-term. The key is to balance risk and reward in order to maximize the value of the portfolio to the organization. Without an accurate assessment of risk, the portfolio manager is treading on dangerous territory.

It should be clearly understood that the large body of well-developed mathematics dealing with the analysis of investment portfolios has only a small relationship to the analysis of a portfolio of projects, programs, and components. The following discussion, while borrowing from that body of science and math, deals with a different type of portfolio. The portfolio discussed here consists of a collection of programs, projects, or operations managed as a group to achieve strategic objectives.

Current State of Portfolio Risk Estimation

The project and program methods of estimating and managing risk are well-trodden ground. The mechanisms of creating risk estimates have been documented, debated, and discussed in many places by many people for many years. Certainly, these estimates can be dressed to have different appearances. They can be made to seem completely factual or represented in a complex and esoteric graphical manner. None of this can be faulted, and all contribute toward the evolution of the art and science of risk management, but the issue addressed in this paper is that of the accuracy of the risk estimation.

However, unlike a project or program, portfolio risk management serves to manage a suite of components as an aggregate pool, adding or removing items as required to maintain a predetermined and acceptable level of organizational risk tolerance and preserve gains made to organizational equity. Portfolio management includes the ability to allocate funding to protect organizational equity and to take actions to cut component funding in order to defend against excessive unanticipated losses. This is a very complex problem to resolve, and it exists in a complex environment.

By simply extending and expanding current project and program methods, it is extremely difficult, if not impossible, to accurately estimate the degree of uncertainty contained within a portfolio. We should not carry our project ranking systems and matrixes into portfolio management as our best tools. The complexities that adversely affect our ability to judge and assess risks accurately at the portfolio level do not exist at the project or program level. Once this constraint in estimation accuracy is accepted, how can we feel confident in obtaining a balanced portfolio of programs and projects as a means to implement an organizational-level strategy? We cannot! This leaves us in a frightening position. The organization is counting on our expertise to carry the strategic goals forward in an intelligent and controlled manner. Yet, there is a very large "unknown, unknown" segment of our risk estimations, and we have only the most rudimentary of methods at our disposal.

This is not to say that the efforts at the project and program level should not be conducted! There is absolutely no substitute for a thorough project and program level risk assessment. These provide critical and important information to the portfolio manager. But, another ingredient must be added into the risk assessment recipe at the portfolio level—complexity. Specifically, the complexity of the challenge the portfolio was created to resolve, and the complexity of the environment in which the portfolio exists must be examined.

Complexities abound. Even the word complex itself, especially regarding portfolio and program management, has needed further explanation (Ireland 2007). Complexities exist in every facet of everyone's life. However, it is safe to say that some undertakings are more complex than others. The project of installing a new pencil-making machine does not have as many or as difficult a set of complexities as a portfolio that the pencil manufacturer may have in order to meet its strategic goals.

This seems to be an insoluble problem. Because of the diverse nature of corporate organizations and their portfolios, arriving at a set of complexity parameters that suits each situation is not possible. In addition, it appears a significant portion of the duty of a portfolio manager is directly related to accurate risk assessments in order to balance risk versus reward. Furthermore, we have decided that we have an inadequate collection of risk assessment tools at the portfolio level. To make matters worse, it appears that we cannot arrive at a complexity solution that satisfies the majority of portfolios! It is time for a rethink.

Risk Assessment Accuracy

It is clear that the accuracy of any portfolio-level risk assessment is determined, to a large degree, by the complexity that can be attributed to the portfolio itself. The more complex a portfolio is, the less certainty or accuracy can be assured for its risk estimates; conversely the less complex a portfolio is, and the more accuracy can be assumed in the risk estimates. This leads us to a fundamental statement regarding complexity and risk estimation accuracy (Figure 11.1).

$$\text{Risk Estimate Accuracy} \propto \frac{1}{\text{Complexity}}$$

Figure 11.1 Fundamental risk estimate vs. complexity statement.

The proposal being made here is to not concern ourselves with mapping high, mediums, and lows, reds, yellows, and greens. What is being recommended is to use a unique computerized model to arrive at complexity values. We should use the power of the very common personal computer to help us estimate portfolio risk. For example, the program evaluation review technique (PERT) was developed in the 1950s and by the late 1950s was solved using computers (Fazar, 1959). Let us take the same step with portfolio risk analysis. Let us use the "learning" and solving power of everyday computers to assist us with our risk complexity problem.

Proof of Concept

The author decided to continue this line of research, including not only desktop analysis but also coding and limited testing of such a software system as a "Proof of Concept." However, more research, design and modeling were required before any Proof of Concept software was implemented.

Fundamental Design

What types of complexity factors might we be concerned with at the portfolio level? There exists a large body of literature dealing with complexity factors in the project, program and portfolio environment that were utilized as a starting point (see references). These were mind mapped, sorted through, duplicates eliminated, and project-specific complexities removed until only a trial set of portfolio complexity factors remained. These complexities were further categorized as "dimensions" and "factors."

Complexity Dimensions and Factors

In the categorization scheme chosen, dimensions are at the highest level and are composed of many factors. For the purposes of the Proof of Concept, eight dimensions were settled upon as determining overall portfolio complexity (Figure 11.2). These eight were chosen as the most reasonable extraction from the available literature and personal experience. No representation is made that this is the final or complete solution for the complexity dimensions of a portfolio. Without doubt, each organization and perhaps each portfolio will to examine its own environmental dynamics and determine what influences are at the highest level of the complexity hierarchy. The result may be less or more than the eight identified here. One of the "charms" of the penultimate complexity/risk assessment tool design will be to allow the portfolio

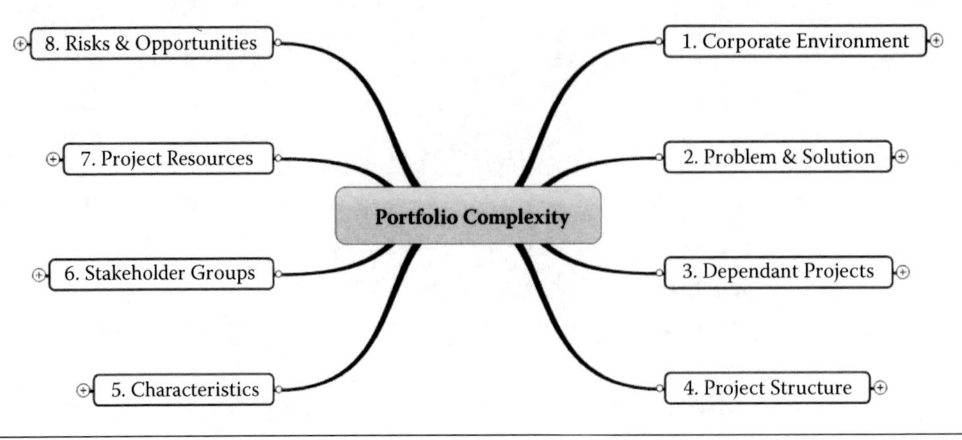

Figure 11.2 Selected dimensions of portfolio risk complexity.

management team to add, subtract or modify dimensions and factors. There exists no single solution to the portfolio complexity question.

While interesting, and a good first step, this level of detail is not a practical aid in determining either the relative complexity of a portfolio or our end-goal, the accuracy of our risk assessments. Beneath each dimension, further decomposition was conducted to include items that appeared to have the greatest affect upon them. For instance, the first dimension, "Corporate Environment" (Figure 11.3) was subdivided into three categories, and they themselves were subdivided.

Every attempt was made to not include "absolute" values. For example, failure of the portfolio's performance was roughly categorized as "no impact, moderate, or large impact." Instinctively, percentages were wanted here. Perhaps something like 10% or

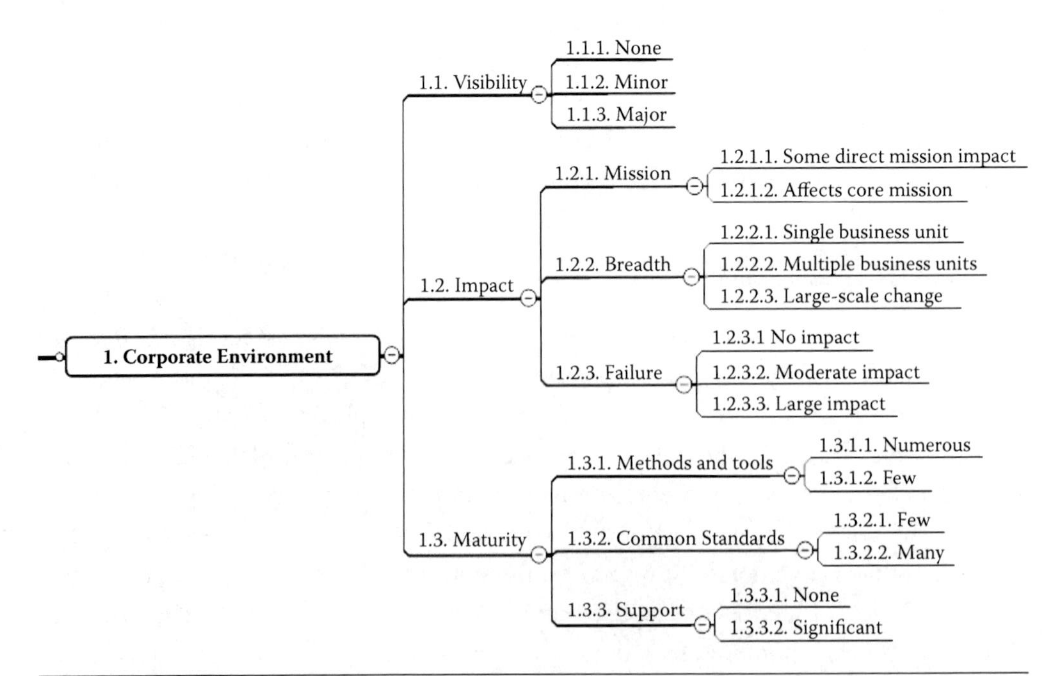

Figure 11.3 Factors contributing to the corporate environment.

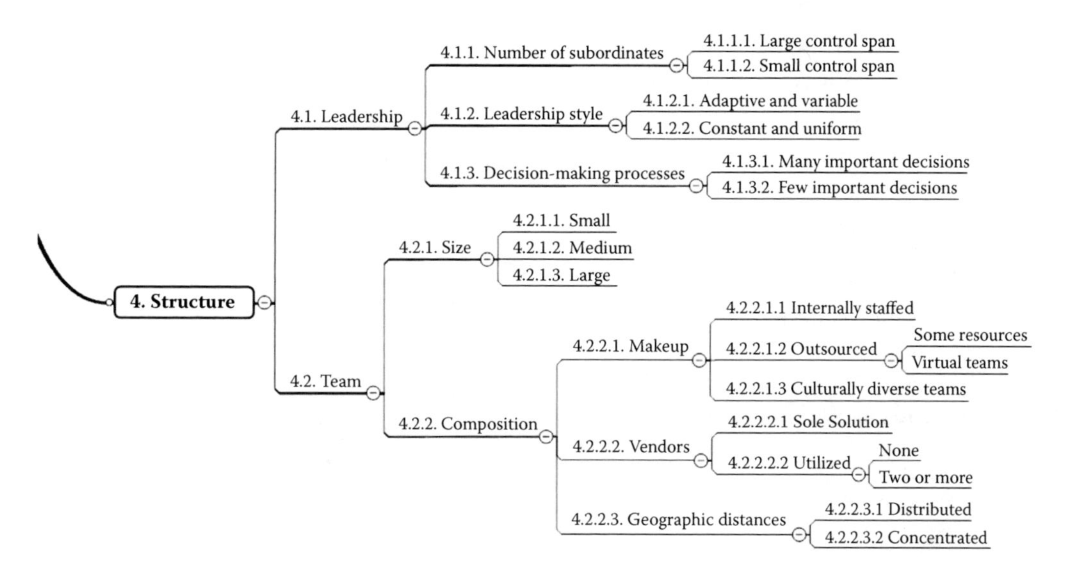

Figure 11.4 Factors contributing to the "structure" dimension.

15% impact to a quantifiable organizational goal. However, incorporation of absolute quantities and measures would disrupt any future scalability of the model within an organization. A failed portfolio may or may not have a large organizational impact, but it is still a failed portfolio! Resources have been spent, and equity lost. Perhaps, because of poor risk assessment accuracy. To be most useful, the complexity/risk assessment model must be applicable to the gamut of the organization's portfolios

A brief look at another of the eight dimensions shows this ideal of "non-absolute" measurement. Dimension 4 was chosen to be the structure of the portfolio within the organization. At the simplest level, the factors were "leadership" and "team." Attempting to quantify those to be useful for our approximation of complexity while avoiding absolute measurements became difficult. For instance, what is the leadership style? Are there defined decision-making processes? Are there a large number of decision-making processes? This dimension and its factors are in Figure 11.4.

While this determination of dimensions and factors is certainly not exhaustive, it can be easily seen that implementing a model with all eight dimensions will rapidly become burdensome for a user. And, the goal was to create a useful model that would help deal with portfolio complexity—not increase it!

How Can This Model "Learn?"

Of course, the word "learn" is misused here. The computer application will not actually learn, rather it will complete some clever math and remember the user's previous responses in order to continually "sharpen the pencil" when determining a complexity factor for the portfolio.

The model conducts a regression analysis to determine a value or values that can be used to modify risk predictions. As stated in the relationship above (Figure 11.1), the higher the complexity the less likely the estimates are to be accurate. Or, put another way, the tolerance of the risk estimates needs to be opened to accommodate the complexity.

Once the portfolio risks have been experienced or perhaps not experienced, the resultant information is fed back into the complexity model. The model then corrects the internal weightings for future predictions. It is this feedback from the portfolio team that is essential to the tool's viability. The more it is used and "corrected," the more accurate its predictions of complexity becomes.

Proof of Concept Development

For reasons of economy and ease of development, several decisions were made regarding programming language and environment. The Proof of Concept was written in Perl 5. Perl is a highly capable, feature-rich programming language that has been continually improved for over 26 years. Perl 5 runs on over 100 platforms from portables to mainframes and is suitable for both rapid prototyping and large-scale development projects. User entry screens were presented in standard HTML forms. Figure 11.5 is the entry form for dimension 6—stakeholder groups. Again, these are only preliminary dimensions and factors, the system allows for on-the-spot user tailoring since each application area is certain to have unique aspects.

All entered data and computed results are stored in an SQL database along with several tags including the original weight, the normalized weight, and of course the applicable complexity dimension and factor description.

The Proof of Concept system then produces a rough calculation of the overall portfolio complexity. This occurs in several steps. The first is the inputting of raw data of the factors for each dimension. The factor input is essentially a weighting done by the portfolio team. The selection of high/medium/low or perhaps 1, 3, 7, and 10 are in

Figure 11.5 Dimension six entry form.

essence weighting factors. These factors are then normalized to provide for an accurate comparison between them and aggregated to the dimension level. Each dimension, in turn, can be weighted for the portfolio's environment.

Several "alpha" testers were recruited to help validate the underlying concepts of dimensions, factors, weighting, regression analysis, correction, and prediction. Comments and suggestions from each of these testers were carefully considered and incorporated where possible.

Proof of Concept Usage

The first step in utilizing this tool would be to determine the dimensions and factors that apply within the specific organization. They can be anything, there are no limits, and the eight proposed dimensions are not firm or fixed. New or modified dimensions and factors would be added to the model as part of its data collection "front-end." It need not be correct and more than likely will never be completely correct. Things change, markets drift, organizations re-align, so shall the model will need to be re-calibrated.

Presentation of the Results

Different graphical representations were reviewed tried and tested. It was difficult to choose one that represented the complexity information in a format that could be readily understood and easily presented to the user. Eventually, a simple bar chart was decided upon, see Figure 11.6. This area of the Proof of Concept will most certainly need future development.

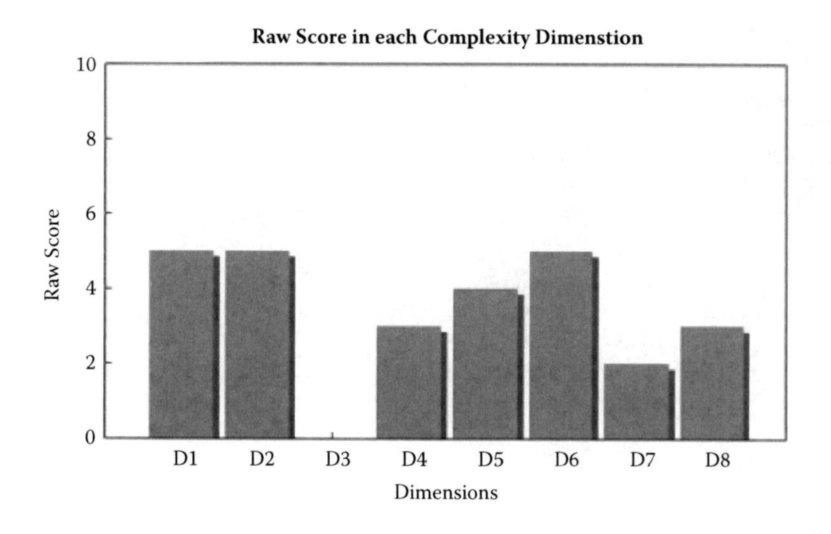

Figure 11.6 Graphical model results.

Applying the Results

Each dimension of the model is normalized, meaning that each resultant score is modified to represent a percentage of a maximum possible score. At this stage in the model, that uniqueness is demonstrated. From the normalized and aggregated dimensional data, a multivariate linear equation is derived with stated confidence intervals that can, in turn, be used to "moderate" the risk assessments.

If a portfolio risk analysis was conducted and it resulted in, perhaps one of the resultant risks with a value of $10,000 and a confidence level of 20%, by combining the results of the complexity model to this analysis, it may be found that the confidence level is too narrow, or perhaps the value stated is considered to be too high. Again, what is suggested here is that the fundamental techniques of risk analysis are conducted, and the results moderated by an estimated level of complexity for the portfolio. And, that estimated level of complexity is established by the portfolio management team and continually updated, pruned, and sharpened so as to improve future risk estimations.

The Future

The concept that complexity inversely affects the accuracy of risk estimates appears clear. Yet we, as practitioners, do very little to incorporate the complexity facet into our thinking. Much work remains to be done in this area. To date, the idea has been conceived, a proof of concept coded, and some testing done by individuals in several different application areas. This is far from being a fully tested model, or one whose value is proven. Yet the concept is very tempting!

References

About Perl. (n.d.). *The Perl programming language*. Retrieved February 27, 2014, from http://www.perl.org/about.html

Curlee, W. &. Gordon, R.L. (2011). *Complexity theory and project management*. Hoboken, NJ: Wiley.

Evans, E.. (2004). *Domain-driven design: tackling complexity in the heart of software*. Boston: Addison-Wesley.

Fazar, W. (1959). "Program evaluation and review technique." *The American Statistician 10* April asd. Web. 17 Dec. 2013.

Hass, K. B. (2009). *Managing complex projects a new model.* Vienna, VA.: Management Concepts.

Ireland, L. (2007). Project complexity: A brief exposure to difficult situations." ASAPM. N.p., n.d. Web. 17 Nov. 1013. <http://www.asapm.org/asapmag/articles/PrezSez10-07.pdf>.

Jackson, K. (2012). *Winning the complexity battle*. Newtown Square, PA.: Project Management Institute.

Merkhofer, L. Technical terms used in project portfolio management (Continued). *Product Portfolio Management to Q-Sort*. Lee Merkhofer Consulting, n.d. Web. 23 Feb. 2014. <http://www.prioritysystem.com >.

Project Management Institute. (2013). *The standard for portfolio management.* Third Edition. Newtown Square, PA: Project Management Institute.

Stacey, R. Griffin, D., & Shaw, P. (2000) *Complexity and management: fad or radical challenge to systems thinking?* London: Routledge.

"Taming complex projects." (2005). projectcomplexity.de. N.p., n.d. Web. 29 Aug. 2013. <http://www.projectcomplexity.de/pdf/Paper_Florian_Prange_NFF_2005.pdf>.

Thiry, M. (2013). *Ambiguity management the new frontier*. Newtown Square, PA.: Project Management Institute.

"Underestimating project complexity. (2013) "ProjectManagement.com -. N.p., n.d. Web. 5 June 2013. <http://www.projectmanagement.com/presentations/219130/Underestimating-Project-Complexity->.

12

Marketing the Project Portfolio

RODNEY TURNER, PhD AND
LAURENCE LECOEUVRE, PhD

An essential, but under recognized, part of communication on projects, programs, and portfolios is marketing of projects and programs. To win the support of the various stakeholders to a project, program, or portfolio, you need to sell them the benefits, and persuade them that the commitment they are going to make, a price they pay, is worth the benefit to them. There are three organizations involved in the management of projects, programs, and portfolios: the project or program itself; the investor, comprising all of its portfolios; and contractors undertaking projects and programs for the investor. All three organizations have to do marketing, and they have to market to a range of different stakeholders. The different stakeholders are effectively different market segments, and need the messages sent to them, the promotion, tailored in different ways. They will all support the project or program in a different way, and so it will have a different cost to them, and they will all perceive different benefits. So the project or program needs to be promoted to them in different ways and in different places.

Project, program, and portfolio marketing is a new concept, and in this chapter we will describe some of the current ideas. In the first section we will describe the three organizations involved in project, program, and portfolio management, and the different types of marketing they do. Two of the organizations, the investor and the contractor, do their marketing at the portfolio level. The third, the project or program, does the marketing at that level. We then describe the two traditional elements of marketing, market segmentation and the 4Ps (product, price, place of sale, and promotion). A model of project results suggested by Turner (2014) is used to suggest a range of different stakeholders involved in projects, programs, and portfolios and to show how it suggests they need different forms of communications. We then consider how the different stakeholders perceive the product, price, promotion, and place of sale. We next describe the marketing that needs to be done by the three organizations: the project or program, the contractor, and the portfolio or investor.

The Three Organizations

Winch (2014) suggests that there are three organizations involved in the management of projects, Figure 12.1:

1. *The investor:* is a permanent organization that initiates the project or program, provides the resources for its execution, and receives the benefits from the operation project's output. The investor will manage the projects and programs it does through a set of portfolios, and so undertakes its marketing through the portfolios. This type of marketing we call marketing OF the project.
2. *The contractor:* is a permanent organization that undertakes a portfolio of projects or programs, doing work for a range of clients. Contractors need to do marketing to win new business, and this marketing also is done at the portfolio level. They need to identify where clients have a need for potential new projects and sell their capability for the project, they need to maintain communication during the project, and then after the project they need to maintain communication with the client, to win new business. This type of marketing we call marketing FOR the project.
3. The project or program is a temporary organization. It is the vehicle through which the investor makes the investment to deliver beneficial change, that is builds a new asset, (the project output), which through its operation, (the project outcome), will provide benefit. The project needs to engage various stakeholders and so will communicate with the stakeholders to win their support. This type of marketing we call marketing BY the project. (For simplicity, for the rest of the chapter we will refer only to "project" rather than "project or program".)

As you see, two of the organizations involved in project marketing are permanent organizations and do their marketing through the project portfolio. The other is

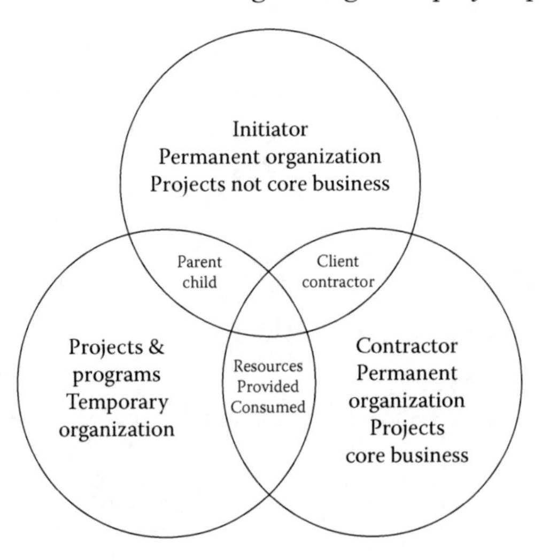

Figure 12.1 Three organisations involved in the management and marketing of projects, (© Turner, see Turner, 2014).

temporary, and the marketing is done at the project level. We will discuss the marketing by the three organizations in the reverse order: project, contractor, and investor; BY, FOR, and OF.

Marketing

We now consider two elements of traditional marketing as they relate to projects, programs and portfolios:

- market segmentation
- the 4Ps

Case

We are going to try to illustrate the ideas that follow by suggesting what it means for a project in the United Kingdom in an early stage of feasibility: the so-called HS2. This is a high speed railway line that has been proposed connecting London, Birmingham, and Manchester. It is called HS2 because HS1 was the high speed line that connected London to the Channel Tunnel. HS1 was the first new mainline that had been built since the late 19th century. All lines built in the 20th century were light rail. Current estimates suggest that HS2 will cost £50 billion, although previous experience on similar projects suggests the final cost may be double that amount.

Being a mega project, the project is in reality a program, consisting of many lines, bridges and tunnels, stations, and other projects. A program consists of a portfolio of related projects, but also the project is part of a portfolio of rail and other transport projects the government is undertaking.

Market segmentation

Market segmentation is a marketing concept whereby prospective buyers are placed in groups (or segments) that have common needs and will respond to a marketing action in a similar way. Market segmentation enables companies to target different types of consumers who perceive the benefit of certain products and services differently from one another and so will buy them in different ways and pay different prices for them. It is suggested that each market segment should obey three criteria:

1. Homogeneity: each segment should have common needs
2. Distinction: each segment should be different from the others
3. Reaction: the members of a given segment should react in a similar way and differently from other segments

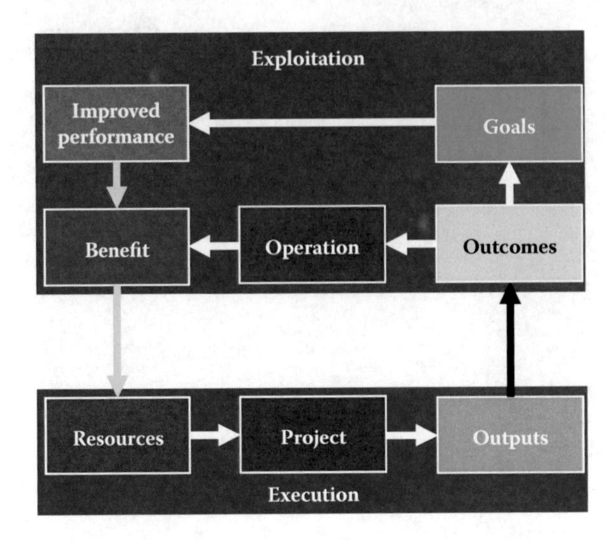

Figure 12.2 Results-based view of projects, (© Turner, see Turner, 2014; Turner, Huemann, Anbari and Bredillet, 2010).

Figure 12.2 illustrates different levels of project results, (Turner, 2014). There are different stakeholders associated with each element of this model, and they all represent different market segments.

Resources Resources provided to a project will take many forms: money, materials, people, and data. Different stakeholders will be associated with all of them. Bankers may provide money, suppliers materials, and contractors the people. The people actually working on the project are other stakeholders.

On HS2, the project will be paid for by the government out of tax revenues. So the country's tax payers are the ultimate financiers. The government is trying to persuade the tax payers that the project is worthwhile because it will enable them to get to Birmingham 10 minutes faster than now is the case. People say that they do not think it is worthwhile to pay £50 billion to get to Birmingham 10 minutes faster. It may be worthwhile paying that amount to get back 10 minutes faster, but what tax payers may be more interested in is reduced congestion, not having to stand the whole way to Manchester, and economic development in the North of the country.

The Project There are many stakeholders involved in the execution of the project. But most of them will be covered under the other areas. For instance we have already met the stakeholders providing resources. So we focus here on just the investor, the client, and the main contractor.

The investor: This is the organization proposing the project and so will primarily be marketing to other stakeholders trying to win their support for the project. But the main and other contractors will also be trying to win the investor's attention to give them the work of the project. For HS2 the main investor is the government, (though tax payers provide the actual money).

The client: The investor is the main client, but in a portfolio or program of projects, there are by definition projects which make up the portfolio of program. Each of those projects may have a client separate from the investor. HS2 will in reality be a program made up of a portfolio of projects. It may be broken into several stretches of line, and bridges, tunnels, and stations, which may be separate projects. There was a high speed line in the Netherlands, where the track foundation was done as a re-measurement contract, with the investor as the client; bridges, tunnels, and stations were done as design and build, again with the investor as the client, but the construction of the track was done as a Public-Private Partnership project, using Build-Own-Operate, (BOT), with the client as the Special Project Vehicle that owned the concession agreement. The client needs to market to the investor to win the concession and to the main contractor to encourage the contractor to bid for the work.

The main contractor: We met contractors providing materials and labour under resources. The main contractor(s) will manage the design, procurement, construction, and commissioning of the projects that make up the program or portfolio. There may be a different main contractor for each project. They need to undertake the marketing described in Section 4 of this chapter to win this and future projects.

The Output The output is the new asset the project constructs. In the case of HS2 it is the track foundation, the track (with signalling), bridges, tunnels, and stations. Again many of the stakeholders described under other elements will have an interest in the project's output, so we mention here only the IMBYs, (In My Back Yard). There will be NIMBYs, (No! Not In My Back Yard), and YIMBYs, (Yes, Please In My Back Yard). For HS2 the NIMBYs include people whose houses will be demolished or damaged by the construction of the line. YIMBYs may be people who see the project as providing economic opportunity in their community. The investor needs to tell the YIMBYs about the project and the economic opportunity it provides and may do that through television, newspapers, and social media. NIMBYs need to be told how they can claim compensation and be persuaded that the project is of great benefit to the country, and some collateral damage is unfortunately necessary. They can be contacted in similar ways.

The Outcome The project's output, (the new asset) will be operated to provide new competencies. The people who make use of those new competencies are the consumers. In the case of HS2, the outcome is reduced congestion, greater capacity, and faster travel times, and consumers will primarily be passengers and freight companies. Freight companies may be particularly interested in better access to European markets. At project commissioning the investor needs to contact them to make them aware that the new services are there and persuade them to use it. This may involve traditional marketing.

Operation To provide the outcome, the output is operated. The people doing this are called users or operators. They need to be communicated with throughout the design and construction stages of the project to get their input to the design of the output and its ease of operation. They may also need training. On HS2 they will include train drivers, signallers, station managers, and employees and their trade unions.

The Benefit What the consumers pay to use the service, or the project outcome, provides a benefit to the investor. The consumers also get benefit they are willing to pay for from the operation of the project's output. The company that was the investor during the project phases usually becomes the owner of the asset during operation. But sometimes the investor sells the new asset to a new owner who receives the income stream. That is particularly the case for the construction of office blocks or rental housing. An investor identifies the opportunity and constructs the building and then sells it to a financial institution who will receive the rental income. As we said for HS2, the track may be built as a BOT project, and so the owner of the concession agreement will receive income from the operation of the track. Stations may be sold to financial institutions. Traditional marketing may need to be used to identify them and persuade them to buy the stations.

Goals With time the operation of the project's output and outcome may enable the investor to achieve higher level goals and performance improvement. In the case of HS2 this may be economic development and higher tax revenues for the government. But as we said above, the ultimate investor in the project is tax payers, and they need to be persuaded the economic development is such that it makes the project worthwhile. It is the increased capacity and faster journey times to Europe for freight and the economic development that accrues from it, not getting to Birmingham 10 minutes more quickly, that makes the project worthwhile and tax payers need to be persuaded of this goal.

The 4Ps

Traditional marketing theory talks about 4Ps of marketing: product, price, place of sale, and promotion. Some people now identify seven or even eight Ps, but we want to focus here just on the traditional four.

Product Marketing theorists will tell you that people do not buy physical products, they buy benefits. You do not buy a computer mouse, you buy the ability to control a computer; you do not buy a bottle of water, you buy the ability to quench your thirst. Project managers need to learn project communication is not about telling people about Gantt charts and critical path networks, but it is about telling investors, consumers, users, YIMBYs, contractors, and others about the benefit they will get from the project. As we saw above, the types of benefits they will get will be quite different, and so the messages to the different stakeholders need to be framed in different ways.

Price Each of these stakeholders will make some commitment to the project, and that commitment is a price they pay to receive their benefit from the project. Some provide financial support, some provide their time, some provide emotional support, and some provide political support. The price people are willing to pay of course depends on the level of benefit they perceive they are getting. So when communicating with project stakeholders, you need to understand what benefit they are getting and how they perceive it, and understand what level of support they are willing to give, and phrase the messages to them accordingly.

Place of Sale Next you need to understand where the project impacts on the stakeholders and their lives. That is where you need to communicate with them. There will be different ways of communicating with different stakeholders. Contractors will be contacted through the trade press; YIMBYs and NIMBYs through television, newspapers, and social media; consumers via advertising literature; and users at their pace of work or through their trade unions.

Promotion So you need to tailor the message to the stakeholder, the benefit they perceive, the commitment they will make, and where the project impacts on their lives. You will do this through the communication plan.

Communication Plan

When developing a communication plan to promote the project to the stakeholders, there are a number of key questions you need to ask yourself.

Who is the target audience? We saw above that there are many different stakeholders. You need to segment the market for the project, understand what the objectives of each stakeholder are, what benefit they perceive, and what contribution they are going to make.

What are the objectives of each communication? Recognizing the objectives will then enable you to understand what you are trying to achieve when you communicate with each stakeholder.

What are the key messages? Once the objectives are known, you can identify what information needs to be communicated to each stakeholder.

Who will do the communicating? Different stakeholders need to be communicated with by different stakeholders. Who does the communicating may also depend on the information being communicated. The communication may be done by the investor, the main contractor, the portfolio manager, the program manager, the project manager, or technical experts. To be credible, some information needs to come from senior managers, other information from technical experts, and yet other information from people who understand project progress.

When will the information be given? Different stakeholders need to be contacted at different times. The YIMBYs and NIMBYs may need to be first contacted before the feasibility study starts to win their support from an early stage. Financiers and the main contractor need to be contacted during the feasibility study. The first contact with the consumers needs to be made then to demonstrate the commercial viability of the project and to the users to show technical feasibility. Suppliers of labour and material are first contacted during design, and the consumers and users are re-contacted during commissioning. Communication with financiers needs to be maintained during operation to show their investment is being repaid. The National Aeronautics and Space Administration (NASA), for example, continues to market Project Curiosity, the Mars rover, during its operation to maintain support of the public and politicians.

How will feedback be encouraged? Communication is two way. Communication is not transmission of information. You do not send out a message and hope somebody hears it. Do not scream in space. Under communication you send a message, someone hears it and sends a response, and you hear the response and know your message has been understood. So you must encourage feedback. Ask people to respond, and show you are listening, through your body language, by answering their questions, and by acting on their suggestions. If a stakeholder makes a suggestion that increases the value of the project, do it. If they make a suggestion which reduces the value of the project, avoid it. But if they make a suggestion that has no impact the value of the project, it may still be worthwhile implementing it to show you are listening.

What media will be used? You can choose from:

- Social media: including Facebook, Twitter, LinkedIn, and You-tube
- Project web pages
- Newspapers and radio
- Seminars, workshops, focus groups, and control meetings
- Videos, CDs, and newsletters
- Bulletins, briefings, and press releases
- Project, program, and portfolio lunches
- Open days and project exhibitions
- ....

Marketing BY the Project

The first of the three organizations we consider is the project itself. This is marketing by the project to engage stakeholders. Figure 12.3 is a stakeholder engagement process, (Turner, 2014).

Emotional Intelligence

But before we describe the steps of this process, and how they relate to project marketing, we would like to introduce a model of emotional intelligence. The engagement of stakeholders and marketing by the project requires emotional intelligence. A model of emotional intelligence is suggested by Müller and Turner (2010). There are four major competences of emotional intelligence, comprising 19 competencies. The objective of emotional intelligence is relationship management, and this is the purpose of stakeholder engagement, building relationships with the stakeholders to win their support for the project and by persuading them that the benefits they will get from the project are worth the commitment they are going to make. The four competences of emotional intelligence are:

Self-awareness: The leader needs to be aware of his or her own emotional responses to situations. When he or she is self-aware, two more competences follow.

Self-management: When the leader is aware of his or her emotional responses, he or she can manage the responses.

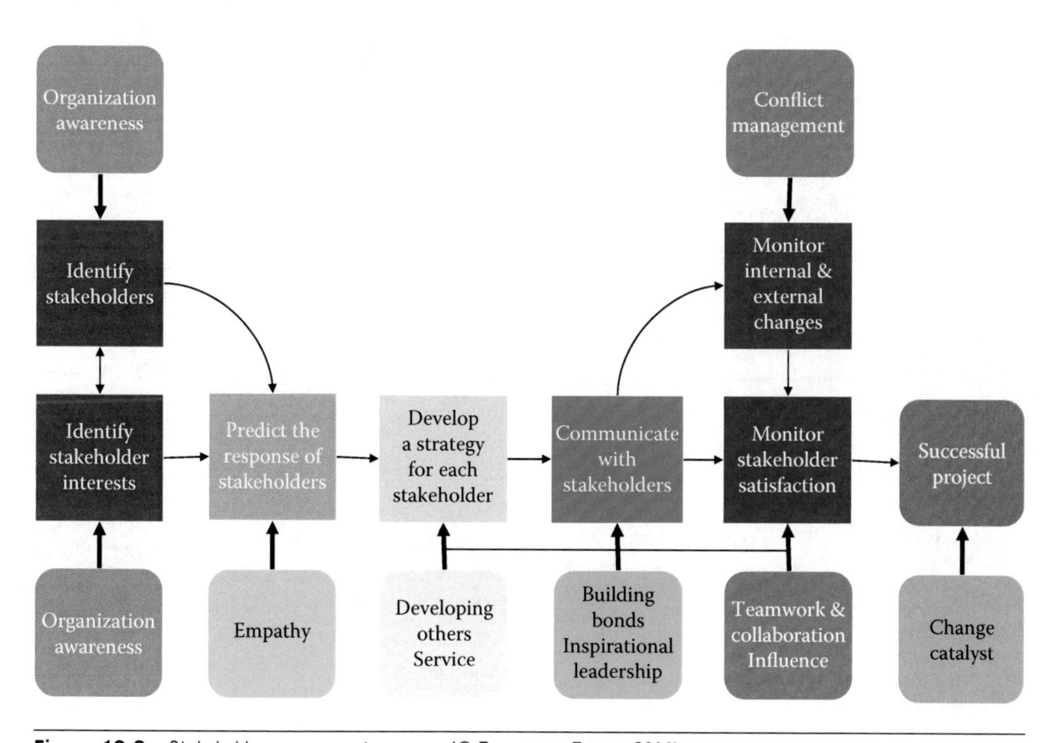

Figure 12.3 Stakeholder engagement process, (© Turner, see Turner, 2014).

Social awareness: He or she can then be aware of emotional responses in other people. From this and self-management flows the fourth competence.

Relationship management: When the leader can manage his or her own emotional responses and is aware of responses in other people, he or she is better at building relationships with the other people.

The 19 competencies associated with these four competences are shown in Table 12.1. We mainly focus on those associated with social awareness and relationship management.

The Stakeholder Engagement Process

Identify the stakeholders and their interests The first step is to identify the stakeholders and their interests. Figure 12.3 and the ensuing discussion should help in this step. This requires organizational awareness: you need to be aware of who the potential stakeholders are and their interests. You are trying to identify what commitment they have to make to the project (price) and what benefit they will perceive from the project (product). You also need to be aware of how the project will impact on their lives, (place of sale).

Predict the response of each stakeholder You need to try to understand how each stakeholder will respond to the project. What benefit, if any, do they perceive from the project, what commitment do they have to make, and if so how will they respond. This continues to require organizational awareness but also requires empathy; you need to

Table 12.1 The 19 Competencies of Emotional Intelligence

COMPETENCE	ASSOCIATED COMPETENCIES
Self-awareness	Emotional self-awareness
	Accurate self-assessment
	Self confidence
Self-management	Emotional self-control
	Adaptability
	Initiative
	Optimism
	Transparency
	Achievement
Social awareness	Empathy
	Organizational awareness
	Service
Relationship management	Building bonds
	Team work and collaboration
	Inspirational leadership
	Influence
	Developing others
	Conflict management
	Change agent

be able to predict their responses. There are three questions you can ask about each stakeholder to help in this step:

1. Is the stakeholder for or against the project or ambivalent?
2. Can the stakeholder influence the outcome?
3. Is the stakeholder aware of the project and its outcomes?

Different stakeholders will have to make different levels of commitment, perceive different benefits, have different levels of influence, and different levels of awareness. So the answers to these questions will identify different market segments for stakeholders in the project.

Develop a communication plan for each stakeholder So we come to the fourth P, working out how we are going to promote the project. This continues to require organizational awareness to understand what motivates the stakeholders, and empathy, to be able to predict how they will respond to your communication. However, you also need skills of developing others and service. You need to develop people so that they buy into the project and can make a positive contribution to it, but at the same time you are providing them with a service, the benefit the project gives them, and you need to be aware of that and develop you own sense of service. We described above the components of a communication plan.

Communicate with each stakeholder Then you need to enact the communication plan with each stakeholder. Through this active approach, you will be building bonds and providing inspirational leadership.

Monitor stakeholder satisfaction As the project progresses, you will monitor stakeholder satisfaction. This requires organizational awareness and empathy. You must have empathy to sense how stakeholders are reacting. You will also need to continue to build others and provide inspirational leadership. However, you will also need to influence team work and collaboration to encourage stakeholders feel they are working on a single team toward a shared goal. Through that you will also need to influence their behaviours. If stakeholders are not behaving as expected, you may need to revisit your communication strategy.

Monitor internal and external changes It is also possible unexpected changes occur, which change the stakeholders' positions and thus their response to the project. These need to be monitored, and if this means stakeholders are not behaving as expected, then you may need to revisit your communication strategy. It is possible that the changes will create conflicts and so draw on your conflict management skills.

Successful project Hopefully, the stakeholders engage with the project as desired, leading to a successful project and proving your skills as a change agent.

Marketing FOR the Project

The second organization we consider is the contractor. The project-based organization needs to win the next and future projects, and marketing at this level needs to be conducted as part of the portfolio of projects it is doing for all its clients, because as we shall see some marketing activities take place before, after, and between projects.

Marketing Models

An early model was developed by Cova, Ghauri, and Salle (2002). They suggested three phases of project marketing:

Independent of any project: Before any project has been actually identified, the contractor should be in contact with potential clients so the clients are aware of the nature of their business and the types of projects they can do. They should make their potential clients aware of their competencies, and make the client aware of how they (the contractor) can satisfy the client's potential needs. The contractor also needs to be aware of any projects the client may be considering, and try to influence the client's definition of those projects so the projects match their competencies and not those of their competitors.

Pre-tender: Once the contractor has identified a potential client, the contractor must liaise with the client to determine whether it is worth bidding for the project. The contractor must truly understand the client's requirements and whether they match its (the contractor's) competencies. The contractor may continue to try to influence the definition of the project to make it more closely match its competencies. The contractor also needs to understand the chance of winning the bid, determine whether the potential profit will be justified by the cost of bidding, and the chance of winning. As we describe below, throughout this stage and the next, the contractor needs to market to three types of decision makers in the client organization: the strategic decision makers, the operations managers, and the technical managers. They may require three teams of people to market to each of these three groups.

Tender: If the contractor decides to bid for the project, the contractor must prepare the tender, make the offer, engage in the negotiations, be successful as the contract is awarded, and start the work of the project.

Throughout these three stages, the emotional competencies identified above are important, but particularly consider organizational awareness, empathy, service, building bonds, inspirational leadership, and influence.

Lecoeuvre-Soudain and Deshayes (2006) developed a four-stage model. They effectively combined independent of any project and the initial stages of pre-tender into pre-project. They combined the latter stages of pre-project, tender preparation, and mobilization into project start. They then introduced two new stages, project delivery and post-project. During the project, the client's decision makers are project stakeholders, and they need marketing as we described above. But the key insight is the post-project stage. The contractor must remain in contact with the client looking forward to future projects. This quite clearly raises project marketing to the portfolio level, since contact with clients will be maintained through the portfolio of projects the contractor has done. If the contractor is providing logistical support to past projects, it will help the contractor to maintain contact with the operational managers, but the contractor must also work to maintain contact with the strategic decision makers and technical managers.

The Focus of Project Marketing

Lecoeuvre-Soudain and Deshayes (2006) suggest six areas of focus of project marketing, see Figure 12.4:

1. Relationship management (Rel)
2. Trust (Tru)
3. Collaboration (Col)
4. Communication (Com)
5. Training (Tra)
6. Going with (providing mentoring, coaching and support) (Gwi)

Relationship management, trust, and collaboration we mentioned previously as emotional competencies, and communication we mentioned as a key element of marketing. Training and going with are closely related to the emotional competency of developing others. Figure 12.5 shows how the emphasis on these six areas of focus varies during the four stages of project marketing.

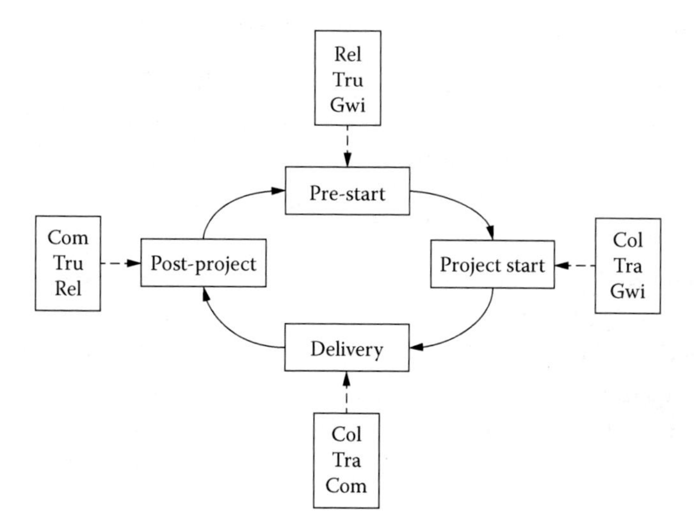

Figure 12.4 A four-stage project marketing process as viewed by the project marketing literature, (© Lecoeuvre and Turner, adapted from Lecoeuvre-Soudain and Deshayes, 2006).

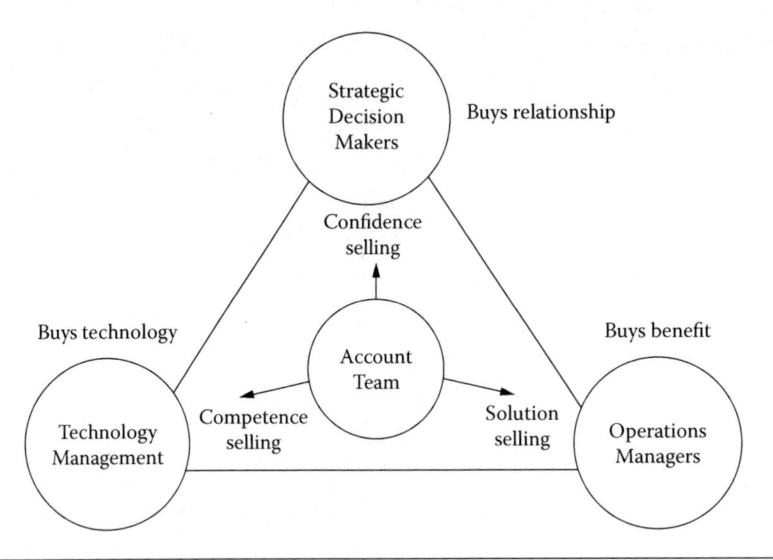

Figure 12.5 Three customers for the contractor's account team, (© Turner, see Turner, 1995).

Pre-project: The emphasis is on building relationships and trust and on providing mentoring, coaching, and support.

Project start: The emphasis is on collaborating and training and continuing to provide mentoring, coaching, and support

Project: The emphasis is on collaboration, training, and communication

Post-project: The emphasis is on communication to maintain relationships and trust.

Who Is the Focus of the Contractor's Marketing?

Figure 12.5 suggests the contractor needs to aim its marketing at three sets of decision makers in the client organization, (Turner, 1995). The contractor should identify different groups of people to target these three groups of people.

The strategic decision makers: These are the people who will ultimately decide to do the project and determine which contractor will be awarded the contract. These people are interested in the project's goal, Figure 12.2. The contractor's board of directors should target these people, with the help of the marketing department.

The operations managers: They are both the users and the consumers. There may be one set of operations managers who will operate the project's output, (the users), and another set who will make use of the project's outcome and obtain the benefit, (the consumers). These people are not interested in the technology. The consumers want the project's outcome to satisfy their requirements and provide them with adequate benefits. The users want ease of operation. It will usually be the role of the sales and marketing department to communicate with these people, although the project manager may also be involved. It is essential to make them comfortable that the project's outcome will satisfy their requirements and provide them with the benefits they want and an output that will be easy to operate.

The technical mangers: These are the people who will judge the contractor's technical solution and will be able to determine whether the project's output will work to provide the outcome. The contractor's technical managers must communicate with these people to persuade them of the contractor's technical competence.

Marketing OF the Project

The third organization involved in the management of projects is the project initiator, the organization that identifies the project opportunity, that an asset can be built, or a change undertaken that will provide benefit, initiates the work of the project to build that asset, and provides the finances. Very little has been written about this element of project marketing. Figure 12.6 shows the project as part of an investment process, (Turner, Huemann, Anbari, and Bredillet, 2010). The project builds the project output, the investment, a new asset that will be operated to provide benefit to repay the cost of its construction. The project is just part of the investment process. Figure 12.7 is a typical project process, with six stages: concept, feasibility, front-end design, execution, close-out, and exploitation.

Also, little has been written about the marketing of project's output and outcome. Early work on marketing during the pre-project and early project phases was done by Foreman (1996). However since she was looking at internal marketing, she mainly focused on one part of the initiating organization marketing the project to other parts of the same organization, and so to an extent it fell under stakeholder engagement. Marketing of the output itself, or the product made by it during commissioning and early operation, can be covered by standard marketing theory. However we believe

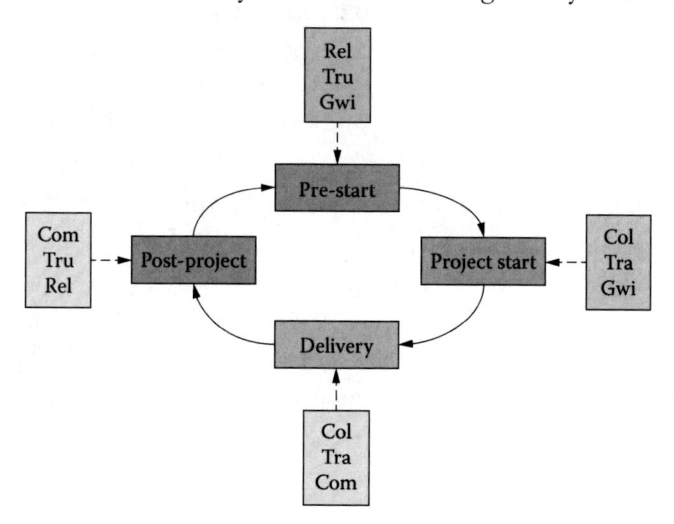

Figure 12.6 The investment process, (after Gareis, 2005; see Turner, Huemann, Anbari and Bredillet, 2010).

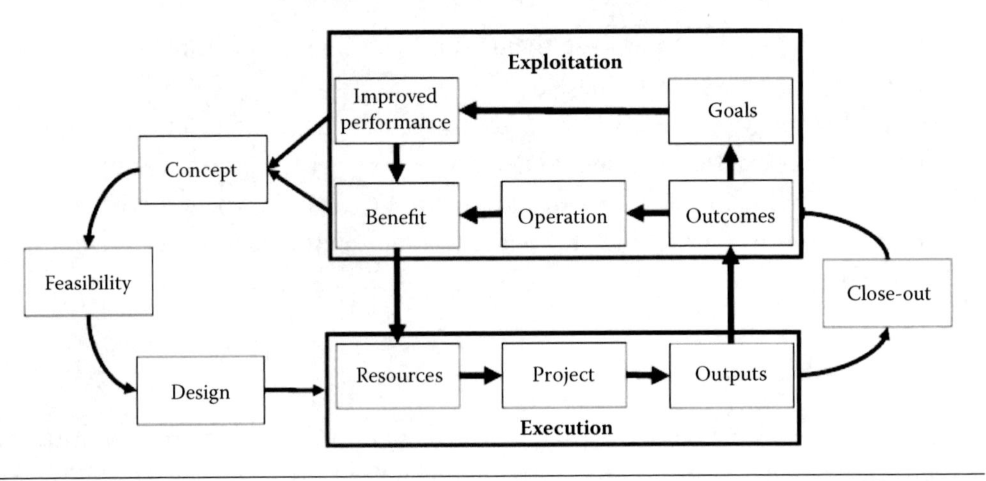

Figure 12.7 The project and investment process and overlaid on the three levels of project results, (© Turner, see Turner, 2014).

different marketing approaches need to be used during the pre-project and early project phases, and other marketing during the operation phase to win political or public support for future projects, as shown in Table 12.2.

Concept: The sponsor identifies a problem or opportunity and suggests a new asset can be built to solve the problem or exploit the opportunity so that the expected benefit will justify the expected cost. A potential investor needs to be persuaded to support the concept by showing the investment will provide adequate returns. They need to be persuaded to provide finances for the feasibility stage against the potential benefits. Political or public support needs to be obtained for infrastructure projects. Recall the example in the United Kingdom is the HS2 project, to build a high speed line to the North of the country. The wrong message has been sold. People do not think it is worth £50 billion to reduce the journey time to Birmingham by 10 minutes. They might be better persuaded it is justified by the potential economic development and to save having to stand all the way to Manchester.

Feasibility: The feasibility of the concept is shown, and the initial business plan is developed. The interest of the investor needs to be maintained, and the investor needs to be persuaded to provide finances for the front-end design by being shown the potential returns justify the cost. Political and public support needs to be maintained for infrastructure projects.

Front-end design: The business plan is further developed. The investor needs to be persuaded to provide the finances for execution by being further persuaded of the efficacy of the proposal to provide adequate returns. Political and public support needs to be maintained for infrastructure projects.

Execution: As money is spent the investor may waiver in its commitment. It needs to be maintained. If the project does not progress as expected, especially if it becomes overspent, political and public support must be maintained.

Table 12.2 Marketing of the Project's Output by the Investor

PROJECT STAGE	AIM OF MARKETING	EXAMPLES
Concept	Win support for the initial project concept Obtain political and public support Gain the interest of the potential investor Obtain funding for feasibility	Winning public support for infrastructure projects. For instance with HS2, persuading that: 1. Travelling to Birmingham 10 minutes more quickly is worth £50bn 2. Economic regeneration of the north of England is worth £50bn
Feasibility	Maintain support for the project concept Maintain political and public support Obtain funding for front-end design	
Front end-design	Convince interested parties in the efficacy of the proposal Maintain political and public support Obtain funding for detailed design and execution	
Execution	Maintain political and public support	
Closeout	Sell the project's output Establish a market for the outcome	Sell a newly built building to an investment company or to the owner occupiers Develop a market for a new product
Operation	Establish a market for the product made by the project's output Prove to the public or political sponsors the value of the product and obtain support for future projects	Use standard marketing For example NASA's marketing of Project Curiosity, the Mars rover

Commissioning: A developer will try to sell the asset. Otherwise an internal or external market needs to be developed for the outcome. In the case of the developer or external marketing, it will be covered by standard marketing theory. In the case of the internal market it will be covered by the work of Foreman (1996).

Operation: If the outcome is being sold, the market for that product must continue to be developed. Also the value of this new asset may need to be demonstrated to win political or public support for future projects.

Closing Remarks

Communication is an essential part of project, program and portfolio management. Müller and Turner (2010) showed that project and program managers are not very good at communication, but it is correlated with success on all projects. A primary focus of marketing for the project and its parent organization is to win support for the project, its output and outcome, and we have shown that is covered by the concepts of marketing. We have also shown that support for the project is an emotional commitment by the stakeholders, and so the emotional intelligence of the person doing the communication is critical. The contractor is not trying to win support for the project, its output and outcome, but its involvement in the project, and the focus of marketing are different.

References and Further Reading

Cova B., Ghauri, P., & Salle, R. (2002). *Project marketing—beyond competitive bidding.* Chichester: Wiley.

Foreman,S. (1996). Internal marketing. In Turner, J.R., Grude, K.V, & Thurloway, L. (eds) *The project manager as change agent.* London: McGraw-Hill.

Lecoeuvre-Soudain, L. & Deshayes, P. (2006). From marketing to project management. *Project Management Journal. 37*(5):103-112.

Müller, R. & Turner, J.R. (2010). *Project oriented leadership.* Aldershott: Gower.

Turner, J.R. (1995). *The commercial project manager.* London: McGraw-Hill.

Turner J.R. (2014) *The handbook of project-based management* (4th ed) New York: McGraw-Hill.

Turner J.R., Huemann, M., Anbari, F.T., & Bredillet C, N. (2010). *Perspectives on projects.* New York: Routledge.

Winch, G. M. (2014). Three domains of project organising. *International Journal of Project Management. 32*(5): 721–731.

13

Strategic Portfolio Management through Effective Communications

AMAURY AUBRÉE-DAUCHEZ, PMP, PgMP

Introduction

Challenging economic, financial, and social conditions force organizations to gain competitive hedge optimizing operations and implement change. Such change must be orchestrated through portfolio management to ensure its high alignment to business strategy, mission, goals, and objectives; it must genuinely put emphasis on effective governance and compliance while leveraging best practices and managing risks. However, to get value for money, organizations must place communication on the forefront enabling controlled, secured, and seamless flows with whomever is concerned, whether collaboration relates to normal work or project portfolio, whether communication is internal, or involves external entities such as partners, vendors, customers, authorities, bloggers, journalists, etc.

Among those organizations considered highly effective communicators, 80 percent of projects meet original goals, versus only 52 percent at their minimally effective counterparts, according to the Project Management Institute's (PMI) *Pulse of the Profession In-Depth Report: The Essential Role of Communications* (2013). The same report revealed that US $135 million is at risk for every US$1 billion invested in the project portfolio. Further research on the importance of effective communications uncovers that 56 percent (US $75 million of that US $135 million) is at risk because of ineffective communications.

This chapter defines the concepts involved and how portfolio management as well as communication, including collaboration aspects, have evolved during the last decades both from a process and an enabler standpoint including a review of some of the literature in the field. Next, it describes the importance of communication to strategic portfolio management. It is followed by a discussion to present how communication should be implemented around strategic portfolio management in order to maximize its impact and relevance throughout the organization and the key considerations during the implementation process. A summary concludes the chapter.

Evolving Concepts and Processes

Portfolio Management Becomes More Strategic

There are many factors impacting the rise of portfolio management as a business discipline including the propagation of Markowitz's portfolio theory (1952), a growing focus on cost visibility following various financial crises, corporate scandals, project failures, and the importance given organization wide to accountability and governance relying on compliance and standardization mechanisms. Portfolio management attempts to answer the 'Catch 22' problem according to which strategy without action is only a daydream, but action without strategy is a nightmare.

As a matter of fact, strategy must have a vision which implies change; changes are orchestrated through ad-hoc works, projects, programs, and sub-portfolios. However, experience shows (Royer, 2003) that, whether or not they deliver on their promises, these activities tend to justify their own existence if they do not loop back to the strategy. While project and program management focus on "doing the work right," the purpose of portfolio management is "doing the right work".

While the first edition (2006) of the *Standard for Portfolio Management* from PMI defines portfolio management as "an approach to achieving strategic goals by selecting, prioritizing, assessing, and managing projects, programs, and other related work based upon their alignment and contribution to the organization's strategies and objectives" (p. 5), the third edition (2013) simplifies it to "the centralized management of one or more portfolios to achieve strategic objectives" (p. 178).

Since inception, portfolio management is an integral part of the organization's overall strategic plan. This said, for many years, the discipline was rather perceived just as information technology (IT) portfolio management with outputs and benefits belonging to the support side of the organization, and the Chief Information Officer (CIO) responsible for the oversight. More recently, other members of the Chief Executive Officer's team (CxOs) have started to take the lead on portfolio steering. First, change is daily on company agendas; it has a bigger stake than before, and corporate mistakes have a much lower level of acceptance. Second, technology itself is changing the nature of the relationships between employees, managers, consumers, and partners with traditional border lines getting fuzzier: operations and support, internal and external, clients, and shareholders. Last, methodologies and supporting platforms are more extensive at bridging the gap between strategy and operational execution.

As outlined by Levin, Artl, and Ward (2010) in practice, however, few organizations have a complete understanding of all the work that is under way in the organization, even if an enterprise-wide portfolio management office is in place. While the Chief Executive Officer (CEO) and Chief Portfolio Officer (CPO) focus on the large programs and complex projects, they probably are not aware of many of the smaller projects that are in progress and do not follow the portfolio management process. What is key to understand is that strategic portfolio management communication involves third parties. Communication is instrumental in uniting people, organizational units, vendors, and partners to overarching mission, goals, and objectives.

Communication: from Linearity to Simultaneity

The first major model for communication was introduced by Shannon and Weaver (1949) for Bell Laboratories. The original model was designed to mirror the functioning of radio and telephone technologies and consisted of three primary parts: sender, channel, and receiver. In that model, communication is the process by which a person, group, or organization (the sender) transmits some type of information (the message) to another person, group, or organization (the receiver).

In later extensions of such linear model (Berlo, 1960), communication becomes social interaction where at least two interacting agents share a common set of signs and rules. In that respect, collaboration, i.e., working with each other to do tasks and to achieve shared goals (Collins English Dictionary, 2012), becomes tightly imbricated with communication. Indeed, when collaboration goes across organizational fences it is not only a means to optimize resources and share knowledge among portfolios, programs, and projects, it is also one of the most important strategies applied to alleviate conflicts; when parties in conflict each desire to fully satisfy the concerns of all parties, there is cooperation, and a search for a mutually beneficial outcome. In collaborating, the parties intend to solve problems by clarifying differences rather than by accommodating various points of view.

Beside formal and informal lateral communication, the portfolio management ecosystem combines upward and downward communication. Robbins & Judge (2013) define the former as a flow to a higher level in the group or organization used to provide feedback to higher-ups, inform them of progress toward goals, and relay current problems, whereas the latter is a flow from one level of a group or organization to a lower level used to assign goals, provide job instructions, explain policies and procedures, point out problems that need attention, and offer feedback about performance. However, despite its wide acceptance derived from its simplicity, generality, and quantifiability, these approaches had a fundamental limitation: they assume that communicators are isolated individuals.

In light of these weaknesses, Barnlund (2008) proposed a transactional model of communication. The basic premise of the transactional model of communication is that individuals are simultaneously engaging in the sending and receiving of messages. Stacey (1995) indicates that effective management focuses on ever-changing agendas of strategic issues. These agendas consist of multiple challenges, stretching aspirations, and ambiguous issues. When managers deal with the issues on their strategic agendas, they are performing a real-time learning activity.

There is no overall framework to which they can refer before they decide how to tackle an issue. Through discussion with each other and with customers, suppliers, and even competitors, they are discovering what objectives they should pursue and what actions might work. The communication is spontaneous in the sense that it is not directed by some central authority. The communication that occurs depends upon the individuals involved. And that depends upon the boundary conditions, or context, provided by individual personalities, the dynamics of their interaction with each other, and the time they have available, given all the other issues requiring attention.

While still operating within separated parts, communication and collaboration are now holistic and span across interconnected systems. Effective portfolio management must find an ever-moving balance between structured and unstructured flows of information. Brafman and Beckstrom (2008) explain that the decentralized "sweet spot" is the point along the centralized-decentralized continuum that yields the best competitive position. The organization must enable enough decentralization for creativity but requires sufficient structure and controls to ensure consistency and compliance. The forces of centralization and decentralization continue to pull the sweet spot to and from one another. In any industry that is based on information – whether it is music, software, or banking – these forces pull the sweet spot toward decentralization. The more security and accountability become the norm the more likely it is that the sweet spot will tend toward centralization.

Embedded Complexity

The evolution of portfolio management as well as communication as described above actually reflects changes both in our society and organizations. For executives, managers, employees, and partners, it is sometimes even perceived as a revolution. First, formal communication flows still exist; they are basically formally prescribed patterns of interrelationships between various elements of the organization often dictating who may and may not communicate with whom; however, they coexist with informal exchanges. Then, Davis and Meyer (1999) coined the term "blur" to describe the combined effect of speed, connectivity, and intangibles on the business world, speed being how fast change occurs, connectivity is how networked businesses are linked to one another to complete a value chain, and intangibles are the ratio of non-physical to physical assets. On that basis, Evans and Roth (2004) developed the notion of collaborative knowledge networks, which underline several topics addressed hereunder.

Through cross-cultural differences, diversity, and gender mainstreaming as well as new work practices, globalization, and social shifts contribute to increase to an even-higher portfolio management complexity:

- Cross-cultural difference. Effective communication is difficult under the best of conditions. Cross-cultural factors clearly create the potential for increased communication and collaboration problems. A gesture that is well understood and acceptable in one culture can be meaningless or lewd in another. Only a few companies have documented strategies for communicating with employees across cultures, and not many more require that corporate messages be customized for consumption in other cultures.

- Diversity and gender mainstreaming are the public policy concept of assessing the different implications for women and men of any planned policy action, including legislation and programs, in all areas and levels. While linked gender balance with mainstreaming essentially offers a pluralistic approach that values the diversity among both women and men, some organizations tackle these changes by making bold individual appointments to strengthen their commitment to this theme (Burston, 2012).
- New work practices. Shifts in company structure lead toward greater and freer communication and sharing of information, outsourcing, offshoring, global teams, mobile workforce, flexible work arrangements, less hierarchical organizations, and the influx of the Millennial generation. About remote work, while not all companies allow employees to work off site, data from Gallup's State of the American Workplace report (2013) show that nearly four in 10 (39%) of the employees surveyed spend some amount of time working remotely or in locations apart from their coworkers. And, Gallup finds that companies that offer the opportunity to work remotely might have some advantages when it comes to hours worked and employee engagement.

Evolving Supporting Platforms

Background

In the early days, project management software ran on big mainframe computers and was used only in large projects. These early systems were limited in their capabilities and, by today's standards, were difficult to use. In May 1957, the Remington Rand Corporation and the DuPont Corporation started a joint venture to develop the Critical Path Method (CPM) mathematical technique for managing plant maintenance projects, which allowed DuPont to save 25% on its plant shutdowns. This technique was popular but expensive and was later dropped by DuPont after a management change took place; it lost traction but came back even overtaking the Program Evaluation Review Technique (PERT) developed by the U.S. Navy until the "Precedence" technique used by many scheduling systems was developed in the 1970s. Nevertheless, according to Rich (2011), project management became the first commercial software program, which ran on a computer using a "stored program".

As Aubrée-Dauchez states (2005), in order to implement effective communications and enable proper collaboration among stakeholders, project portfolio management platforms shall provide much more than dashboards and monitoring reports. A discussion on supporting tools must be integrated into a broader context including, inter alia, the exchange of information between different stakeholders, document management systems, the different methodologies that can inspire or directly influence the activities or platform coding, and last but not least, the mandatory commitment of senior management. In this review, the technical layer as such is less important than the application layer (LEADing Practice, 2014); whether portfolio management users access information via a file server, the Intranet or the Internet, it does not affect the business logic per se.

Bottom-up

With the acceleration of micro-computing in the 1970s and 1980s, organizations started to manage projects on personal computers with individuals or groups gathering in separate files summary information on progress, risks, and issues. This approach is sometimes referred to Enterprise Portfolio Management (EPM) since it is geared to supporting day-to-day project portfolio management.

This bottom-up solution also accommodates spreadsheets, text files, and even sometimes hand-written notes that have to be consolidated at the program and portfolio management levels. Such an approach proved to be cumbersome and prone to errors hindering collaboration and communication and making portfolio management difficult and unable to perform well in achieving strategic goals.

Top-Down

As noted by Levin et al (2010), the emphasis therefore has changed to ensure that only those programs and projects that support the organization's strategic objectives are selected and pursued. Therefore, alongside of formalization and standardization of portfolio management practice, platforms started to emerge in the late 1990s to provide a top-down approach to managing, tracking, and reporting on enterprise strategies, projects, portfolios, processes, resources, and results.

This approach, however, faces inherent limitations; most top-down solutions tools fail to accommodate work other than projects such as unstructured work or ad hoc requests. Moreover, collaborative project elements and lower-level communication are not noise anymore in a business environment embedded in complexity. There is nothing more outdated than a static spreadsheet, even received a few minutes ago. With that in mind, most of the top-down solutions offer, at cost, integration connectors to synchronize, to the extent possible, data with other applications.

Integrated

In order to overcome the shortfalls of the unidirectional solutions and sometimes in parallel to the development of top-down solutions, integrated platforms have been made available on the market around the year 2000. Basically, project information is stored in a central database and protected from unauthorized access and corruption.

Project managers can drill down into project details, while portfolio managers can apply business logic to select, prioritize, and assess projects working dynamically on the same data sets and accessing the same document and discussion repositories. The integrated solution has organization-wide default and custom fields allowing comparison across the project portfolio.

About communication, it is worth noting that Aberdeen (2013) has termed Integrated Communications (IC) as the approach that enables organizations to integrate solutions from a variety of vendors. Unified Communications has long appeared as the mythic 'Holy Grail' of business communications, the seamless integration of all communication channels: voice, voice messaging, email, Short Message Service (SMS), text, chat, or instant messaging (IM), presence, video, and fax with the ability to move a conversation effortlessly from one channel to the next. This promise has yet to be fulfilled for many organizations, in part because of a lack of a compelling business case, and until recently a lack of industry standards that would make agile, cross-channel business conversations viable. The relatively recent emergence of standards has driven down the cost and increased the adoption of advanced communications capabilities.

Social and Agile

Organizations often try to address business challenges by relying on the traditional managerial tools to pursue operational excellence: establishing well-defined roles, best practice processes, and formal accountability structures. However, Cross's research (2010) shows that such tools, though valuable, are not enough. The key to delivering both operational excellence and innovation is having networks of informal collaboration. Within departments of large global companies, Cross notes innovative solutions often emerge unexpectedly through informal and unplanned interactions between individuals who see problems from different perspectives. Additionally, successful execution frequently flows from the networks of relationships that help employees handle situations that do not fit cleanly into established processes and structures.

Organizational leaders, who learn to harness and balance both formal and informal structures, can create global organizations that are more efficient and innovative than organizations that rely primarily on formal mechanisms. However, even though individual employees may be able to identify local patterns of collaboration, broader configurations of informal collaboration tend to be far less visible to senior leaders. In the face of this reality, Cross found that organizational network analysis offers a useful methodology to help executives do two things: assess broader patterns of informal networks among individuals, teams, functions, and organizations and then take targeted steps to align networks with strategic imperatives.

Some project portfolio management (PPM) platforms acknowledge that communication and collaboration are much more than meeting and talking about work; it is about connecting workers to each other and to information whenever they need it, wherever they are. They have embraced Mobile First principles (Wroblewski, 2011), and social network user experience, they provide agile project management features out of the box, and they extend portfolio management to an end-to-end organization work life cycle. Forrester (2010) states that social networking software that enables people to post their activities and receive feedback from their peers has taken off like a rocket, particularly with consumers. This type of peer collaboration, exemplified in consumer-oriented social networks, represents the new design metaphor needed to make PPM more broadly accepted and productively used.

Importance of Communication to Strategic Portfolio Management

According to Locker (2001), experts consider communication to be a key process underlying all aspects of organizational management. Contemporary scholars have referred to organizational communication as "the social glue . . . that continues to keep the organization tied together" (Roberts, 1984, p. 4) and "the essence of organization" (Weick, 1987). Writing many years earlier, Chester (1938, p. 91) said, "The structure, extensiveness and scope of the organization are almost entirely determined by communication techniques".

Since portfolio management provides key capabilities for achievement of an organization's strategy, there may be a major focus at the executive level both to assemble and to communicate detailed information on the progress of major objectives and impacts of the components, as well as any changes to previously communicated plans.

Ongoing and well-targeted communication is a key requirement for maintaining stakeholder confidence in and support for the objectives to be achieved and the approaches being implemented. In addition, in order to ensure coordination and effective teamwork, communication among the teams responsible for the various components need to be planned, formalized, and managed in concert with best practices. Various portfolio events or milestones need to be communicated both inside and outside the organization. This could include achievement of a major objective, elimination of a component, and other matters requiring corporate communications.

The challenges are that manual project requests through different tools make it difficult to stay on top of opportunities, squeaky wheels demand attention, but distract from more important priorities, and manually collecting the status of a portfolio of projects is like "herding cats". The organization's leaders must validate, align, and optimize the right portfolio investments and communicate about them.

The strategic portfolio management system put in place must ensure that every project aligns with organization's goals and objectives by standardizing project requests; quickly validate, align, and prioritize projects in portfolios; and make the right decisions on trusted data. Gartner (2013) reports that PPM will evolve to cover the project life cycle from a bright idea to a post-project review. Because projects move businesses and products forward, increased attention to request and service desk management has become more important to the end-to-end project portfolio management model.

To reflect the importance of communication, the third edition of the *Standard for Portfolio Management* (PMI, 2013) added a new knowledge area on Communication Management which outlines two processes, namely:

- Develop portfolio communication management plan
- Manage portfolio information

Implementing Strategic Portfolio Management Communication

Recognize Maturity Levels

The Ancient Greek aphorism "know thyself" indicates that each organization must become conscious of where it stands in terms of communication around portfolio management in order to be in a position to move to the next maturity level, if deemed necessary and affordable. The level of maturity of portfolio management in the organization will influence the way communication is organized. Bayney & Chakravarti (2012) outlines five maturity levels in the People competency domain suggesting levels of maturity pertaining to communication and presentation (p. 156):

- Level 1 or the Ad-hoc level is characterized by using ineffective communication that may result in duplication of efforts across divisions. Decisions are not communicated effectively.
- Level 2 or the Basic level recognizes the need for a communication plan for portfolio management. The organization is working toward standardizing communication.
- Level 3 or the Standard level implies that a communication plan has been prepared and is used for portfolio management.
- Level 4 or the Advanced level has decisions communicated throughout the organization, with open communication being the norm.
- Level 5 is known as the Center of Excellence (CoE) level in which the organization's culture is collaborative and communicative. People are encouraged at all levels to submit ideas to foster continuous improvement.

Likewise, Bayney & Chakravarti outlined five maturity levels in the Strategy competency domain suggesting levels of maturity pertaining to collaboration (p. 158):

- Level 1 or the Ad-hoc level is characterized by projects initiated based on division needs and without regard to the impact on other divisions.
- Level 2 or the Basic level recognizes the need for identification of cross-functional synergies to achieve an enterprise focus.
- Level 3 or the Standard level implies that synergies, overlaps, and conflicts in each division's portfolio are discovered and resolved. Moreover, cross-functional opportunities are identified.

- Level 4 or the Advanced level has projects proactively reviewed to identify synergies, overlaps, and conflicts. Divisions collaborate to resolve them.
- Level 5 is known as the CoE level in which portfolio planning, execution, and review are performed as a partnership.

Standardize and Simplify Communication Processes

Whatever methodologies and enablers are chosen to streamline portfolio management processes, they must be tailored to the organization and properly communicated to the various parties involved. In highly competitive markets, the need for trustworthy information becomes critical. The challenge for stakeholders is not obtaining more information but obtaining the right information. The system put in place must allow people to access, discuss, and share reliable and up-to-date information. An important step is to create and disseminate a simple formalized process for evaluating potential and current projects based on alignment to corporate strategic and financial goals. However, it is crucial to establish a process that requires every potential project to meet pre-determined criteria for acceptance giving executives confidence that they are making data-driven decisions on the greatest business value brought in and are not just appeasing influential stakeholders.

There are common questions that should be discussed upfront:

- What are the high-level objectives of the project? It is not uncommon for a project to morph into something very different from what was originally intended. Specifically identifying the desired outcome of every project helps project teams, sponsors, and stakeholders remain focused.
- What are the estimated costs of the project; what are the anticipated rewards? Without the answers to these questions, it is difficult to determine if the proposed project will provide any business value, let alone the greatest business value.
- Does the proposed project align with the mission, vision, and values of the organization? Individual projects must represent the execution of strategic direction and financial goals if the desired result is to maximize the return on investment (ROI).
- What risks are associated with pursuing the project under consideration? If risks can be identified and evaluated while the project is still in the consideration process, actions can be taken to mitigate risk and increase a project's probability of success.

In order to increase productivity and reduce costs, communication and collaboration around portfolio management will be eased by:

- Embedding rules and procedures in the process
- Developing and configuring user-friendly, natural, and enjoyable artifacts

- Creating repeatable work processes using enterprise templates
- Providing users with a means to update status with minimal impact on their work day
- Giving business leaders immediate access to business intelligence they need to make strategic decisions
- Gathering lessons learned geared toward continuous improvements
- Using cloud-based solutions whether these are public, private, or hybrid

Connect and Engage with People

The communication system deployed shall reduce the information overload by cutting out excess tools and connecting collaboration with work whether partners are or are not within corporate fences; in a nutshell, more context, less meetings. Connecting and engaging with people implies removing or at least reducing barriers that can retard or distort effective communication. Common barriers are technical and security layers, filtering, selective perception, emotions, language, silence, communication apprehension, etc.

- If the level of user involvement depends on the type of project or activities (Wysocki, 2009), being the ultimate communication piece between customers and implementers, the program and project managers shall be trained and possibly certified in using similar terminology, most often a blended version of a globally recognized methodology. They shall be invited to actively engage in topics of interest in dedicated forums.
- In order to avoid a mismatch between the classical paper-based, non-disclosure agreement (NDA) suitable to asynchronous communication channels, such as e-mail and the access to modern synchronous environments such as the professional social networks, it is crucial to have a process in place to safeguard the organization for improper use of information. Ideally, NDA signatures shall be embedded in a two-step authentication process eluding paperwork.
- Most enterprise work scenarios require documents of various shapes and sizes at many different stages of work, for collaboration, approvals, and sharing. Documents shall be an integral part of the business and portfolio management life cycle. Teams should not waste time and resources inefficiently managing the creation of documents and digital assets from beginning to end. Documents shall be organized and accessible from a central location by managers, team members, external stakeholders, or any third party contributors keeping all discussions, questions, comments, and versions in the context of the document throughout its life cycle.

Increase Visibility and Collaboration

An effective portfolio management communication system shall provide all levels of the organization with visibility and insight into the truth about workloads, dependencies, and when things will really be done. Independently of the channel, information is conveyed through managers, and senior managers shall get real-time visibility into who is working on what and how to justify the resources used, and the ones still needed. In order to make informed decisions, executives, managers, and project teams with the data shall be able to securely access standard dashboards as well as configurable reports whether they are on company premises or traveling for business.

To maximize impact and relevance as a CoE, the enterprise portfolio management office shall dispose of a properly designed communication room composed of wide screens displaying live portfolio information 24/7; a specialized library gathering references pertaining to portfolio, program, and project management; as well as video facilities to record and disseminate achievements and lessons learned.

Summary

Once a senior manager working for a huge global corporation was asked what was the most important topic to address in portfolio management. His interviewer was indeed expecting him to answer the adherence to processes. He opted, however, for communication. According to that leader, despite the plethora of methodologies, frameworks, platforms, tools, and techniques that can be used or not depending on the context, especially in an organization so big, diversified, and globally distributed; communication is unique; as such, it is not an option.

Somehow communication transcends its environment by allowing collaborators to address strategy, performance, governance, and risks topics aligning operations and portfolios of programs and projects to achieve overarching business objectives, see Figure 13.1.

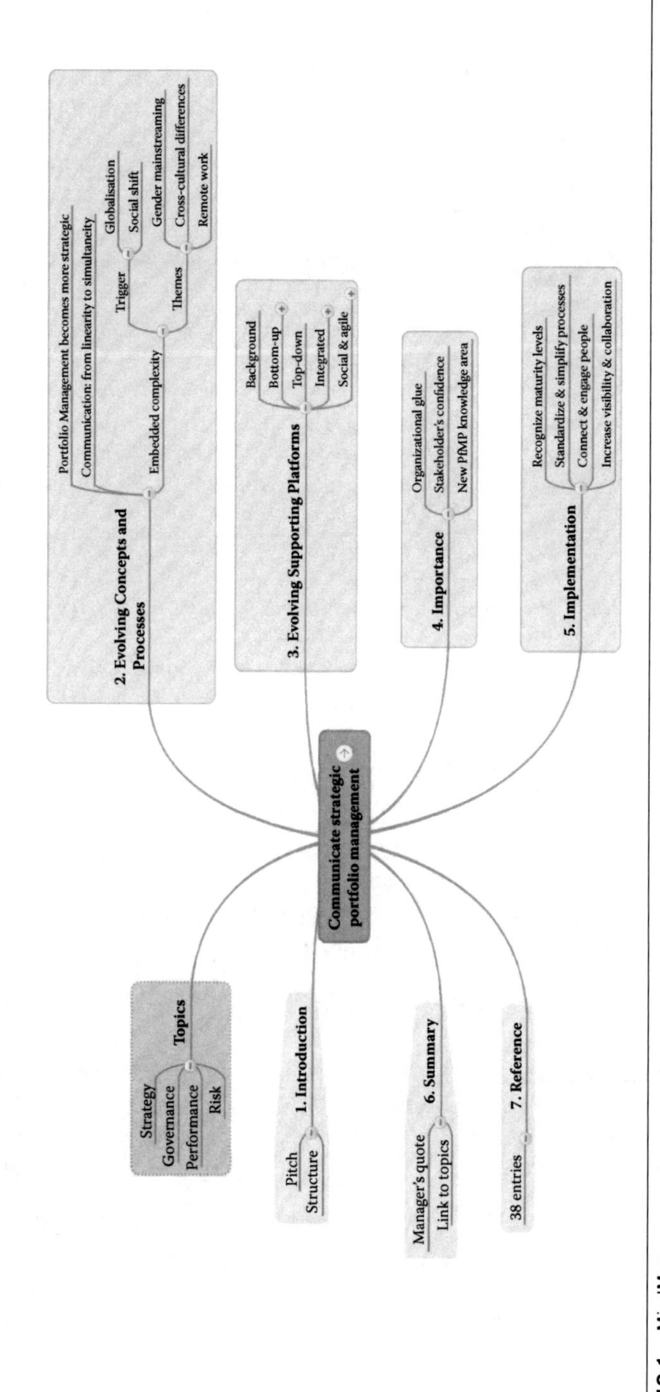

Figure 13.1 MindMap

References

Aberdeen Group Analyst Insight. (2013). *Harnessing the power of next-generation communications.* Boston, MA: Aberdeen Group.

Aubrée-Dauchez, A. (2005). *Portfolio management toolset overview. Memorandum.*

Barnard, C. I. (1938). *The functions of the executive.* Cambridge, MA: Harvard University Press.

Barnlund, D. C. (2008). A transactional model of communication. in Mortensen, C.D., *Communication theory.* (2nd ed.) New Brunswick, NJ: Transaction Publishers, 47–57.

Bayney, R. & Chakravarti, R. (2012). *Enterprise project portfolio management: Building competencies for R&D and IT investment success.* Boca Raton, FL: J. Ross Publishing.

Berlo, D. K. (1960). *The process of communication.* New York, NY: Holt, Rinehart, & Winston.

Booth, C. & Bennett, C. (2002). Gender mainstreaming in the European Union: Toward a new conception of practice and opportunities? *European Journal of Women's Studies, 9*(4), 430–466.

Bowler, N. (2012). *How to align corporate strategy with project execution.* AtTask. Retrieved from: www.attask.com

Brafman, O. & Beckstrom, R. A. (2008). The starfish and the spider. In search of the sweet spot. *Portfolio Trade.* Chapter 8. London: Penguin Books Ltd.

Burston, P. (2012). Google's diversity head: Mark Palmer-Edgecumbe. *The Guardian*, March 15.

Corporate Executive Board Corporate Leadership Council (2010). The role of employee engagement in the return to growth. *Bloomberg Business Week*, August 13.

Cross, R., Gray, P., Cunningham, S., Showers, M. & Thomas, R. J. (2010). The startling cost of inefficient collaboration. *MIT Sloan Management Review 52,* 83–90.

Davis, S. & Meyer, C. (1999). *Blur the speed of change in the connected economy.* New York: Warner Books Edition.

Drucker, P. (2006). *The effective executive: The definitive guide to getting the right things done.* New York: Harper Collins Publishers

Forrester Consulting (2010). *The case for social project management.* June. Cambridge, MA: Forrester.

Gallup Blog. (2013). *Remote workers log more hours and are slightly more engaged.* July 12. Gallup.

Greenberg, J. & Baron, R. A. (2002). *Behavior in organizations: Understanding and managing the human side of work.* Upper Saddle River, NJ: Pearson Education.

Hobart, B. & Sendek, H. (2009). *Gen Y now: How generation Y changes your workplace and why it requires a new leadership style.* Novato, CA: Select Press.

Kerzner, H. R. (2006). *Project management systems approach to planning, scheduling, and controlling.* 9th edition. Hoboken, NJ: John Wiley & Sons.

Krebs, J. (2009). *Agile portfolio management.* Richmond, WA: Microsoft Press.

LEADing Practice (2014). *Value reference fram*ework. LEADing Practice. Retrieved from: http://www.leadingpractice.com

Levin, G., Artl, M. & Ward, J.L. (2010). *Portfolio framework: a maturity model.* Arlington, VA: ESI, International.

Locke,r K. O. (2001).*Business and administrative communication.* Burr Ridge, IL: McGraw-Hill.

Matos-Camarinha, L. M. & Afasarmanesh, H. (2004). *Collaborative network organizations. A research agenda for emerging business models.* Boston, MA: Kluwer Academic Publishers.

Office of Goverment Commerce. (2011). *Management of portf*olios. Norwich: The Stationery Office.

Project Management Institute. (2013). *PMI's Pulse of the profession in-depth report. The high cost of low performance. The essential role of communications.* Newtown Square, PA: Project Management Institute.

Project Management Institute. (2006). *The standard for portfolio management.* Newtown Square, PA. Project Management Institute.

Project Management Institute. (2008). *The standard for portfolio management.* Second Edition. Newtown Square, PA: Project Management Institute.

Project Management Institute. (2013). *The standard for portfolio management.* Third Edition. Newtown Square, PA: Project Management Institute.

Rich. (2011). *Project management software – A brief history (Part 1).* EPM Live. Retrieved from: http://epmlive.com/?s = Project+Management+Software

Robbins, S. P. & Judge, T. A. (2013). *Organizational behavior.* 15th edition. Upper Saddle Ridge, NJ: Pearson Education.

Roberts, K. H. (1984). Organizational communication. in Kast, F. and Rosenzweig, J. (eds.). *Modules in management.* Chicago: SRA Associates, p. 4.

Royer, I. (2003) Why bad projects are so hard to kill. *Harvard Business Review. 81.* February, pp. 48–56.

Shannon, C. E. & Weaver, W. (1949). *The mathematical theory of communication.* Urbana, IL: University of Illinois Press.

Stang, D. & Handler, R. A. (2013). Magic quadrant for cloud-based IT project and portfolio management services. Stamford, CT: Gartner.

Weick, K. E. (1987). *Theorizing about organizational communication. Handbook of organizational communication.* Newbury Park, CA: Sage.

Wroblewski, L. (2011). *Mobile first. A book apart.* Paris: Groupe Eyrolles.

Wysocki, R. K. (2009). *Effective project management: Traditional, agile, extreme. Client involvement vs the complexity/uncertainty domain.* 5th edition. Hoboken, NJ: John Wiley & Sons, Inc.

14
Project Portfolio Management and Communication

WANDA CURLEE, DM, PMP, PgMP, PMI-RMP, PfMP

Portfolio management is a recent phenomenon which has been acknowledged by the Project Management Institute (PMI®). PMI acknowledged project portfolio management when the association published the first standard on portfolio management in 2006. Since the initial standard publication there have been two updates.

Project portfolio communication has some commonality with projects and programs, but strategy is *always* the focus for a project portfolio. The Project portfolio manager (PPM) needs to ensure that leaders of the company or the organization understand the project portfolio. The PPM must ensure the dashboards and the communication strategy are at the correct level of information for executives.

The PPM needs to speak to the project and program managers. This is a different discussion. The PPM is normally extracting information from the project and program managers. The PPM is consistently reviewing and analyzing the data to ensure the project portfolio is meeting the Key Performance Indicators (KPIs) established. The strategic communications and discussions with project and program managers will be discussed in a later section.

The banking, financial, and wealth management industries have used portfolios for many years. The project management environment has recently adopted the concept of portfolios. This discipline of project portfolios is still developing. Some in project management only believe "true" projects and programs may be part of a project portfolio. These individuals discount research and development portfolios as not project oriented.

Over time, the project portfolio discipline will mature, and various industries and types of projects will be included or discounted as part of a project portfolio. Achieving strategic or organizational objectives is the purpose of the portfolio (PMI, 2013b). Communication for the project portfolio will differ from project and program communication.

The sections below provide an overview of the portfolio's various communication paths, relevant stakeholders, suggestions for the portfolio communication plan, ways to integrate communication throughout the portfolio, methods to ensure communication stays at a strategic level, approaches to integrate strategic information for executives, and the necessity of communicating to stakeholders, internal and external.

General Communications and Project Portfolios

The project portfolio manager needs to remember the formula for the lines of communication. The formula is (N * (N – 1))/2 where N is the number of people in the overall group (PMI, 2013a). As the number of people increase the PPM should recognize communication becomes more complex. Relaying the aggregate project portfolio information (PMI, 2013b) begins in the planning phase.

The PPM should realize that organizational communication is complex and important to the organization establishing a project portfolio. Most likely the project portfolio constituents are not fully colocated with all of its stakeholders, customers, suppliers, and employees. The portfolio is no longer housed in a single building or area, and technology is making it less likely that this will ever be the case again. Although some organizations are now having employees move back to a brick and mortar environment (e.g. Yahoo, Hewlett-Packard) to enhance face-to-face communication as some Chief Executive Officers (CEOs) now are of the opinion that having everyone in the office enhances day-to-day communications; however, there are other technology-based communication methods that perform the same function as these face-to-face meetings (Curlee & Gordon, 2013). The continuous monitoring of team members that organizations of the past felt were essential is no longer seen as an organizational priority. Individuals involved with a project portfolios are expected to perform their tasks with others in mind so that the portfolio can achieve the stated objectives.

Technology and poor face-to-face communication skills cannot replace delivering an effective message; in fact, operational communication is one of the most critical elements of a project portfolio as many members may be virtual. Communication is moving away from the point-to-point, face-to-face communication of the past. Face-to-face communication is being replaced by different methods of communication (Curlee & Gordon, 2013). Telephones, electronic mail, over the computer communications, virtual meeting rooms, Wikis, and blogs have increasingly replaced face-to-face communication with regard to passing along stories that support organizational needs. These methods of communication are both formal and informal and may readily simulate an oral mode of communication while decreasing costs.

The contemporary PPM should learn to communicate more effectively both in a non-face-to-face and face-to-face setting so that messages are received and comprehended. Senior leaders do not have the time to spend to clarify or push PPMs in the correct direction. The communication distribution system has emerged as an important force in effective internal and external organizational communications. This is particularly important in both disbursed organizations and those moving back to the brick and mortar environment because one needs to be sure that people involved in project and/or program not only receive the necessary information, but they understand the current and future intent of the KPIs and project portfolio (Curlee & Gordon, 2013).

Organizations continue to evolve, and because of this evolution, there is great speculation about the future direction of merging the virtual with the traditional brick and mortar organizations. The new organizations that can adapt to change will become even more important, and this adaptability is essential for project portfolio management. Project portfolios are not static units that will remain the same throughout their life. Project portfolios must have built in adaptability, and they must be designed with the future in mind, including industry changes, strategic company changes, and changes within the economy. Project portfolio managers who realize technology will be changing rapidly and have a plan to be able to upgrade incrementally will do far better than those PPMs that design project portfolios for a set fixed need without constantly revaluating trends affecting the portfolio (Curlee & Gordon, 2013).

For a virtual organization to be successful, trust must be part of the foundation of the organization. Trust demands boundaries and learning. Trust requires bonding and leaders that have a certain touch to management. Trust includes such factors as competence, integrity, and a concern for others (stakeholders). These elements of trust are necessary to make an organization successful. Trust must be visible at all levels of an organization in order to achieve excellence. No virtual organization can be successful, let alone effective, without the individuals involved having a high level of trust at all levels.

The PPM's leadership style benefits by promoting trust and collaboration in a faceless environment. Creative manners and communication help the PPM establish trust. The effective PPM establishes trust between himself or herself and each team member individually. The PPM should also promote trust within the group. Devoting time to building relationships with each team member is a must for the PPM. The PPM needs to learn how to interact with each team member and to keep track of their individual progress. A PPM needs to be interested not only in the project portfolio but also in the person. This personal approach will make a difference when building trust. Individuals trust those that know them.

Trust does not end with the individual; the PPM must build trust as part of a group. The group must learn to trust each other and to trust the judgment of the group. Trust is created when individuals respect one another. Trust must be visible at all levels of an organization in order to achieve greatness. No virtual organization can be successful, let alone effective, without the individuals involved having a high level of trust between one another.

A PPM should avoid just using strategies that create trust for a group. This approach is a common error made by leaders of project portfolios. For example, a team meeting will not build trust when people just report on their progress. A staff meeting does nothing to build trust if trust is not on the agenda. If a portfolio manager wants to use this time to build trust, then put trust at the top of the agenda and stick to it. Talk with the team about trust and have them talk to each other about what trust means to them. Communicate about how effective trust is (or is not) being built in the project portfolio. The results are surprising once trust is established. A PPM must remember that building trust is not an either or proposition. Trust must be built individually, and it must be built as a group. Studies have clearly shown that without trust, a virtual team is more likely to fail (Johnson & Johnson, 2000; Uhl-Bien & Marion, 2009).

One study (Useem & Harder, 2000) found that executives in eight major U.S. corporations agreed that it was difficult to establish trust in a multi-cultural environment. Many project portfolios are multi-cultural, either internal to the company, the country, or various countries. This lack of trust led to the companies establishing duplicate processes and procedures and different systems, with the result that there were many international companies instead of one cohesive enterprise. If managers do not already trust and respect virtual employees, then they need to build trust and respect for all employees and stakeholders. Many managers need visual clues and interrelationships to be comfortable.

Trust and Communication

Portfolio communication may be complex. Project portfolios normally span across functions and may even span across organizations. This approach may mean the stakeholders, leaders, and others are no longer visible to the PPMs where they can use 'management by walking around' to learn about what is happening since it is no longer a viable option in the evolving business. Employees, whether virtual or traditional, need to learn to better communicate because there is less face-to-face contact. As technology improves and becomes less expensive to implement, the virtual organization will become the norm. Virtual no longer means individuals working from home. The virtual employee may be in a brick and mortar situation working with others in another brick and mortar location. The continuous monitoring of the project portfolio by organizations will become easier on the face because of technology; however, the PPM will be the value-added person to determine the value of the data and to communicate it to the correct parties. Work will be done in a manner where no one can see it happen. To meet this new way of conducting business, a PPM will need to increase the level of communication.

There are three views to communication. They are as follows: communication as interaction, communication as action, and communication as transaction. In the end, all of these styles of communication can be effective (Fox, 1999). PPMs have to decide which one(s) they will use consistently and then use them. Communication, whether virtual or face-to-face, builds trust when there is predictability. Some PPMs feel that if they can tell a story that parallels the current situation with a positive outcome, they can help individuals deal with the current situation. In the past, PPMs would build this trust through face-to-face communication; however, that option is no longer always possible. PPMs must feel comfortable in both mediums (Curlee & Gordon, 2013).

Communication and trust building can be seen as a series of gatekeepers, individuals who control the flow of communication to others. This view addresses the relationship of individuals as mediators in a dynamic communication systems; a relationship of cause and effect. Gatekeepers can either restrict information or mediate communication. Either form can be noble; however, these gatekeepers can be far more insidious as they can be skilled politicians who control knowledge to maintain their expert status. Although a PPM might feel more in control when he or she uses this method, this form of restricted communication will most likely be seen by others as negative. When information becomes a medium of exchange, it can create some relationships that are not founded in trust. PPMs should use this kind of communication carefully.

In contrast to gatekeepers, others believe that the project portfolio is not about the single steward whose actions control the project portfolio's destiny. Leading a project portfolio is about a single individual who harnesses a personal and business network in order to achieve impressive results (PMI, 2013b). The PPMs' teams, virtual or face-to-face, require that the PPM shift from a focus upon personal ability to a focus on group results.

PPMs and their organizations need to learn to communicate better in order to build trust and to ensure that their messages are received and comprehended. The communication distribution system has emerged as an important force in effective internal and external organizational communications. This is particularly important for the hybrid teams or face-to-face teams interacting with other face-to-face teams in other locations and organizations, because of the potential impact upon the project portfolio and ultimately the company/organization.

Trust should be important to the PPM, because he or she may be dealing with several cultures whether the culture is within the various parts of the company/organization, cultures spanning countries, or even cultures spanning within the PPM's own country. The emphasis is for the PPM to foster one cohesive enterprise even if the company has had numerous mergers or "right" sizings. What is the true culture of these organizations? If PPMs do not already trust and respect employees, then they need to build trust and respect with those in the project portfolio. Some PPMs may need visual clues and interrelationships to be comfortable. PPMs need to learn how to move past this limitation and learn to rely upon other cues for security.

Project Portfolio Strategic Communication

Project portfolios are focused on delivering strategy for the organization (PMI, 2013b). Strategic communication normally is centered on senior leadership and driving a message of critical importance to the organization or company. A PPM needs to structure the communications, including dashboards, to meet the needs of the project portfolio but also the needs of leadership.

When developing the project portfolio communications plan, the PPM needs to include the following (among other items):

- Objectives for the communication,
- Stakeholder identification,
- Expectations of the stakeholders,
- Policies and constraints, and
- Methods to deliver the communication (PMI, 2013b).

The PPM must deliver the information in a concise manner. This will include a combination of written (including dashboards and emails) and spoken (including one-on-ones, presentations, steering committee meetings, status meetings, governance meetings, and so forth) communication.

Dashboards need to concentrate on the visuals. A dashboard should deliver the information in pie charts, colored status indicators, bar charts, Gantt charts, and other visually appealing and information rich content. The information would be enhanced when stakeholders, and leadership can drill down into the data. Too many times executive dashboards are static and do not allow the leader to drill deeper. Hyperlinks on the actual dashboard allow leaders to drill down as far as they wish. Some leaders are detailed oriented, while others are not. Having this option frees up the time of the PPM to focus on issues at hand.

The PPM must have confidence in the tool to use hyperlink. The tool(s) used for the project portfolio needs to be customized as much as possible to the culture of the organization. Or, if the organization is not accustomed to project portfolio management, then the PPM should be instrumental in selecting an appropriate tool(s) to deliver leadership's and stakeholder's needs.

The PPM must support the team as well as help members move toward the desired goal. Leaders, such as PPMs, who only communicate tasks and timelines, are not as successful as those that truly lead the team. Leaders not only have to help the team toward the new goal, but they have to identify and garner support of the extended network of experts that support the goal. The PPM identifies these stakeholders and gains their support in order to make the program successful (Curlee & Gordon, 2013).

One successful strategy observed is to the create a stakeholder matrix outlining level of participation, roles, and contact information. This matrix is important to allow project portfolio team members to understand how resources need to be deployed to the various projects/programs. A team that uses positive communication to stakeholders can become powerful in project portfolio success. Furthermore, successful PPMs are skilled at gaining support of customers and stakeholders. The PPM needs to build a stakeholder matrix and communicate this information to appropriate team members. Ask team members if they ever pass along information regarding the project portfolio to these stakeholders. Creating positive marketing of a portfolio will help ensure its success. Other leadership strategies for successful project portfolios are: developing and transitioning team members, developing and adapting organizational processes to meet the team's needs, allowing leaders to transition when appropriate, and ensuring the team receives appropriate training for virtual communications, technology, and enhanced skill sets (Curlee & Gordon, 2013).

A PPM should review the changes that will be required by the project portfolio and then develop a roadmap to the new organizational processes that will be able to meet these new requirements. Leadership transition is important to the success of a project portfolio. The PPM will not always be the person who is best suited to lead the team at all times. There will be times where it would be better to allow someone else, perhaps with more tactical ability, or perhaps an influential stakeholder, to lead the team on an interim basis.

The PPM should also establish metrics for team performance. These metrics should institute high expectations to encourage the extra effort required to overcome the communication hurdles (Duarte & Snyder, 2006). As the team becomes familiar with the expectations and the performance metrics, the amount of undesirable conflict should be reduced.

Project Portfolio Day-to-Day Communication

Experts (Tuckman, 1995; Rigby, 2009) agree that one team-building model is applicable project portfolio teams as the stakeholders, projects, and programs change often. Several experts identified the Tuckman model as a team model that applies (Curlee & Gordon, 2013, Curlee & Gordon, 2011, Duarte & Snyder, 2006). A PPM can apply this model to his or her team (where some may be virtual) in order to minimize disruptions during the project portfolio implementation and then prepare for conflict during those points in the team process.

The Tuckman model may assist the PPM in resolving conflict once it occurs within a team. Since the team process might cause conflict within a team, a successful PPM must be able to resolve conflict when it arises. Negative conflict results in both wasted time and lost productivity for a project portfolio because conflict costs the organization money and is a contributing cause of delays. It is estimated that the cost of conflict, when including ineffective managing of interpersonal situations, conflict avoidance, and lost days accounted for $20,000 per employee, per year in addition to 20-25% of a manager's time spent dealing with team disagreements (Johnson & Johnson, 2000). The PPM needs to identify potential periods of conflict as well as to determine strategies to cope with destructive conflict.

The PPM should consider technology-based communications instead of completely relying on traditional methods (e-mail, push/pull sites, or blogs). Technology-based communication relies upon the complexity based six degrees of separation. This theory is the foundation of the success of social media websites such as FaceBook®, Plaxo®, LinkedIn®, and others that a company may use for personal or professional reasons. The expansions of these networks caused many companies to restrict these sites; however, some organizations are beginning to understand the advantages these sites offer in certain areas. These websites create a unique opportunity to help the forward progress of a project portfolio.

Complexity theory is based upon the management belief that total order does not allow for enough flexibility to address every possible human interaction or situation. The problem is people are inherently skeptical of less order and flexibility because it is believed there is less control. Evidence (Cooke-Davies, Cicmil, Crawford, & Richardon, 2007) suggests a project portfolio with virtual programs/projects/stakeholders is more effective when a PPM uses aspects of complexity theory to lead the project portfolio. Six degrees of separation is the belief that people are connected by no more than six degrees, and so if one were to reach out randomly into one's network, one should be able to reach anyone after about six connections.

Social media has touched many lives in the professional or personal environment. Many have used instant messenger to communicate to friends, colleagues, or loved ones. Most would agree that it is a fast and efficient means to resolve problems or find the right individual to resolve the problem. In fact, many companies have implemented instant messenger system (such as MS Lync) to ensure intellectual property is not compromised. Companies also are more apt to create private and public FaceBook® pages. The private FaceBook® pages may be for employees to collaborate, while the public FaceBook® is normally for commercial or marketing purposes (Curlee & Gordon, 2013).

By harnessing the power of social media sites, a PPM can leverage a meta-network in a manner to expand support and to improve communications. A well run project portfolio, which normally has virtual programs, projects and stakeholders, verges on controlled chaos since there are many lines of communication and lack of visual cues. A social network is the same as one can post or communicate to everyone in the meta-network by leveraging a known normal network.

One practical way to maintain this balance of control is to leverage these social networking sites in a way that supports the project portfolio and its components. A PPM should consider creating a social networking site that distributes information. In the past, organizations looked to distribute this kind of information through an intranet or via the corporate website. However, these sites tend to restrict information in a manner that is not productive. Studies have shown that if people need to sign on to another site to garner information, they are less likely to use those sites (Curlee & Gordon, 2013; Cooke-Davies, et al, 2007).

Conclusions

The project portfolio manager of today most likely will not recognize the project portfolio in 5-10-15 years. PMI's Portfolio Standard (PMI, 2013b) is approximately 202 pages cover-to-cover. The standard has only had three editions. The PMBOK® (PMI, 2013a) in its early stages was approximately 100 pages. Today, the PMBOK® is over 600 pages cover-to-cover. Expectations are the Portfolio Standard will be as large in a comparable time period.

The PPM can expect new knowledge areas, more sophistication, more templates, and more information on how to do various aspects within the project discipline. There will be more books written on the topic to provide various thoughts on the trends within project portfolio. PMI will continue to award grants to study project portfolios and to feed this information into the various aspects, including new versions of the Portfolio Standard.

The project portfolio discipline is ready for increased maturity. PMI has started a new certification for project portfolio managers. With this recognition, there will be those who wish to grow in the portfolio management discipline to increase their value to companies and leadership.

References

Cooke-Davies, T., Cicmil, S., Crawford, L. & Richardon, K. (2007). We're not in Kansas anymore, Toto: Mapping the strange landscape of complexity theory, and its relationship to project management. *Project Management Journal, 38*(2), 50–61.

Curlee, W. & Gordon, R. (2011). Complexity theory and project management. Hoboken, NJ: John Wiley & Sons, Inc.

Curlee, W. & Gordon, R. (2013). *Successful program management: Complexity theory, communication, and leadership.* Boca Raton, FL: CRC Press.

Duarte, D. & Snyder, N. (2006). *Mastering virtual teams* (3rd ed.). San Francisco: Jossey-Bass.

Fox, D. (1999). Views and models of communication. http://communication-theory.freeservers.com/custom.html

Johnson, D. & Johnson, F. (2000). *Joining together: Group theory and group skills.* Needham Heights, MA: Allyn & Bacon.

Project Management Institute. (2013a). *A guide to the project management body of knowledge* (PMBOK® Guide) Fifth Edition. Newtown Square, PA.: Project Management Institute.

Project Management Institute. (2013b). *Standard for portfolio management.* Third Edition. Newtown Square, PA.: Project Management Institute.

Tuckman, B. W. (1995). Developmental sequence in small groups. In Wren, T. (Ed.), *The leader's companion: Insights on leadership through the ages,* pp. 355–359. New York: The Free Press.

Uhl-Bien, M. & Marion, R. (2009). Meso-modeling of leadership: Integrating micro- and macro-perspectives of leadership. *The Leadership Quarterly, 20*(4), 631–650.

Useem, M. & Harder, J. (Winter 2000). Leading laterally in company outsourcing. *Sloan Management Review, 41*(2), 25–36..

15

Addressing Portfolio Information Issues through the Use of Business Social Networks, Stars, and Gatekeepers

ROBERT JOSLIN PhD, PMP, PgMP, CENG

Introduction

The aim of this chapter is to address one of the key issues associated with portfolio management, which is the lack of necessary and relevant information for timely and informed decision making. Information impacts not just the portfolio manager and his or her team, but everyone touched by the portfolio management processes. Address the information problem and it will address many of the questions relating to portfolio management.

The chapter starts by describing the main problem areas that are constraining successful portfolio management and then focuses on one—incomplete and inaccurate information. In the process of addressing the information issue, the chapter covers portfolio success factors, portfolio success criteria including definitions, and the importance of timely and accurate information on success factors and success criteria. Information management (IM) is then introduced and its importance to portfolio success. Passive portfolio controlling risks are discussed including organizational and human factor risks. The concept of proactive portfolio controlling is introduced with the underlying need for reliable information and how to obtain through a detailed understanding information flows through an organization. The mechanics behind business social networks are discussed with a comparison of hierarchical and network structures with their respective information flows and governance structures. This then leads into the core makeup of a network including the description of likely actor types who can address the information shortfall impacting portfolio managers.

The penultimate section describes a practical framework of how to build a business social network tool to support a portfolio manager in his or her daily job and then concludes with a short story of how one portfolio manager who suffered from taking a passive portfolio controlling approach is introduced into a new approach of active portfolio controlling leveraging the benefits and insight generated from business social networks.

Problem Areas in Portfolio Management

Portfolio management has provided many improvements to managing projects such as improved tracking, resource allocation, scheduling, analysis, and governance and oversight. However portfolio management is still not where it needs to be because it still faces several difficult challenges (Bryan, Matson, & Weiss, 2007). A literature review highlights six problem areas associated with portfolio management, see Table 15.1.

The following sections describe the problem categories in more detail.

Table 15.1 Most Cited Portfolio Management Problems (no order to ranking)

PROBLEM AREA	DESCRIPTION	CITATIONS
Portfolio level activities	Too many projects approved with weak-go ahead decisions: resources, value, and priority are not considered properly. Overlapping and non-integrated projects and tasks within one portfolio and between portfolios are not reviewed. Portfolio manager roles and responsibilities are not clear. No feedback is given at the project level. Projects are not killed when priorities or business cases no longer make sense.	(Cooper, Edgett, & Kleinschmidt, 2001)
Portfolio competencies and methods	Methods and guidelines for portfolio evaluation and project planning and management are inadequate. Mathematical portfolio approaches historically have provided inadequate treatment of risk and uncertainty, portfolio managers are unable to handle multiple and interrelated criteria, and people generally fail to recognize interrelationships with respect to payoffs of combined utilization of resources. Managers perceive several techniques to be too difficult to understand.	(Cooper, 2007a)
Human resources	There is a human resource shortage, a lack of commitment, and inadequate competencies at the project level.	(Cooper, Edgett, & Kleinschmidt, 1999, 2000; Elonen & Artto, 2003; PM Solutions, 2013; Prifling, 2010)
Information management across the portfolio, programs, and projects	There is a lack of useful information on projects, and an inadequate flow of information across the organization. The information flow from projects to the other parts of the organization and vice versa is limited. There are issues with the reliability of information and also limited availability of information.	(Cooper et al., 1999, 2000; Elonen & Artto, 2003; Prifling, 2010)
Management of project-oriented business	Projects are given a second priority and are not rewarded. Many lack a defined owner, business, or personnel strategy for portfolio management. Rapid and recurring changes in roles, responsibilities, or organizational structure are the norm. Many bodies are entitled to set up a project and independently own objectives of a unit.	(Elonen & Artto, 2003)
Project-level activities	The pre-project phase lacks improper implementation. Project progress monitoring is infrequent. Poor project management resulting in projects overruns.	(Cooper et al., 2000)

Portfolio-Level Activities

Portfolio management activities are greatly influenced by the culture of the organization. Prifling (2010) identified three dimensions of organizational culture, which result in oversized portfolios: safeguarding culture, consensus oriented, and sustainability oriented.

Organizations exhibiting one of more of these three dimensions are at risk in trying to manage oversized portfolios; therefore, these organizations need to be efficient in killing non-performing and limited-value projects. Portfolio managers who do not have clearly defined roles and responsibilities with an agreed-upon portfolio governance structure should address these issues; otherwise, it will reduce the effectiveness of the portfolio management activities (Cooper et al., 2000).

Portfolio Competencies and Methods

The lack of portfolio competencies required by portfolio managers has been cited as a problem area since the first research papers were written on this topic. Portfolio managers need to have a mix of skills ranging from understanding of the business, markets, projects and programs, impact of interdependencies with cause and effect, management accounting, identifying and managing risks, resource balancing across the portfolio, complex mathematical modelling, statistics, analysis skills, presentation skills, communication skills, and political skills with the ability to guide and influence senior management. This challenging role is beyond many people's management abilities; therefore, the importance and effort to identify and select the right portfolio manager as well as to develop individuals with potential should not be underestimated (Fricke & Shenhar, 2000).

Humans Resources

Organization project resources are invariably constrained; therefore, balancing resources according to skill and priority within and across portfolios is a challenge. This is exacerbated by project managers with assigned skilled resources who are unwilling to share them with other projects especially if the projects are in another division with their own divisional targets (Kerzner, 2010, p. 16).

Management of a Project-Oriented Business

Organizations have been known to give projects a low priority over operations especially in resource-constrained organizations and therefore put projects as a second priority. It is ironic that organizational improvement projects, whose objectives are to address the inefficiencies of the organization, cannot progress because the resources needed are tied up in operations. This `Catch 22' situation increases the risk of project failure (Elonen & Artto, 2003).

Information Management across the Portfolio, Programs, and Projects

A lack of reliable and timely information results in delayed or ill-informed decisions impacting the performance of the portfolio as well as the perceptions of the individuals who are involved in the portfolio management processes (Rajegopal, Waller, & McGuin, 2007, p. 18).

Project-Level Activities

One of the most cited reasons for poor portfolio performance is poor project and program management. If the projects or programs are only partly meeting their goals, this will have a direct impact on the performance metrics of the portfolio (Martinsuo & Lehtonen, 2007). Issues include: insufficient preparation in the pre-project phase; infrequent project progress monitoring, and general poor project management, which results in project overruns.

With the above problems it is not surprising that portfolio managers have a challenging role but does it need to be so? The problems listed in Table 15.1 are most likely related to the fact that one or more portfolio success factors are missing or incorrectly implemented.

The next section describes the importance of defining portfolio success criteria, the role portfolio success factors play in achieving portfolio success, and the underlying need of timely and reliable information.

Portfolio Success Criteria

Portfolios are a collection of projects, programs, and operations; therefore, all directly contribute to a portfolio success. Success is defined according to the context of a project or program and varies within and across industries (Shenhar, Tishler, Dvir, Lipovetsky, & Lechler, 2002). Regardless of how success is defined, it must be measurable, and metrics require the use of success criteria. The following definition of portfolio success criteria has been adapted from a definition of *project success criteria,* Morris & Hough (1988):

> Portfolio success criteria are the measures used to judge on the performance of a portfolio; these are the dependent variables that measure success.

Success criteria are seen on balanced scorecards to determine the progress of a project, a program, or the performance of a portfolio. Portfolios are different from projects and programs since they have a strategic, longer-term view and are ongoing unlike

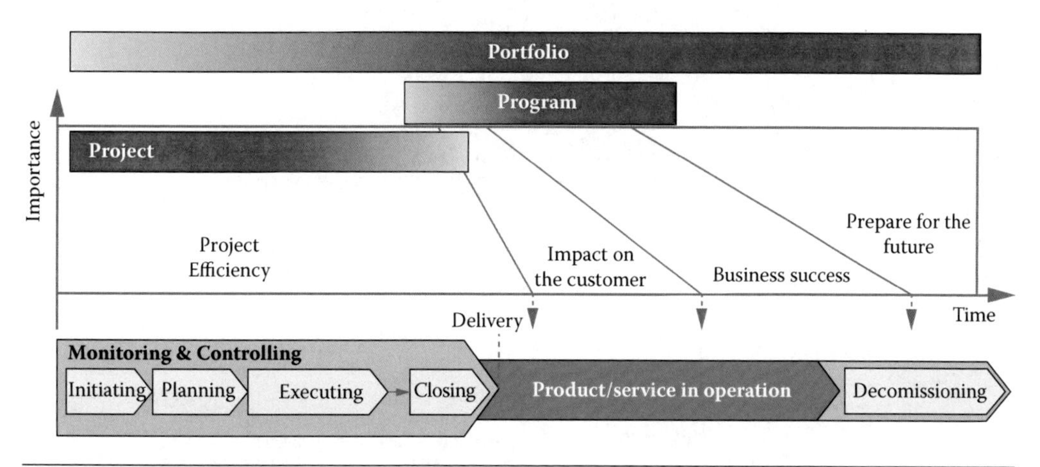

Figure 15.1 Relative importance of success criteria over time

a project or program,* which have finite durations. For this reason portfolio success criteria are more forward looking than their component counterparts as can be seen in Figure 15.1.

Once the appropriate success criteria have been defined and agreed upon, the next step is to ensure the appropriate success factors are in place at the right time within the phases of the project and program life cycles. Portfolios do not have life cycles per se, but they do have success factors as these also need to be present in conjunction with the project and program success factors.

Portfolio Success Factors

Portfolios are typically based on a line of business and once established are ongoing. In theory they do not have a life cycle, but literature references portfolio life cycles, which are mainly based on practitioner literature (Clack, Scott, Graeber, Honan, & Ready, 2012). Portfolio success factors vary according to the portfolio type, industry, culture, and other influences (Martinsuo, 2013; Teller, Unger, Kock, & Gemünden, 2012). This observation is supported in research literature where contingency theory states that no one way or approach is best for managing all types of portfolios (Voss & Kock, 2013). Contingency theory has been applied extensively to projects and programs (Hanisch & Wald, 2012) and more recently to portfolio management. With this concept in mind the need of different success factors will all apply within and across projects, programs and portfolios. The definition of project success factors from Turner (1999) has been extended to include portfolio success factors:

> Portfolio success factors are elements of a project, program, or portfolio which, when influenced, increase the likelihood of short-, medium- and long-term success; these are the independent variables that make success more likely.

* Some programs last for decades (Levin & Ward, 2011), but in theory all have a finite duration.

- Portfolio finance and non-finance related goals aligned to strategic goals
- Effective monitoring, control and alignment

Strategic Alignment*

Stakeholders

- Good communications/feedback
- Effective consultation with key stakeholders
- Support from senior management
- Good customer relationship

Governance

- Formal governance procedure (effective)
- Identifying, prioritizing, authorizing component programs and projects
- Resource balancing depending on component merit
- Clear communication channels
- Delegation of responsibility coupled with governance procedures
- Component initiation and closure according to governance and overall portfolio timing

Complexity

- Risks addressed/assessed/managed at all levels

Leadership

- Manage portfolio according to project type e.g, NPD vs client delivery versus maintenance project
- Leadership quality of portfolio manager and EPPO
- Skilled and suitably qualified portfolio manager and team
- Schedule regularly updated including component schedules
- Well functioning project organization with accountable, dedicated cross functional teams and strong leaders
- Effective use of Enterprise Project Portfolio Office (EPPO)

Adaption

- Flexibility in management
- Environmental influences managed

Figure 15.2 Success factors at the portfolio level.

A great deal of research exists on project success factors including the impact of context (Hanisch & Wald, 2012). Program success factors are less researched than project success factors. The main differences between project and program success factors are the ones relating to adaptability; long-term funding; a detailed understanding of the market; and environmental, legal, and regulation issues. Portfolio success factors are the least researched, but enough literature exists to be able to collect and categorize success factors along six success dimensions.* Figure 15.2 shows the success dimensions with their respective success factors.

A portfolio manager will want to ensure that the success factors within the portfolio including those of the project and program have been identified and implemented because they will have a major impact on the overall performance of the portfolio (Martinsuo & Lehtonen, 2007). To implement a success factor requires a considerable investment, and some success factors require greater investments than others. Therefore, it is important to understand which success factors impact the portfolio and which do not, as the degree of impact depends on the context of portfolio. Implementing success factors that have little or no impact on the portfolio outcome will consume time and resources as well as being a management distraction.

The success factor framework (Figure 15.2) explicitly addresses all of the portfolio issues identified in Table 15.1, except for one which is 'timely and accurate information'—which is implied throughout the framework. Without timely and accurate information many of the portfolio success factors would not be identifiable, implementable, or measureable.

The next section introduces the concept of Information Management (IM) and the impact of IM on portfolio management.

* A success dimension comprises of one or more success factors

Information Management

Background

IM can be viewed from three distinct perspectives: the organization, the library, and the individual (Detlor, 2010). The focus of this chapter is on the 'organization perspective'.

IM is about the control over how information is created, acquired, organized, stored, and distributed so as to achieve an efficient and effective information access, processing, and decision making. Detlor (2010) defines IM as a unit with the responsibility of defining organizational informational requirements, planning and constructing the information infrastructure and information systems applications, managing the system, and manage staff activities. The next section describes the importance of IM to portfolio management.

Importance of IM for Portfolio Management

Portfolio managers need timely, accurate information to be able to take informed decisions to steer a portfolio (Rajegopal et al., 2007). This implies that the effectiveness of portfolio managers is dependent on the maturity of the organization's IM. Organizations, such as the Software Engineering Institute (SEI), Gartner Group, and the Enterprise Information Management (EIM) institute plus individual researchers in the IM field have developed IM maturity models to help organizations determine their information maturity.

Referring to Figure 15.3, there are five levels of IM ranging from unstructured and ad-hoc information to highly structured information, which are aligned to the

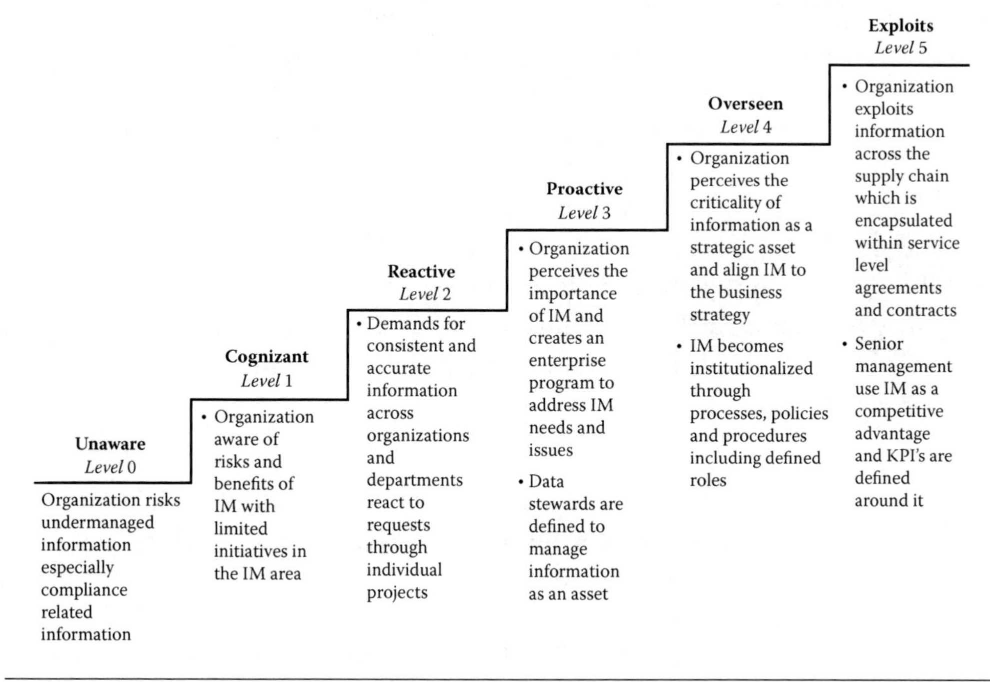

Figure 15.3 Example of an IM model.

business strategy and used as a competitive advantage. Portfolio managers operating within an IM environment at levels 0 to 3 need to be more proactive in obtaining the necessary information for decision making. For portfolio managers operating at levels 4 and 5, there is more structured and easily accessible information, but this creates a risk of adopting a passive portfolio controlling approach by solely relying on the information provided and as a consequence reducing contact with the program and project teams. This behavior is termed 'passive portfolio controlling', and if adopted, will increase the risk of negative events appearing out of the blue impacting performance of the portfolio. The next section describes the risk of passive portfolio controlling in more detail from both a structural and human factor perspective.

Risks of Passive Portfolio Controlling

A portfolio manager should be aware of the risk of relying too heavily on reported information without taking a proactive approach to really understand the basis of the underlying reported information. The telltale signs of a portfolio manager with a passive portfolio controlling approach are when the physical contact with the project and program teams drops during a quiet period that is followed by a crisis mode triggered by an unexpected crisis. This is the start of a spiral of events starting with nasty surprises occurring where project and program statuses change from green to red with no apparent warnings. Then meetings after meetings are scheduled to discover the causes of the issues where hasty solutions are put into place just to see the same problems arising again and again. The portfolio is now unstable, and unless the root cause issue(s) are identified and addressed, the impact on the organization will be severe.

This description is of a portfolio manager who has lost control because of not having accurate and timely information, which helps to provide a holistic picture of what is happening and proactively addresses problems before they impact the performance of the portfolio.

Inexperienced, newly assigned portfolio managers are the most likely to adapt a passive controlling approach because they fall into false sense of security if they perceive the portfolio organization as mature and functioning. One of the reasons they may justify adopting a passive controlling approach is the organization's high benchmarked IM maturity level. The following examples show why this passive controlling approach is extremely risky:

- IM maturity models benchmark the maturity of an organization on the basis of formal roles, metadata definitions, data bases, and data collected, but what they cannot address is the human psychology side of what is reported and if this information is accurate or not.
- IM models give no direct reference to information that flows through formal or informal network structures; therefore, these structures may not be considered as part of the benchmarking. However, with up to 60% of organizational

information flowing through formal and informal networks, they cannot be ignored by the portfolio manager (Bryan et al., 2007).
- IM maturity models do not cover the effectiveness of managing the constantly changing factors (CCFs) within projects such as the ability to manage risks, questions, issues, tasks, decisions, assumptions, change requests, and opportunities within a timely manner without negatively impacting the project (Joslin, 2012).

There are other information traps that an inexperienced portfolio manager may not be aware of which are categorized as organizational structural and human factor related issues, which are described in the next section. For each trap a solution is given to show how an experienced portfolio manager would address each one.

Organizational Structural Issues Impacting Information Quality

Organization Structures Non-conducive to Projects Organizations can be structured in any number of ways and are typically more suited to operational environments than project environments. Solutions exist to reduce organizational barriers by creating projectized and matrix based structures, but both have their strengths and weaknesses (PMI, 2013a). The portfolio manager should pay attention to the way projects are established and proactively influence a proposed structure to reduce the risk induced by unnecessary constraints.

Poorly Thought-out Contracts Hindering Rather than Supporting Projects A common problem across all industries is that many contracts are established by procurement and legal people with little or no experience of complex project management. This lack of experience results in contracts that hinder at best and worst make it impossible for projects to achieve their goals. Goldratt (1990, p. 36) described poorly thought-out polices and hence contracts as bottlenecks similar to production line bottlenecks. Once the bottlenecks are removed throughput increases substantially. If the portfolio manager has no idea of the types of contracts and how they are structured, then there is a risk of poorly thought-out contracts impacting the overall portfolio's performance.

The ideal person helping to establish a new contract is one who has procurement, legal, project management, and systems modeling experience. This person should also understand the essence of the contract(s) associated with the projects he or she is managing. A portfolio manager who has worked his or her way up from being a project and program manager should have such skills. A knowledgeable portfolio manager will take an active interest in the formation of these contracts and gain a detailed understanding of their impact on portfolio performance.

Flaw in the Portfolio Selection Process A theory called 'winner's curse' proves that winners of auctions, always overpay. This theory can also be applied to the portfolio selection process, which highlights a flaw that gives an alternative perspective on low project success rates regardless of industry and country.

Winner curse occurs when the bidders in an auction are unable to estimate the true value of the item for which they are bidding. The winner always grossly overpays compared to the true value because the highest bidder wins (Thaler, 1988). This has been proven in several controlled experiments (Bazerman & Samuelson, 1983).

In the portfolio selection process, sponsors' project and program proposals are bidding to be selected where the uncertainty of all costs and benefits are estimated with some magnitude of error. Therefore it is likely that the projects/programs selected for the portfolio are most likely to have underestimated the costs and overestimated the benefits rather than the average (Moore, 2010, p. 61). The link between winner curse and project/program selection is the 'inability to determine the true value, benefit and cost' of a project/program. If a sponsor is aware of the weakness of the project/program selection process then he or she is likely to exploit this weakness because it is almost impossible for the decision makers to judge if the project or program proposal's underlying assumptions and estimations are accurate.

Now the reader may consider the sponsor's short-term approach flawed because the sponsor's plan will be discovered later on in the project or program life cycle, which is partly true. The project or program is unlikely to achieve its financial and schedule goals because of the distorted values given in the business case. But the project or program will probably be completed especially if an organization does not have a track record of killing non-performing endeavors. A secondary impact of overspend is likely a reduction of funding for other potentially competing projects or programs, which is unlikely to be much of a concern to the sponsor especially if the impacted projects or programs are outside of his or her area. In the event of a review, the results will probably state that the project or program manager was unable to meet the budget and schedule rather than examine the assumptions behind the costs and schedule estimates to determine if they were inaccurate. Therefore, the flaw in the project and program selection process is unlikely to be spotted except by the people exploiting it.

A portfolio manager should understand the concept of winner's curse and take a proactive approach to determine if the selection process is being exploited. To do this will require some investigatory work to review historical project/program outcomes to determine if there is for each sponsor a consistent set of gaps in terms of what was promised versus delivered for his or her projects and programs.

The experienced portfolio manager will review the profiles of each sponsor's project and programs to determine the 'promised values' versus the 'delivered values' over a period of say five years. Referring to Figure 15.4, it shows one project sponsor's projects, called 'Spud', and how his projects have consistently underperformed. This information will raise an alarm to the experienced portfolio manager to dig further into the historical data to determine who the project manager(s) were for these projects, and how well they performed on other projects beside those of Spud's projects.

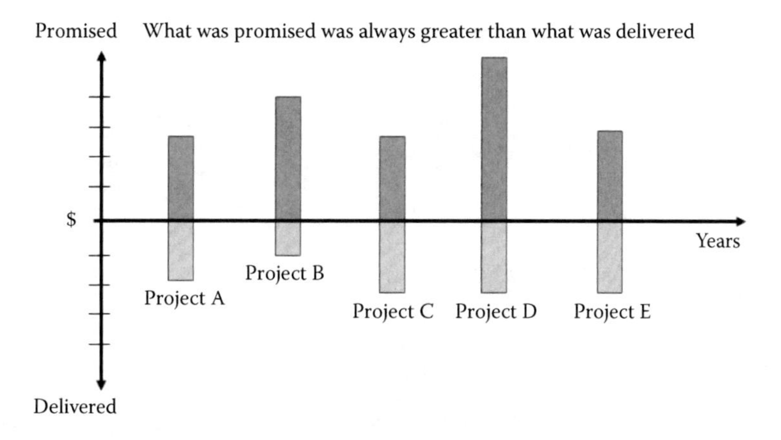

Figure 15.4 Profile of 'Spud's' sponsored projects 'promised versus delivered values'.

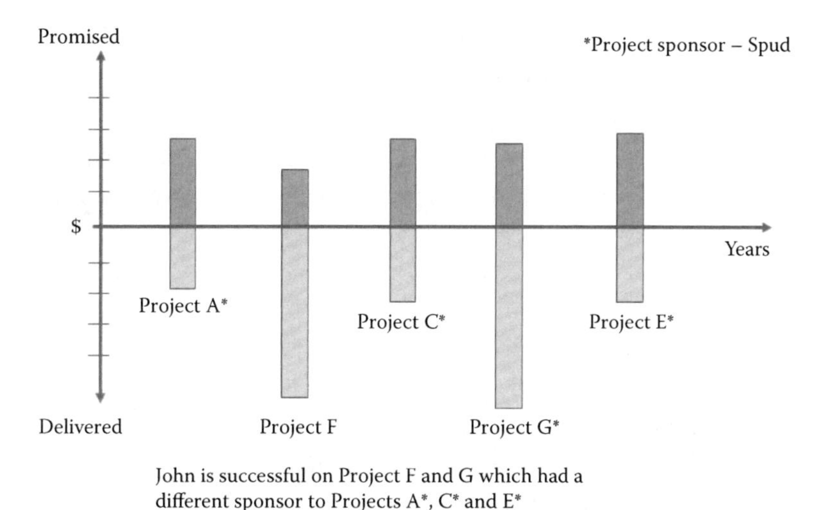

Figure 15.5 Profile of John's (project manager) projects promised (business case) versus delivered results.

Figure 15.5 shows the profile of one of the Spud's project managers who is called John. John was not successful on any of Spud's projects. However, he was very successful for two other projects for another project sponsor where the project goals were exceeded. If one now looked into more detail of the failed projects (A, C, and E) belonging to Spud, John only marginally missed the goals. If Spud's did distort the costs and benefits, then John really did a great job. However, John's efforts were not understood, and therefore, were not rewarded in his performance review because managers were not aware of the winner's curse.

Spud, the sponsor, also used another project manager called Peter on projects B and D. Referring to Figure 15.6, it shows Peter is consistently underperforming on all his projects. Peter was let go after the failure of project D.

If the portfolio manager knew in advance that Spud, the sponsor, had used winner's curse to his advantage i.e., over promises the benefits and under promises of costs,

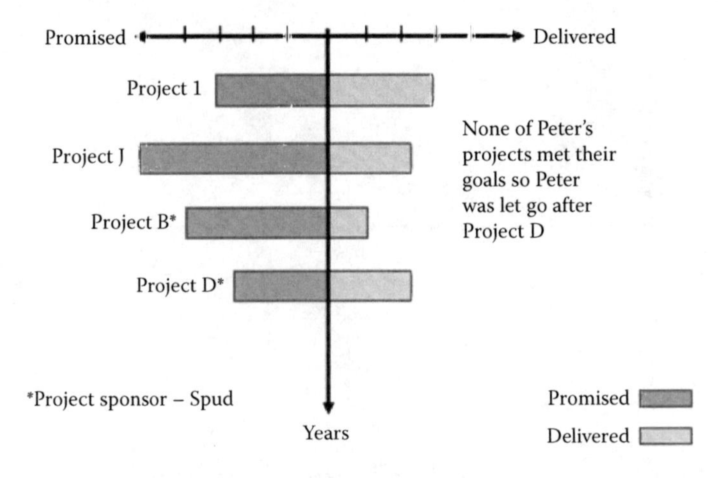

Figure 15.6 Profile of Peter's (project manager) projects promised (business case) versus delivered results.

then Peter would not have been reviewed on project B and D as it would have been impossible to succeed. Therefore reviewing Peter only on projects I and J would show he almost met their stated goals.

The recommendation from the experienced portfolio manager would be to should he should receive additional training and support rather than terminate his employment.

Human-Factor Related Issues

Cognitive Dissonance (self-denial) Cognitive dissonance is the excessive mental stress and discomfort experienced by an individual who holds two or more contradictory beliefs, ideas, or values at the same time (Cooper, 2007b, p. 6).

Project managers and sponsors are both susceptible to cognitive dissonance resulting in distortions of what is discussed and what is reported. This problem is one of the reasons why projects scorecards go from 'Green to Red' without showing 'Yellow'. Cognitive dissonance can also spread across an organization into a corporate culture where project results are changed to meet managers' expectations*.

The state of cognitive dissonance occurs when a project manager believes that two of his or her psychological representations are inconsistent with each other; for example, the belief in himself or herself as a successful project manager (or a sponsor) and the poor state of the project's performance. The project manager (or sponsor) will go into self-denial and continue to report green until the real status of the project or program has been discovered; often when it is too late to recover the project. The situation is exacerbated if both the project manager and project sponsor suffer from cognitive dissonance thus reinforcing each other's self-denial. Cognitive dissonance impacts extraverts more often than introverts. Interestingly, introvert project managers view delays in their projects as ones that are due to environmental impacts and therefore

* Interview findings from a qualitative study (Joslin, 2014)

do not take them personally (Meyer, 2013). This lowers the risk of inconsistent psychological representation and therefore results in more likely reporting of the correct status of a project.

For the interested reader, there are research papers that break down competencies of project and program managers into EQ (Emotional), IQ (Intelligence), and MQ (Managerial) and compares project and program managers (Dulewicz & Higgs, 2005; Müller & Turner, 2007a; Shao, Müller, & Turner, 2012; Turner, Müller, & Dulewicz, 2009). However, none of the papers cover the topic of cognitive dissonance.

The personalities of the project managers who suffer from cognitive dissonance unfortunately drive them to take on greater challenges to prove themselves even with a checkered project management history. If senior managers do not address this issue, or if the project manager has moved to another organization, this problem poses a risk for the portfolio manager. For the experienced portfolio manager he or she will be aware of the risks of unproven senior project managers who appear extraverted and perhaps overconfident and therefore may be susceptible to cognitive dissonance.

Lack of Skills in Project Management In the research community there is an ongoing discussion on whether project management is a profession or an occupation. The main argument for project management not being a profession is that project success rates are stubbornly low. This is unlike any other profession where there success rates are consistently high e.g., dentists, lawyers, architects, and engineers. Getting a project management certification, such as Project Management Professional (PMP) or Projects in Controlled Environments (PRINCE2), does not mean a project manager is experienced and able. It is only the track record of a project manager, as with all other professions, which qualify one as a successful professional.

Lack of skills/competencies in project management is one of the most cited reasons for poor portfolio performance. One simple example is the confusion as to what is the difference between success factors and success criteria, and a lack of basic understanding of which success factors influence project success and hence need to be in place during the project life cycle. All too often projects are started without this knowledge resulting in an inevitable suboptimal conclusion.

An experienced portfolio manager should be able to determine fairly quickly through direct observation of a work of a project manager as to his or her competency. He or she will also get a project 'heart-beat', which is the understanding of the unreported aspects of a project such as team morale, confidence in the project manager of the team members and the sponsor, the impact of stress and deadlines on the team, and unreported short cuts or issues, etc.

Summary of Passive Portfolio Controlling

Passive portfolio controlling has many inherent risks. It is easy to take for granted reported information without understanding the underlying risks. Telltale signs of passive portfolio controlling are constant reactive responses to unexpected project issues that require a great deal of time and effort. A portfolio manager who is in control will better invest his or her efforts carrying out proactive portfolio controlling with the aim to get closer to the projects and program so as to get a 'heart-beat' to complement and cross check the reported information. A project or program 'heart-beat' is more tacit in nature and not necessarily forthcoming but is a key source of information. Heart-beats come from sources within and also outside of the project teams, such as individuals who play an important role in collecting and disseminating information. A portfolio manager first needs to understand how information flows through the organization and then where and how this heart-beat information can be used.

The next section answers the question as to how information flows through an organization.

How Information Flows through an Organization

Decisions based on timely and accurate information in an organization mean people need to share and pass on information. Yet according to a global survey of senior executives only 25 percent of the respondents described their organizations as "effective" at sharing knowledge and information across boundaries (Cross, Martin, & Weiss, 2006). Some organizations are better than others in sharing ensuring there are various channels where information can flow. In extreme cases when information sharing is restricted, peoples' beliefs are so strong they sometimes decide to share information knowing it will be detrimental to their organization's reputation.*

Information is in some ways similar to water where it flows (gravity) to people who need it and/or may benefit from it (need or demand). When water comes to a barrier it builds up until at some point a weakness is found through, over, under or around the barrier. If information flow is blocked for whatever reason then if the need to share and/or the demand for the information is high enough, it will eventually overcome the barrier. However, the time taken to overcome a barrier impacts its timeliness and hence value to the recipient to act on or prevent something as described within the information. If information barriers are understood by the transmitters of the information, then they can take proactive measures to use an alternative channel. Receivers of information who are also aware of barriers within an existing communication channel can look to get quicker access to the sources of the information through an alternative communication channel e.g., e-mail, phone, meetings, and bulletin boards.

There are two types of organizational structures that support communication channels:

- Hierarchical-based structures
- Network-based structures

* These are typically whistleblowers who take risks to make public information that may be contravening local and international laws

In hierarchical-based structures the information that flows across the communication channels is typical command and control type information; whereas in network-based structures, information flow is based on people's work needs and interests.

Network-based structures are broken down into formal networks and informal networks.

- Formal networks are known by management and handled formally where every formal network has a network owner and other defined roles (depending network objectives). The people with assigned formal network roles may have network performance targets, which may or may not be incorporated into their overall personal work objectives.
- Informal networks arise out of a need or interest within an organization, which often bypasses communication and management bottlenecks. These informal networks have no controls and may not be known to management.

Research has shown that organizations need both types networks especially as some of the types of information that flow through network structures is not present in hierarchical structures (Kotter, 2011).

Referring to Figure 15.7, the first column, 'information', lists types and sources of information that are of interest to the portfolio manager. The following two columns show a hierarchical and a network structure and the respective information that flows through them. The first observation in the comparison of hierarchical and network structures is that the majority of organization information flows through network structures. The second observation is that the soft or tacit information flows mainly through both formal and informal networks and little within a hierarchical structure.

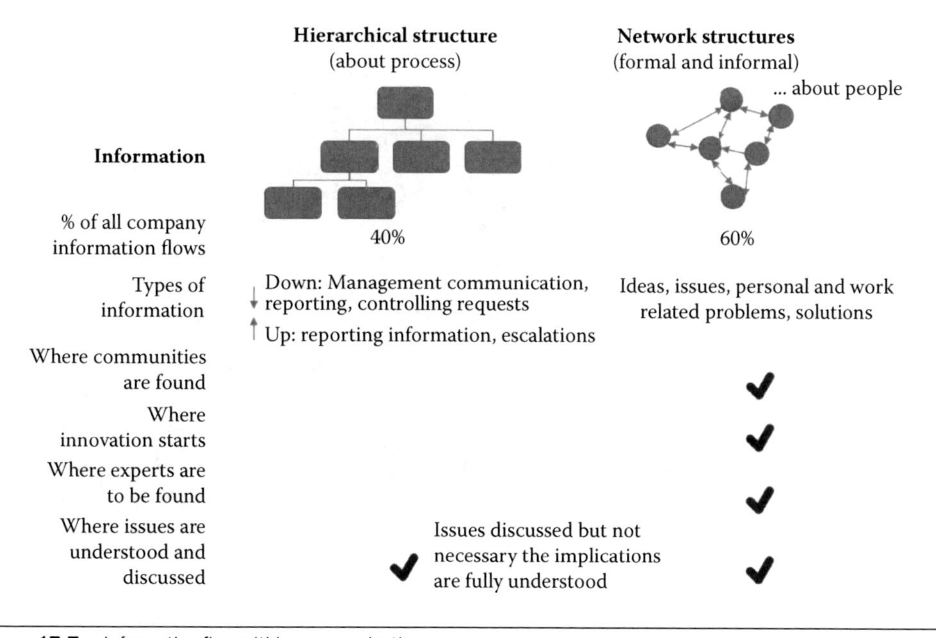

Figure 15.7 Information flow within an organization.

Communities grow by developing networks, innovation flourishes within communities, and problems are resolved by people in communities who have the experience and knowledge about how to fix things. There are many benefits resulting from network structures, but these structures also have risks associated with using them. The next section describes the benefits and the risks of informal networks and how to effectively mitigate the risks through migrating at the appropriate time to formal networks.

Value and Risks of Harnessing Informal Networks

Referring to Figure 15.8, organizations with a low IM level will have many more informal networks than formal ones as talented employees find ways to get around information and decision impasses.

Without informal networks many things would not get done in a timely manner (Cross, Parise, & Weiss, 2007). As organizations with a low IM level start to understand the value of network structures and the information that flows through them, they will start to formalize the structures to reduce the risks of associated with serendipity. This appreciation of network value should trigger the need for IM, which, if implemented effectively, will put the organization on track to becoming a high-performing and potentially innovative organization.

Employees within an organization that has a high level IM are unlikely to set up long-term work related informal networks knowing that they will become formalized structures as the value of the information on the network increases. If work still continues within an informal structure when part or all of the work-related activities have been formalized, then this structure will quickly lead to a risk of information duplication (e-mails, documentation, and discussions), confusion, and ultimately adversely affect the improvements that they were trying to progress to at the outset.

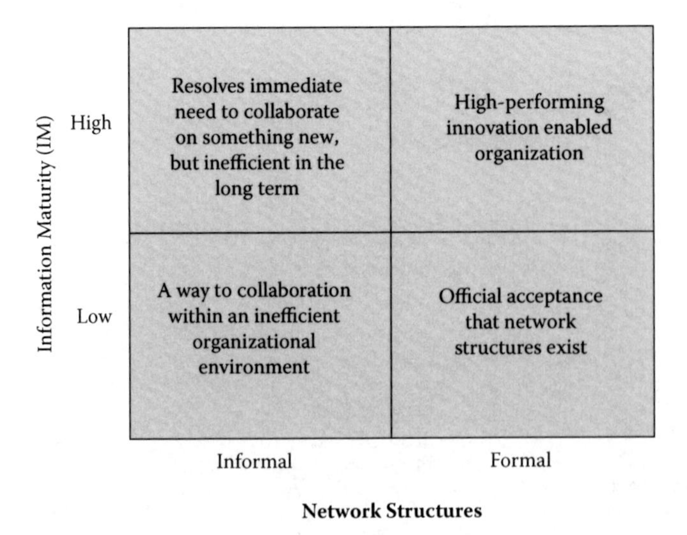

Figure 15.8 Influence of IM maturity on the use of network structures.

In a project management environment, project managers need to be well connected to project stakeholders regardless if their organization has a low or high IM maturity. Organizations that are highly outsourced, engaging several outsourcing partners in a production and/or service process, and complex in structure with numerous service contracts greatly increase the challenge for the portfolio managers to just to stay connected and to determine if their project managers have their projects under control. Pragmatic project, program, and portfolio managers within these types of environments will recognize the value of informal networks and how they will increase the chances to meet their goals. They should also understand the risks of being too reliant on informal networks in the event of a key person leaving or moving on to another organization.

Informal networks are unlikely to fit into an organizational chart especially in organizations with a low IM level. So what does one look like and how can one see the risks of relying on an informal network structure? The next section is written as a case study to show the importance of understanding the impact (positive and negative) of informal networks.

Informal Networks—A Case Study

Referring to Figure 15.9: consider the portfolio where the newly instated portfolio manager (Khan) is managing several project managers, and everything appears to be

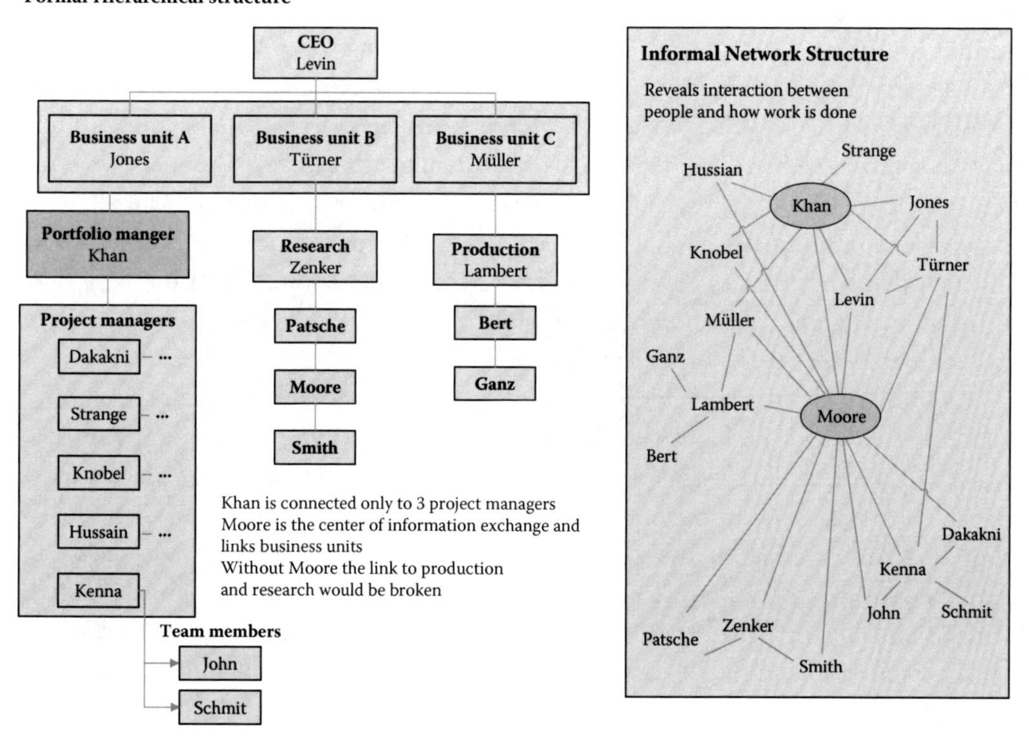

Figure 15.9 Deficiencies in formal structures and information compensated by an informal network structure.

running as one would expect in a large organization. Khan has limited connections mainly to his managers and a few direct reports. Khan is unaware that the majority of his projects are dependent on one person, Moore, who is a 'star', which is a type of connector in ensuring communication flow between research and production.

Khan is unaware that Kenna, one of his direct reports suffers from cognitive dissidence. Kenna has promised Türner, who is the sponsor of his project, he can succeed in an innovative but highly risky development. The project status reports do not show any major issues; however, the truth is far from the reported information. Khan's network is limited being new and is relying on reported project information. Taking this approach introduces the risk of nasty surprises in project status reporting that would be mitigated if he was connected into the business social network where he would find out really what is happening. If Khan was connected then he would discover that Moore would be the obvious person to discuss some of the project heart-beats. He would quickly learn that Kenna's project is in red and risk of failing, with a potential knock-on effect of his own position being questioned.

The scenario describing Khan, Kenna, and Moore happens all too often in organizations where stars* (connectors) ensure information flows across networks (unknowingly to management) while doing their own jobs. It is the stars that ensure projects and work activities progress without being adversely impacting through inefficient hierarchical structures.

Now going back to the description of Khan in his portfolio management position, Khan would soon learn that Moore is considering leaving but is unaware that the majority of Khan's projects are dependent on Moore as the information connector between his project group and the two business units of research and production.

Referring to Figure 15.10, as long as Moore is present everything is working, information is flowing albeit less efficiently than if direct relationships existed. However, as soon as Moore leaves, the network is broken resulting in communications breaks. The result is communications having to resort back to the hierarchical structure with its inherent delays, or someone needing to take Moore's network connection (star) role. Either way, there is a disruption to the projects' performance during this transition time as well as longer term implications.

In summary, informal networks are a workaround for silos, poorly performing hierarchical structures, and low levels of maturity in information management, but they are reliant on serendipity and risk at any time breaking down to key players leaving the network. Well-intentioned, informal networks can quickly consume a lot of time in terms of effort of the networked individuals where research has shown in extreme cases nearly half of the interactions of information networks were not central to decision making (Bryan et al., 2007). Informal networks have their role, but senior managers need to ensure there is a way to formalize informal productive networks to ensure these networks remain formal.

---

* Refer to Table 15.3 for the full definition of a Star.

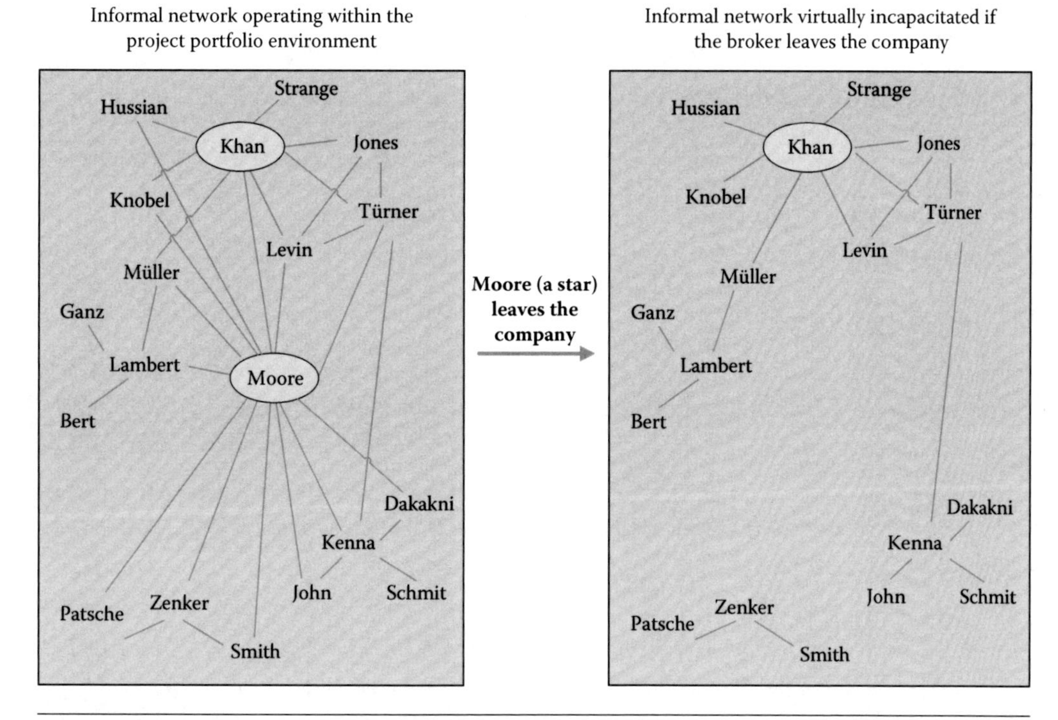

Figure 15.10 A broken network which is due to Moore (star) leaving the organization.

Formal Networks

The main objective of designing new formal networks or to formalize informal networks is to decrease the costs of collaboration and information flow and increase the value to the employees and organization. Formal networks do not replace hierarchical structures as both have their place in an organization. The next section describes the characteristics of hierarchies and formal networks.

Characteristics of Hierarchical versus Formal Network Structures

Most organizations will have coexisting hierarchical and network structures. Organizations that are at a low IM level will likely have more informal networks than formal networks. Organizations at a higher IM maturity will have a higher ratio of formal networks, which should complement their hierarchical structures.

For the portfolio manager it is important to understand the characteristics of both hierarchical and formal network structures in the quest to achieve a high-performing portfolio.

Reviewing the attributes of hierarchical versus formal network structures in Table15.2 shows the importance of formal network structures for project and program environments. The implications for the portfolio manager are the necessity to understand what types networks exists (informal and formal), what information flows across

Table 15.2 Characteristics of hierarchical versus network structures.

	HIERARCHICAL STRUCTURE	NETWORKS (FORMAL)
Underlying principle	Process driven where work is done through authority.	People driven where work is organized through mutual self interest.
Degree of influence	Management hierarchy.—Influence is the level in the hierarchy.	Collaboration and leadership. Influence is the number and type of effective network relationships.
Career progression	Achievers progress in their careers in the hierarchy.	Achievers widen their network.
Number of reporting lines	One or more depending whether it is single or multiple reporting lines	Only one manager per person, one network leader per network, which may also be a star if he or she links internal network clusters together, or a gatekeeper links internal and external network clusters
Traits	Complex hierarchical structure and multiple roles	Flat structures
	Intensive and sometime duplicate and redundant interactions	Point-to-point interactions for developed networks
	Decision-making bottlenecks	Direct decision making
	Command and control information	Information flows on interest and need
	Difficulty in finding knowledge assets and/ or subject matter experts	Networks designed around knowledge and topic area making it easy to find knowledgeable people
	Conflicts between direct reports	Single direct report reduces potential conflicts

these networks, and why. Expanding on the why question—are the use of networks in the portfolio environment addressing communication deficiencies imposed by structural and/or cultural issues, or are they addressing one or more ad-hoc needs? With this information the portfolio manager can develop a strategy on how to best leverage the value of existing network structures within the context of his or her environment. The process of developing a strategy should also include how to reduce the risks of simultaneously using both informal and formal networks.

The next section describes how a portfolio manager can maximize the value of the network structures.

Maximizing the Value of Network Structures

Networks have more to do more to do with people than processes, and as organizations outsource and partner both formal and informal networks will continue to expand. For the portfolio manager and his or her team it is essential to tap into and perhaps drive the expansion and value that can be obtained from networks. The question is where to start?

Steps to Identify, Map, Understand, Optimize, and Extend Networks

For a portfolio manager to gain the maximum from structural networks it will require time, knowledge, and discipline. Every person except a hermit has a network of some

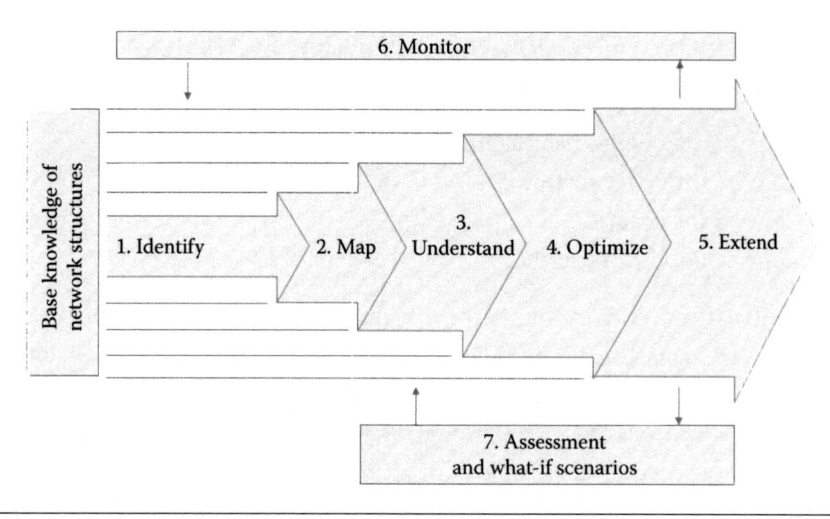

Figure 15.11 Framework to optimize the benefits of business social networks.

type of network whether one is conscious of it or not. It is unlikely than many people manage actively their networks in a structured way to achieve the maximum benefit. Social networks sites such as LinkedIn help connect people and provide simple tools for showing relationships between people. However, there is an absence of theory and functionality to understand how to build a focused network, the types of relationships that can contribute to one goal, and why and how information flows across networks. Portfolio managers may understand to varying degrees the power and limitations of networks, but they are unlikely to employ a structured approach to use networks as a strategic tool to address the problems of timely and accurate information flow. Figure 15.11 is a framework that addresses these issues using a seven step process that if followed should ensure the portfolio manager is in control of the information he or she needs to obtain the highest level of portfolio performance.

Before describing the seven process steps, a 'base knowledge' of the core makeup of a network structures is required, which follows in the next section.

Core Makeup of a Network This section describes the theory behind network structures, which is the core makeup of both formal and informal networks. With a good understanding of the theory a portfolio manager will be able to design a network in such a way to achieve the objectives whilst minimizing the effort and maximize the benefits.

Networks are made up of nodes (sometimes called actors, egos, and units), and the relationships between them (links). The type of content that flows over the links is called transactional content, see Figure 15.12.

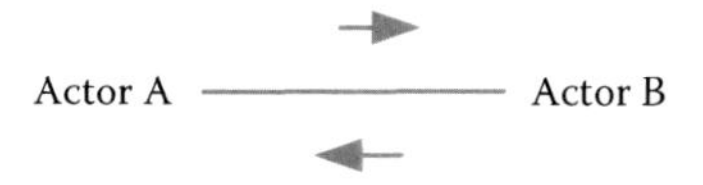

Figure 15.12 Bi-directional transactional content.

Transactional Content When two actors are linked, four types of transactional content are exchanged (Tichy, Tushman, & Fombrun, 1979).

1. Exchange of affect e.g., friendship
2. Exchange of influence or power
3. Exchange of information
4. Exchange of goods and services

'Exchange of affect' and 'exchange of information' are more related to social networks than hierarchies, while 'exchange of influence or power' is typical in a hierarchical structural. But the latter is also present in with formalized networks. Formalized networks may be centralized with mediation by an assigned or assumed network owner e.g., probably a supervisor. In an immature portfolio environment 'influence or power', 'exchange of information', and 'exchange of goods and services', are of the most interest to a portfolio manager. If the portfolio manager is responsible for a mature portfolio the last two transactional types; 'exchange of information' and 'exchange of goods and services' are of greater interest.

Link Attributes The link between any two actors can be described in terms of characteristics as shown in Figure 15.13.

These include:

1. Intensity—the strength of the relationship in terms of honoring obligations and the willingness to help the other by forgoing personal costs
2. Reciprocity—the degree and balance of mutual exchange of information
3. Expectations—the degree to which two actors agree on how they behave in the relationship
4. Multiplexity—the degree of which two actors are linked by multiple roles, e.g., husband and wife or manager and employee. The more roles through which two actors are linked to each other the stronger the relationship

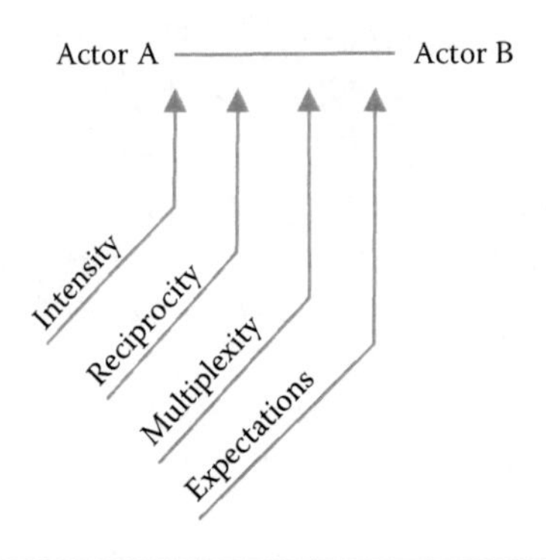

Figure 15.13 Link attributes.

Structural Characteristics The structural characteristics of a network include:

1. Size—number of actors within a network
2. Density (links)—the number of links in a network as a ratio of the theoretical maximum number of links
3. Clusters—grouping of actors where there are more links than elsewhere in the network
4. Network actors—the types of actors and the roles they play in a network

The first three structural characteristics are not described further except to say there is extensive material written about structural characteristics of networks, all outside the field of project management (Aalbers, Koppius, & Dolfsma, 2006; Kratzer, Leenders, & Van Engelen, 2010; Tushman, 1977).

Note: The PMBOK® (PMI, 2013a) contains information on the size of a network but does not cover the other three structural characteristics of a network.

The last structural characteristic requires further investigation because of the potential impact of the actors' type on the performance of a portfolio.

Actor Types in a Network Recent literature on business social networks has reduced actor types from six actors types (described in Figure 15.10) down to three—which the latter are termed broker, central connector, and network member (Burt, 2004, p. 7; Cross et al., 2007). The reason for the reductionism may be to make it easier for management to understand the concept of network structures. However, in doing so, it has also lost the importance of the full set of roles, which are needed for the people looking to understand, build, and leverage the power of these networks. The portfolio manager is one of the people who will need to understand the roles people play in a network and their motivations before a network can be fully understood (its purpose, strengths, and also potential weaknesses).

Referring to Figure 15.14, there are two network clusters and six actor roles. The roles used in Figure 15.14 were first described in a research paper published in the 1970s *Academy of Management Review* (Tichy et al., 1979). Some of the terms originated from even earlier papers, which described stars and gatekeepers within research environments and their impact on innovation and performance within the research and development (R&D) environment (Allen & Cohen, 1969; Tushman, 1977).

Note: For the interested reader who intends to do further reading on this topic the terms brokers and connectors are often used: a broker is equivalent to a star if the broker does not have network connections to external entities and is equivalent to a gatekeeper if he or she does have internal and external connections. A central connector is the modern literature is equivalent to a star i.e., someone within an organization connecting a cluster.

A description of the network roles is given in Table 15.3.

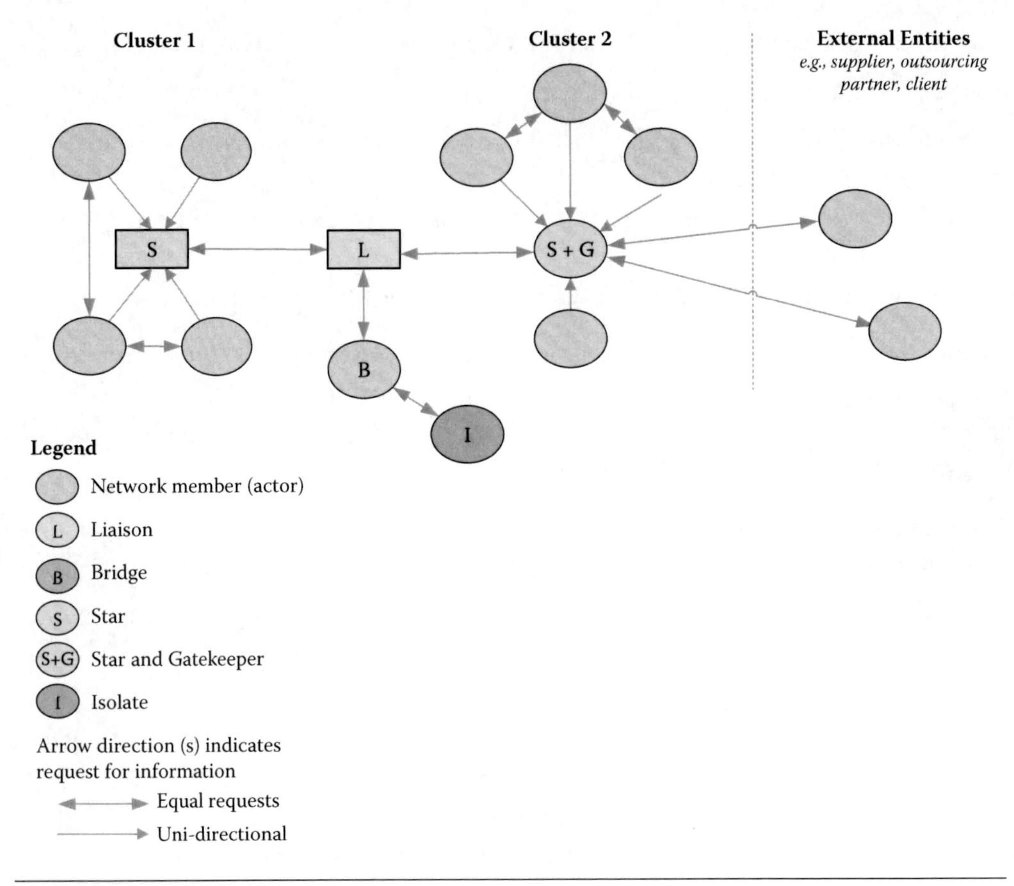

Figure 15.14 Actor types in a business social network structure.

Table 15.3 shows a seventh role called a network owner, which was not originally described in the original papers on social networks in the 1970s. This term was coined in management consulting articles within the last ten years as a way to put some formal rigor to networks to leverage their potential. Network owners can also hold the

Table 15.3 Description of the Business Social Network Roles

NETWORK ROLES (ACTORS)	DESCRIPTION
Network member	A member of a network who uses information to do his or her job, provides new information,. and acts as a conduit for information
Liaison	An individual who is not a member of a cluster but links two or more clusters
Bridge	An individual who is a member of multiple clusters in a network and acts as a connector
Star	A star who is a member of one or more internal clusters who has significantly more communication than non-stars (all other roles except the gatekeeper)
Gatekeeper (includes star)	A star who is a member of one or more internal clusters and extends to external entities (organizations) *Note:* Figure 15.14 only indicates one Star and one Gatekeeper each linking one cluster. This is just simplification for understanding.
Isolate	An individual who is uncoupled from the network
Network owner	Formal networks are assigned a network owner who may have performance or other metrics to meet.

gatekeeper or any other role although most of the other roles beside gatekeeper or star would imply a less than optional choice of network owners.

Stars and non-Stars All network roles have an impact the effectiveness of a network, and therefore, are important for the portfolio manager to understand. The most interesting of the roles for the portfolio manager are the star and gatekeeper, as the people that reflected in these roles are pivotal in ensuring that the networks operate as intended. In Figure 15.14, Moore's role fits the description of a 'star' as he is purely internal focused. The term 'star' were originally given to the high performing R&D people who were the most successful in their teams (Allen & Cohen, 1969). They were smart, well-educated, articulate, and above all well connected and managed their connections to achieve their assigned objectives. Stars have significantly more communication across technical as well a professional departments than non-stars.

Gatekeepers and Their Importance with Regard to External Entities e.g., Outsourcing Partners Gatekeepers are stars with connections to external entities (Tushman, 1977). They are excellent communicators and will ensure that if poorly thought-out processes and service agreement are impacting the objectives they will find ways to workaround the constraints using their networks and, if necessary, extend their networks.

For the portfolio manager it is interesting to see that the attributes of a star/gatekeeper can easily be mapped to a project and program manager profile to determine which people on the portfolio manager's team fit the profile, see Table 15.4.

Some readers may point out that project and program managers have temporary roles whereas the role of a gatekeeper/star implies a more permanent role. This is correct but do not overlook the fact that project and program managers networks will still be useful for the future assignments especially they are in the same area and are for the same sponsor. Therefore, with a new assignment a project or program manager will need to understand what networks exist and how will they impact the new

Table 15.4 Comparison of Attributes of Star/Gatekeeper and Project and Program Managers

STAR/GATEKEEPER	PROJECT MANAGER	PROGRAM MANAGER
Exceptional communicators	90% of their time communicating (PMIa, 2013)	Excellent communicators (PMIb, 2013)
Cited as sources of information (Allen & Cohen, 1969)	Understands the project's technical and business domain[a]	Understands the domain of the program including external factors[b]
Highly educated and high intelligence	Educated and intelligent (Turner et al., 2009)	Highly educated and high intelligence (Shao, 2010)
Focused on meeting assigned goals	Deliverable oriented	Benefit oriented
Early adopters of change (Cross et al., 2007)	Implement change through projects	Drive and promote change through their programs

[a] There is a division in the project management community if effective project managers need to understand the business area that the project is in to be effective

[b] Program managers are more likely to be business focused being more senior and experienced than project managers. As programs comprise on two or more projects and have longer duration, external factors are likely to have a major impact on the success of a program and therefore need to monitored and responded to accordingly

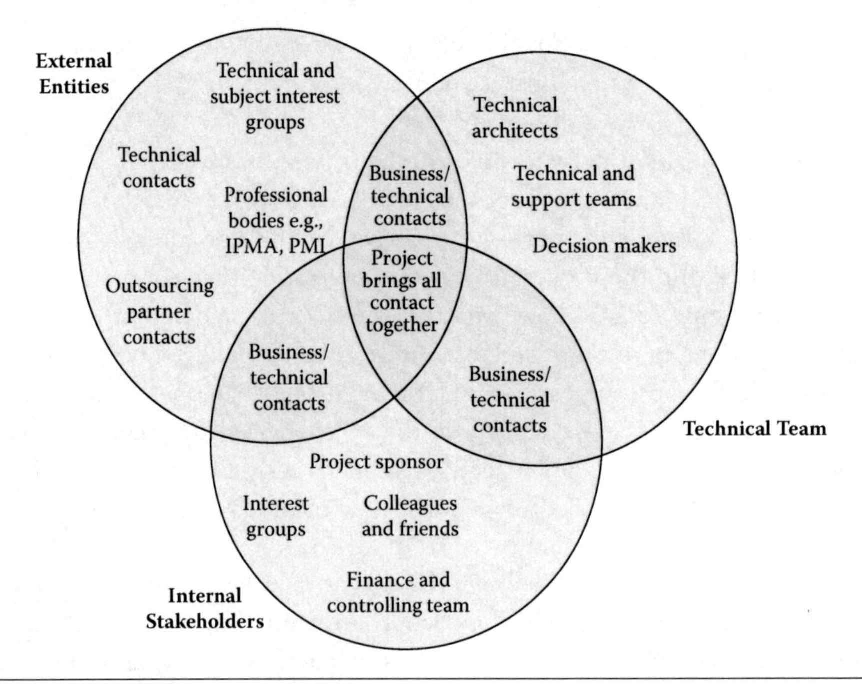

Figure 15.15 Project manager's network per job role bridging departments and external entities.

assignment(s). Then they should determine how to connect into these network(s) and start to change and leverage them for the good of the organization.

Figure 15.15 shows a project network per job role extending across departments and external entities. Business/technical contacts should be stars and gatekeepers helping the project manager connect and stay connected with the stakeholder communities. Is the acceptance into a network or the ability to create a cluster within an existing network dependent on culture or experience or time in an organization? Müller & Turner (2007) show local project managers are more effective than foreign project managers because of their understanding of local culture. Their research however did not investigate if local project managers were more effective because of their existing network rather than foreign project managers who are unlikely to have an extensive network at least from outside of their foreign engagement. Research has not been done to determine if foreign project managers become comparatively more effective over time compared to local project managers. From the author's observations working in foreign international environments for the past 20 years, foreign project managers do understand the need to quickly increase their networks and are also selective about the people who are in their networks.

Project and program managers' profiles match well with the gatekeeper profile. However, there is one gatekeeper attribute that may not be present in many project or program managers—work colleagues consistently approach the gatekeeper for advice on related topics. A gatekeeper is cited as being a source of information within his or her domain (Allen & Cohen, 1969). For a project and program manager, their own domain is the field of project and program management, but this field is only a set of tools and techniques that are applied to a business function or area to achieve something, e.g., development of a product or service. One could argue that for project and program managers to be accurately recognized as gatekeepers they need to also understand the business area they are working within (for the assignment), and the impact of the change their endeavor will have on it. The implications being a detailed understanding of the business and the technical project management aspects are required. This argument of project and program managers necessity to understand the business is also supported in the findings of a research literature review on project success factors by Khan & Turner (2013) showing that having past relevant experience of the business area is a success factor (Fortune & White, 2006; Shenhar et al., 2002; White & Fortune, 2002). Program management research is a relatively new phenomenon and less researched. To date, there is little research on program success factors relating to previous experience of the business area, but it does not mean previous experience is not a success factor. It has just not been extensively researched. Perhaps taking a practitioner perspective of the importance of previous experience may provide additional insight? If the reader reviews job postings for program managers, experience of the industry or area is invariably requested, which shows the importance of understanding the business area. With the above information one can conclude that project and program managers' roles fit into the stars and gatekeepers profiles and therefore like the gatekeepers should have all the characteristics to enable them to flourish in their positions and enjoy a high level of success.

Note: Academic research on the topic, project success factors show the need for previous project experience, and second, projects are contingent on context, and therefore, the project manager needs also to adapt to the environment, which also means understanding of the business (Shenhar & Dvir, 2002). Could this be one of the reasons why project and program success rates are so low that many certified project and program managers may not be stars or gatekeepers because they do not have the knowledge, the network, and the profile, which is required for successful project and program outcomes?

Framework Process Step 1: Identify The first of the eight steps that must be completed before a portfolio manager can address the information issues associated with portfolio management is to identify the networks that exist within the portfolio community.

The first step is to understand where are the structural networks, who are the actors, what roles they have, what is the main purpose of each network, and if its purpose is to overcome an organizational structural weakness, etc., see Figure 15.16.

The challenge is to collect data on a constantly changing network. The easiest way is to identify the core contacts of a person's network and use the structure of the relationships with and among the core contacts to make an inference about the kind of network that person maintains (Burt, 1995). With on-line surveys and intelligent logic it should be relatively easy to collect the necessary data including the professional

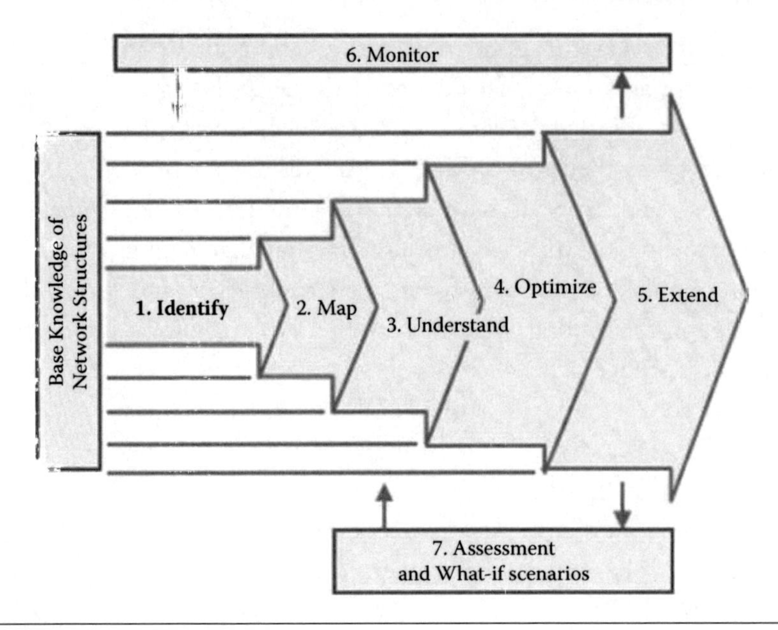

Figure 15.16 Step 1: Identify.

relationships with the contacts. Once all the people have completed the on-line survey then the process of mapping can start.

Note: Data protection laws need to be reviewed to ensure compliance. All the information contained in the networks should only be business related. Therefore, on-line surveys should be structured to ensure compliance. If structural network data are intended to be viewed across countries then compliance must be ensured in all countries.

Framework Process Step 2: Map Once the data are collected then a suitable mapping tool needs to be identified, which is similar to an advanced form of layered network diagrams used in project scheduling, see Figure 15.17. There are several tools available for mapping business social networks such as NetDraw, Social Networks Visualizer (SocNetV), and connected action.

Figure 15.18 shows the results of the data collected by a portfolio manager using an online survey from his or her program and project teams including their stakeholders. The figure shows a network by job title, which is one of many views to visualize a structural network. Each network view provides a different aspect of the network e.g., by role, by job title, by relationship, by network interest, by department by region, and by project or program and therefore will provide different unique insight.

Once the network has been mapped and visualized, the next step to understand how to read it and recognize what information can be obtained.

Framework Process Step 3: Understand Network diagrams contain a lot information. It is important to know how to select a view of a network and interpret the information based on what one is looking for or is looking to answer, see Figure 15.19.

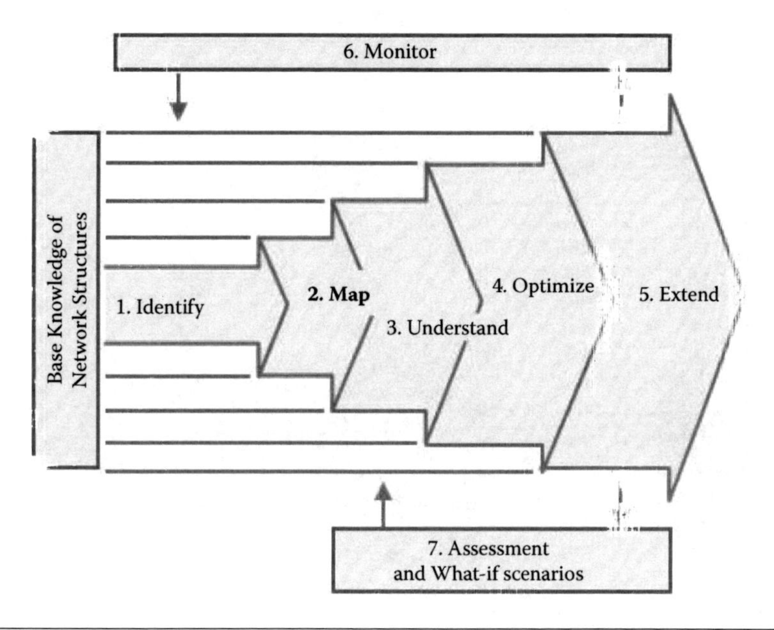

Figure 15.17 Step 2. Map

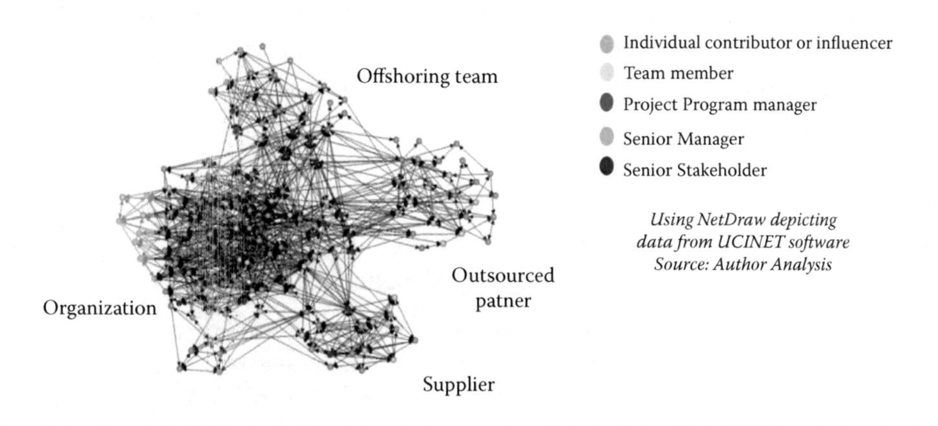

Figure 15.18 Structural network of portfolio.

The portfolio manager can use the business social network to address existing problems such as information issues as well as any other problems where the network can provide a level of insight. Another use of the network is for the portfolio manager to discover if there are stars and gatekeepers that may or may not be a project or program manager who can give a heart-beat on one or more projects and programs. If the portfolio manager is concerned about a particular project manager who is reporting green where the project reality is different, then the project manager's network can be analyzed to determine if there are signs that the project manager is not working effectively e.g., operating as an isolate, bridge, or a liaison rather than the preferred star. The more the network is used the greater understanding will be obtained from the network.

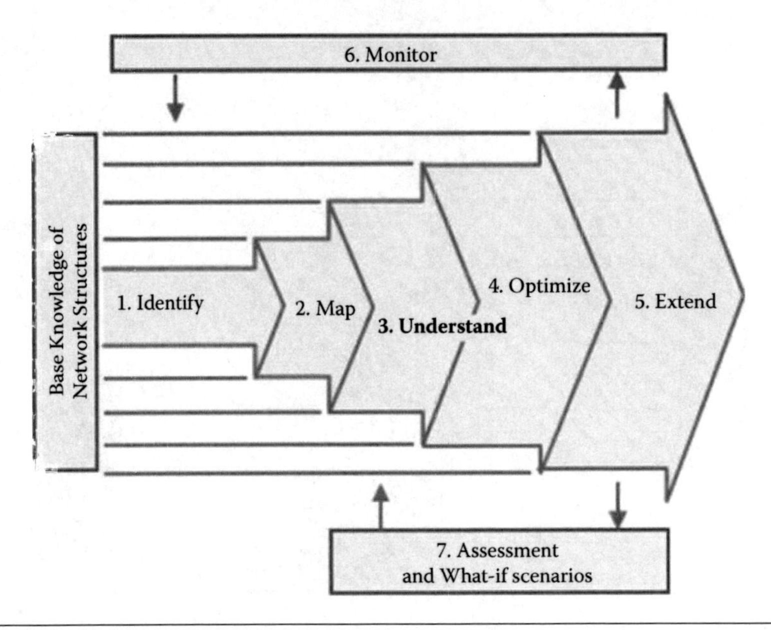

Figure 15.19 Step 3: Understand.

Framework Process Step 4: Optimize Network diagrams have so much information it is easy to interpret them in different ways depending on the desired information. The portfolio manager is looking to address his or her information problems and therefore would be looking to see if the network can give insight into particular problem(s), see Figure 15.20.

A portfolio manager may see bottlenecks in the network structures for current or future portfolio needs. For example if the organization is planning a merger or outsourcing a part of the IT function, is the portfolio manager's team prepared for

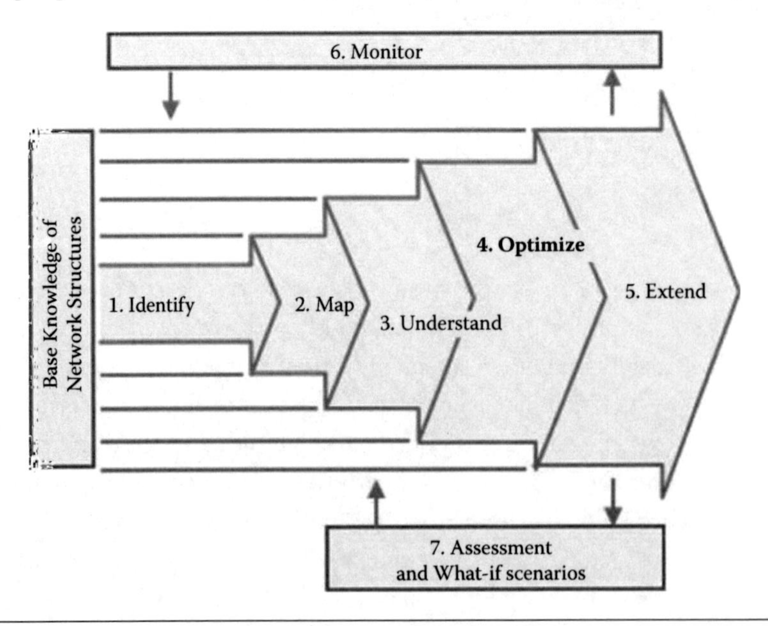

Figure 15.20 Step 4: Optimize.

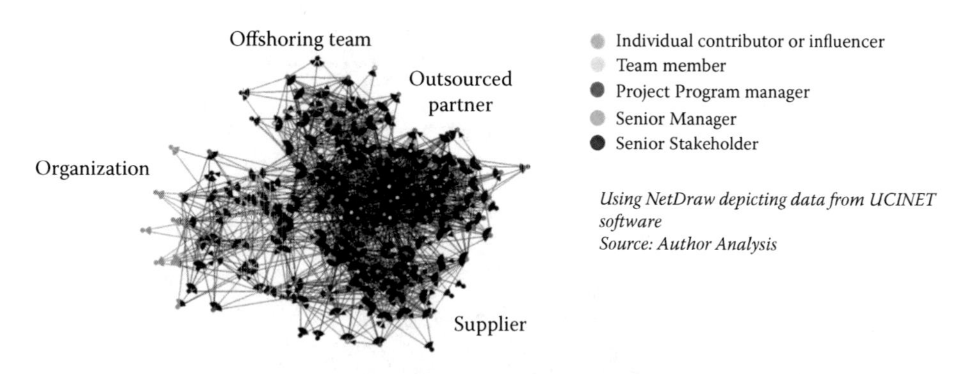

Figure 15.21 Integrated network of four entities with a low risk of bottlenecks.

this change? Every change to an organization invariably disrupts ongoing activities including projects and programs. One of the questions the manager instigating the organizational change should ask is "How well prepared are the existing structural networks to support the organizational change?" To answer this question the portfolio manager, who has already done the mapping work, will be able to answer this question. Referring to Figure 15.21, the network shows there are potential bottlenecks of communication from the outsourcing partner to the main part of the portfolio's manager's organization, which is the off-shoring team. The bottlenecks exist because the main links to the outsourcing partners is through senior managers and not directly into the project management team. To reduce the risk of a poor organizational integration, the structural deficiencies associated with the proposed outsourcing partner can be resolved by finding ways to integrate the partner proactively. The can be achieved by encouraging the existing stars and gatekeepers to build relationships and then motivate project team members to do the same. The portfolio manager then can make these recommendations and carry out a new survey to see if the integration has improved. Figure 15.21 shows the results of the efforts by the stars, gatekeepers, and other network members. The network structure is far more integrated thereby reducing the risks of communication bottlenecks that existed prior to this proactive effort.

In summary networks can be optimized for current or future scenarios to ensure potential negative impacts that are due to an organizational, contractual, or any other change can be reduced and absorbed by the network.

Framework Process Step 5: Extend The extend process step goes into more detail how to extend a network from a personal perspective and what considerations should be taken into account in deciding how to do it, see Figure 15.22.

A network takes time to develop and maintain; therefore, it is important for the person extending his or her network to carefully select the people to extend it to without wasting time in duplicating links or not fixing the information shortfalls.

Referring to Figure 15.23, Network A is an efficient construct where there is only one link to each actor, but the value of the network is limited because it is sparely populated. This means that none of the contacts are linked to anyone else. Network B shows

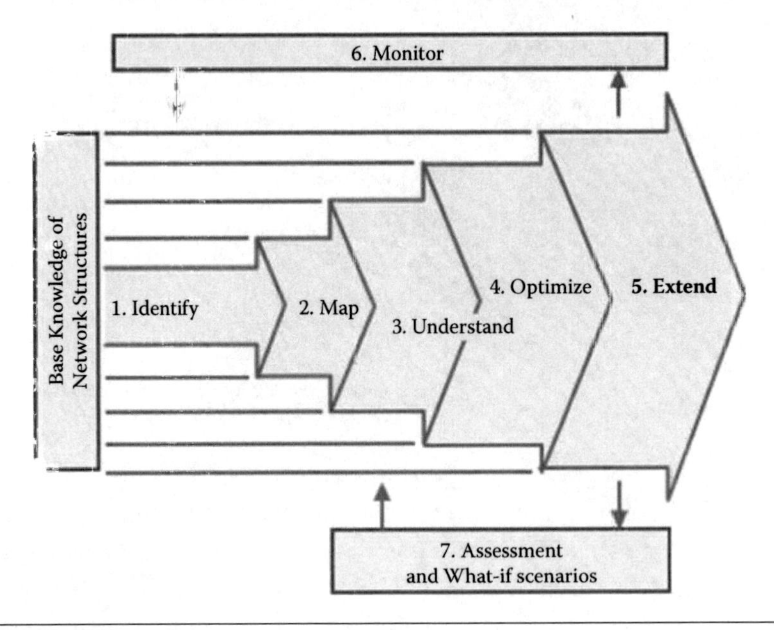

Figure 15.22 Step 5: Extend.

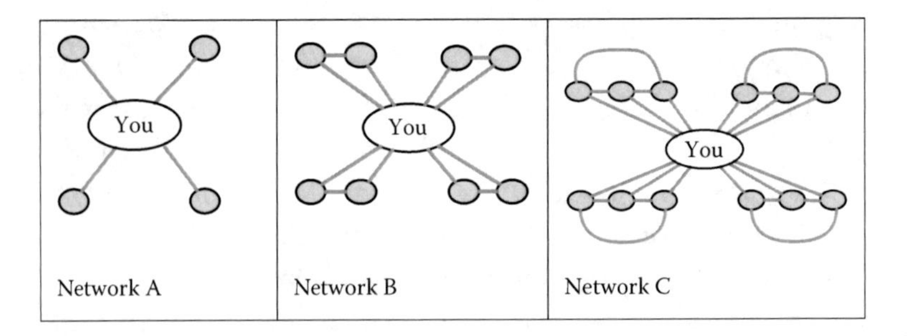

Figure 15.23 Network options.

a larger network with four clusters,* but "you" in this network is linked to every actor within each cluster, therefore, requiring extra effort to maintain the relationships when one link per cluster would provide a similar amount of information with less effort. Network C shows four clusters with three actors per cluster and "you" in this network is linked to everyone in all four clusters. One can quickly see the redundancy; hence the cost of maintaining redundant links where the ideal linkage would be one per cluster. This redundancy is called lost opportunity cost where the saved time could be used to develop and maintain other network clusters that can influence your ability to succeed.

There are two types of redundancies in networks. Referring to Figure 15.24, the first form of redundancy is by cohesion. The same information could be obtained by maintaining one link and not three. The second form of redundancy is by structural equivalence. It could be that your direct reports, or first level of network contacts, are duplicating their efforts. There are various options to reduce structural redundancy

* Two actors per cluster

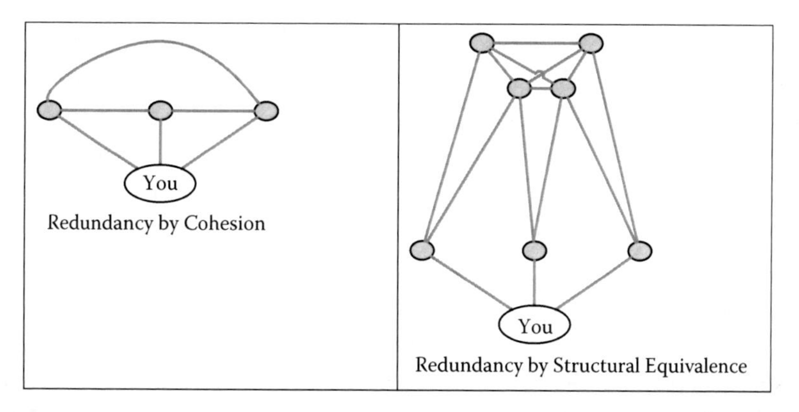

Figure 15.24 Structural indicators of redundancy adapted from (Burt, 1995).

such as reducing two of the three direct links or to discuss with the first direct links to reduce their duplicate links. The way this reduction is done, and the implications of doing it, depends on the actual situation and therefore are context sensitive.

With the basic knowledge of the above the portfolio manager can now determine from the network structure how to extend his or her own structure as well as how to influence effectively his or her direct reports to improve the effectiveness of their networks.

Framework Process Step 6: Monitor At this point in the framework the portfolio manager has an operational tool, which has been populated with network information. However, networks are dynamic and are changing; therefore, a monitoring process is required, which includes identifying new networks or network clusters and changes to the existing network(s), see Figure 15.25.

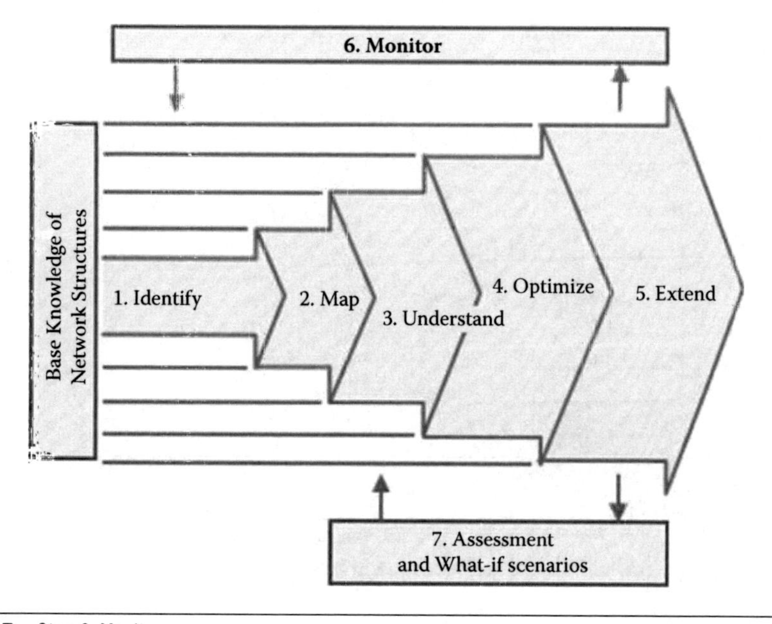

Figure 15.25 Step 6: Monitor.

If the network tool is intended to help support the business, then the tools and underlying processes should be institutionalized into the organization and available to allow access to a greater community.

The decision of whether to give limited access or open access will vary from organization to organization. There will be advocates that support the tool and others that oppose it because the information it contains will bring new insight that was not previously possible. The advocates will be driven by opportunities to improve efficiency and effectiveness, where the non-proponents will be concerned about providing transparency into probably what constitutes transparent and probably inefficient units. Providing transparency also provides dangers by highlighting who are true stars and gatekeepers are as they can be targeted by senior managers who like to keep their silos as silos. Having a star or gatekeeper link into a silo is a potential risk to the silo manager as stars and gatekeepers may break down an information barrier which was specially erected. A star or gatekeeper can quickly understand the actual workings of a section, department, or outsourcing partner, which means if the problems are known, they can be addressed. All too often senior managers keep out of each other's divisions so that any structural problems may be kept secret within their divisions. Stars are crucial in all organizations, but their roles are not necessarily liked or appreciated by some managers.

Framework Process Step 7: Assessment and What-if Scenarios The final process step is to use the new tool to help assess the current situation of a project or program i.e., problems, concerns, risks, additional resource assignment, and reporting audits/checks, see Figure 15.26. The tool can also be used for planning new projects including resource selection (team, steering committees, and project sponsor).

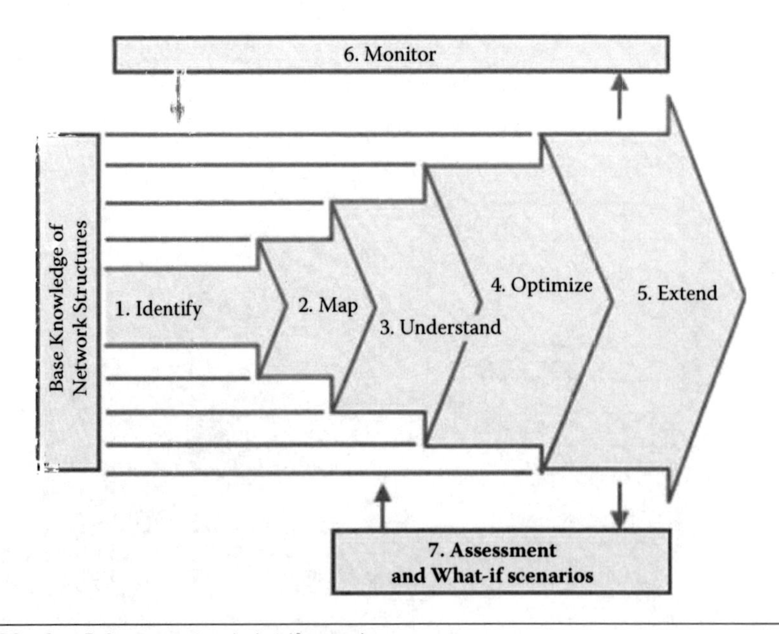

Figure 15.26 Step 7: Assessment and what-if scenarios.

What-if scenarios are useful for planning new projects but also to determine the impact of external factors such as organizational checks on the performance of the portfolio.

Examples of what-if scenarios are shown in Table 15.5.

Table 15.5 Examples of What-If Scenarios Applied to the Business Social Networks.

SCENARIO	DESCRIPTION	BENEFITS
What is the likely impact on project performance if the project sponsor's proposed project manager resource is assigned?	Project sponsors like to select their own trusted direct reports as their project manager. However, project sponsors are invariably in operations, and their trusted persons are typically operational people with little or no project experience. Using the network tool, the project sponsor's proposed project manager can be analyzed to see if there are additional risks in this case associated with the proposed project manager's network.	A case can be made why the project sponsor's proposed project manager should or should not be accepted where the network tool gives a new perspective in helping make the right decision
What will be the impact of the proposed organization change on planned and existing projects/programs in the portfolio?	Organizations are continually changing structures and reporting lines without necessary giving consideration to the impact on structural networks and the ability to get things done.	The information provided by structural networks may help to influence whether a proposed structure makes sense from the perspective of achieving targets. Structural networks may also provide insight into how organizational structures should be selected where the information flows without constraints.
What is the impact on the performance of the portfolio team of the proposed multi-vendor outsourcing deal?	Outsourcing contracts rarely consider a holistic impact on the performance of a function that is being outsourced. Skills, roles, processes, and tools are outsourced with little consideration to how they fit together in context of planning, setting up, running, and handing over of one or more projects. The tools typically used to determine if outsourcing make sense or not are mainly from a finance and strategic perspective but with little consideration of the impact to the mechanics of the function in question or the soft factors of the motivation of individuals. System dynamics (process modelling) is one tool which provides a detailed view if the contracts and Service Level Agreements (SLAs) support or hinder projects and where the bottlenecks are. On the soft factor side structural networks will show the importance and extent of the current relationships and the potential impact of the proposed outsourcing deal.	Modeling the proposed outsourced environment using both systems modeling and structural networks will give an understanding of whether the proposed organization's contracts will perform as expected, and the structural network analysis will show the level of structural network integration and where the gaps exist.
With a problem project, what is the impact if the incumbent project manager is replaced with one with better network connections?	A project manager may perform better when assigned to one project type than another, and/or assigned to a project where he or she is better integrated into a network cluster. The network analysis of every project manager will show their clusters, which is a good indication of what they like to do which may be different to what they are actually assigned to do.	The structural information on a low-performing manager could be used to find this person a new and better-suited position as well as to find potential replacements who are better connected.

(continued)

Table 15.5 Examples of What-If Scenarios Applied to the Business Social Networks (continued)

SCENARIO	DESCRIPTION	BENEFITS
What is the impact on portfolio performance if project managers with limited networks have personal performance targets to increase their networks according to the needs of their individual projects?	If there is a correlation between an efficient and effective structural network and project success, then project managers who are new to an organization can use the network tool to help them develop their network in an efficient and effective way.	Faster ramp-up times for new project and program managers tend to occur. Higher project and program success rates and customer satisfaction tend to result.
What is the impact on the overall performance of the portfolio if the project managers' networks are optimized?	This question requires Key Performance Indicators (KPIs) taken from successful project and program managers and a profiling of their personal networks. It should be possible to extrapolate the increase in performance of the portfolio using the top quartile of successful project and program managers.	If not addressed, the cost of projects not meeting their stated objectives therefore provides motivation to put corrective actions in place.

The above examples relate to the impact of project and programs. What about operations could this tool also help? The structural network tool can be applied to operations to determine if operational processes are efficient or not. If the structural networks are configured differently to the operational processes, then it suggests that processes are ineffective and need to be redesigned. Operational KPIs should also show this problem, but what they can do unlike the structural networks is to provide insight to design solutions based on information flows (need and necessity).

Summary of Framework Section

The framework describes a structured approach to understand and develop the value of the existing assets within an organization through better communication and information flows. A portfolio manager using the framework approach will have a better understanding of how well networked his or her project and program managers are into the stakeholder community plus he or she will also see who are his or her stars and gatekeepers. Few organizations today have a systematic approach to develop their own networks, where probably the most advanced are the management consulting companies who build practice knowledge data bases. The internet gives a glimpse of what is possible with social network sites such as like LinkedIn and Facebook, but this is the tip of the iceberg to possibilities and potential especially in the complex work of portfolio management. The framework described in this section should help stimulate interest within organizations to develop the ideas further and perhaps pilot projects to determine the potential of managing and extending network structures.

Who would be a likely project sponsor for such a pilot?... the portfolio manager looking to find ways to resolve his or her information issues.

The last and concluding section in this chapter is a short story of how Khan, a portfolio manager described in the first case study, addressed his information issues.

Scenario of How the Portfolio Manager Addresses the Information Needs in a Networked Advanced Organization

Khan has been retained for a two year assignment into rapidly growing R&D division, which is partly owned by his parent organization. This division has been so successful that it started a second line of business which required a second portfolio manager. The portfolio manager of the first line of business, called Nahk, has been successfully managing his portfolio worth hundreds of millions, and his project and program success rates are 90% +, something unheard of in the industry. Nahk is an unusual person who talks about stars, gatekeepers, isolates, and other terms, and Khan is totally unaware of their meanings nor are they typically used in his day-to-day portfolio vocabulary. Khan is invited to meet his new peer, and the two discuss experiences and approaches. Khan starts by explaining his experience in the parent organization and explains one of the problems he encountered was he thought the reported information was reliable, but for some reasons nasty surprises came up again and again, which required him to attend emergency meeting after emergency meeting. Khan went on to describe how difficult it was to determine if there was a core problem, which was putting his portfolio on the verge of being out of control. Khan described how project managers and sponsors reported and also said how well their projects were progressing, but he had a feeling from the reports that the KPIs and some comments were giving less rosy picture. When Khan challenged them there was always a reason or excuse to explain the meaning behind the numbers. Khan wished he find a way to proactively determine if the information reported was accurate, and who he could approach for missing and unreported information. Khan went on to say that some project managers always seemed to succeed even though there were mixed reports of how effective or ineffective they were, whereas other project managers never met their goals. In this group there was also a group of project managers that promised a lot but disappeared or found other positions before anything was delivered.

Nahk listened to Khan, and said he had the same experiences as a portfolio manager but found a way to address the information issues that seem to plague their role. Nahk told Khan he had a theory that successful people work with successful people, and they know how to achieve their goals. So Nahk explained to Khan that he created a proposal for a pilot project to understand the structural networks of successful project managers, and how their networks and profiles compared against project managers who were struggling in their positions. Using open source software and an on-line questionnaire, Nahk identified the networks of the top and bottom quartile project managers and visualized their networks. The first observation he made was that the successful project managers had extensive networks, whereas the struggling project managers did not. Nahk then assigned one successful project manager and one struggling project manager to areas where they had few network connections. What he noticed is successful project managers created more new connections compared to the struggling project managers. Khan was listening to Nahk's description of his experiences and how he was building up information on this team. As Nahk was describing how he solved the various problem, Nahk showed Khan a series of network diagrams with people (or actors) as nodes, with different colors and the links showing types of information flow. Nahk then went into more detail by showing how his software program could show overlays of networks building up a picture of a project manager network project by project. After one hour Khan realized why Nahk was so successful in portfolio management. Nahk knew where to go and who to ask those difficult 'heart-beat' questions to check if a project was progressing as reported. He was proactive and instrumental in the set up and development of his project managers' structural networks to ensure they had the best chances of success. Nahk then took Khan on a tour of the organization during which Khan noticed how people approached Nahk with questions that to Khan would require time to think and analyze. Instead Nahk seemed to have answers to most of the questions, and the ones he felt required more investigation he promised a response within a day.

Khan realized that Nahk had not only full control of his portfolio but also many aspects of the business knowledge of the projects and program through the knowledge he had at his fingertips, which was something Khan never had. Nahk had truly solved the issues with poor information management by understanding the strengths and weaknesses of his people using structural networks and using this information for helping his people become better integrated into the organization.

Khan, not to be outdone, asked how structural networks can work in global projects where there is no mapped network or contacts in a remote location. Nahk responded with a smile that he employed an old concept from the 18th century Boar War – called scouting (Begbie, 1900). Nahk continued to explain that the founder of Sir Baden Power was a Lieutenant-General in the British army and invented scouting, which was originally a form of reconnaissance but later formed into a boy movement called scouts. Nahk went on to say that using proven reconnaissance techniques linked with interpersonal skills and a modern approach of mapping structural networks, the relationships helped seed the project network of relationship in the country offices, which were then used as a basis to map, understand, optimize, and extend.

Khan knew after this first meeting with Nahk that his new assignment would be for more productive and pleasurable compared to his previous assignment.

Khan can thank an approach that originated from the 1950s but was never was fully exploited because of a lack of tools and technology at that time. This approach has been largely overlooked in portfolio management until now.

References

Aalbers, R., Koppius, O., & Dolfsma, W. (2006). *On and off the beaten path: Transferring knowledge through formal and informal networks* (No. 2006/08). *CIRCLE Electronic Working Paper Series* (pp. 1–33). Lund, Sweden.

Allen, T. & Cohen, S. (1969). Information flow in R&D labs. *Adminstration Science*, *14*, 2–19.

Bazerman, M. & Samuelson, W. (1983). I won the auction but don't want the prize. *Journal of Conflict Resolution*, *27*(4), 618–634.

Begbie, H. (1900). *The story of Baden-Powell: 'The wolf that neversleeps'*. London, UK: Grant Richards.

Bryan, L., Matson, E., & Weiss, L. (2007). Harnessing the power of informal employee networks. *McKinsey Quarterly* (pp. 1–10). New York.

Burt, R. (1995). *Structural holes: The social structure of competition*. Boston, MA: Harvard Business Press.

Burt, R. (2004). *Brokerage and closure: An introduction to social captial.* Oxford, England: Oxford University Press.

Clack, A., Scott, F., Graeber, M., Honan, D., & Ready, J. (2012). *Insights and trends: Current portfolio, programme, and project management practices*. PricewaterhouseCoopers: Zurich, Switzerland.

Cooper, R.G. (2007a). Managing technology development projects. *IEEE Engineering Management Review*, *35*(1), 67–77.

Cooper R.G, Edgett, S., & Kleinschmidt, E. (1999). New product portfolio management: practices and performance. *Innovation Management*, *6782*(99), 333–351.

Cooper, R.G, Edgett, S., & Kleinschmidt, E. (2000). New problems, new solutions: making portfolio management more effective. *Technology Management*, *43*(2), 18–33.

Cooper, R.G, Edgett, S., & Kleinschmidt, E. (2001). Portfolio management for new product development: Results of an industry practices study. *R&D Management Management*, (4), 1–39.

Cooper, J. (2007b). *Cognitive dissonance: 50 years of a classic theory. Vasa*. London, England: SAGE Publications.

Cross, Martin, R. & Weiss, L. (2006). *Mapping the value of employee collaboration. McKinsey Quarterly*. Washington, DC.

Cross, R., Parise, S. & Weiss, L. (2007). *The role of networks in organizational change. The McKinsey Quarterly*. Washington, DC.

Detlor, B. (2010). Information management. *International Journal of Information Management*, *30*(2), 103–108.

Dulewicz, V., & Higgs, M. (2005). Assessing leadership styles and organisational context. *Journal of Managerial Psychology*, *20*(2), 105–123.

Elonen, S., & Artto, K. a. (2003). Problems in managing internal development projects in multi-project environments. *International Journal of Project Management*, *21*(6), 395–402.

Fortune, J., & White, D. (2006). Framing of project critical success factors by a systems model. *International Journal of Project Management*, *24*(1), 53–65.

Fricke, S. & Shenhar, A. (2000). Managing multiple engineering projects in a manufacturing support environment. *Engineering Management, IEEE 47*(2), 258–268.

Hanisch, B, & Wald, A. (2012). A bibliometric view on the use of contingency theory in project management research. *Project Management Journal*, *43*(3), 4–23.

Joslin, R. (2012). Information structuring "methodology": Tools and techniques for effective program management. In Levin, G. (Ed.), *Program management: A life cycle approach* (pp. 319–362). Boca Raton, FL: CRC Press.

Kerzner, H. (2010). *Project management best practices: Achieving global excellence* (2nd ed.). Hoboken, NJ: John Wiley & Sons Inc.

Khan, K., & Turner, R. (2013). Factors that influence the success of public sector projects in Pakistan. In *Proceedings of IRNOP 2013 Conference, June 17–19, 2013, Oslo, BI Norwegian Business School,* Oslo, Norway.

Kotter, J. (2011). Hierarchy and network: Two structures, one organization. *Harvard Business Review*. Retrieved March 10, 2014, from http://www.forevueinternational.com/Content/sites/forevue/pages/1494/16_3_Hierarchy_and_Network_Two_Structures_One_Network.PDF

Kratzer, J., Leenders, R. T. A. J., & Van Engelen, J. M. L. (2010). The social network among engineering design teams and their creativity: A case study among teams in two product development programs. *International Journal of Project Management, 28*(5), 428–436.

Levin, G. & Ward, J. L. (2011). *Program management complexity: A competency model.* Boca Raton, FL: CRC Press.

Mabin, V. (1990). Goldratt's" Theory of Constraints" thinking processes: A systems methodology linking soft with hard. Retrieved from: http://www.systemdynamics.org/conferences/1999/PAPERS/PARA104.PDF

Martinsuo, M. (2013). Project portfolio management in practice and in context. *International Journal of Project Management, 31*(6), 794–803.

Martinsuo, M. & Lehtonen, P. (2007). Role of single-project management in achieving portfolio management efficiency. *International Journal of Project Management, 25*(1), 56–65. doi:10.1016/j.ijproman.2006.04.002

Meyer, W. G. (2013). *Termination of troubled projects and the critical point theory.* SKEMA Business School, Lille, France.

Moore, S. (2010). *Strategic project portfolio management: Enabling a productive organization.* Hoboken, NJ: John Wiley & Sons, Inc.

Morris, P. & Hough, G. (1988). *The anatomy of major projects: A study of the reality of project management.* New York: John Wiley & Sons Inc.

Müller, R., & Turner, R. (2007a). Matching the project manager's leadership style to project type. *International Journal of Project Management, 25*(1), 21–32.

Müller, R., & Turner, R. (2007b). The influence of project managers on project success criteria and project success by type of project. *European Management Journal, 25*(4), 298–309.

PM Solutions. (2013). *The state of project portfolio management (PPM)* (pp. 1–12). PM Solutions: Glen Mills, PA.

Project Management Institute. (2013a). *A guide to the project management body of knowledge. (PMBOK® Guide).* Fifth Edition. Newtown Square, PA: Project Management Institute.

Project Management Institute. (2013b). *The standard for program management.* Third Edition. Newtown Square, PA: Project Management Institute.

Prifling, M. (2010). IT project portfolio management–A matter of organizational culture? In *PACIS 2010 Proceedings* (pp. 761–772). Retrieved from: http://aisel.aisnet.org/pacis2010/18/

Rajegopal, S., Waller, J., & McGuin, P. (2007). *Project portfolio management: Leading the corporate vision.* Basingstoke, UK: Palgrave Macmillan.

Shao, J. (2010). *The impact of program managers' leadership competences on program success and its moderation through program context. PhD Paper.* SKEMA Business School, Lille, France.

Shao, J., Müller, R., & Turner, R. (2012). Measuring program success. *Project Management Journal 43*(1), 37-49.

Shenhar, A. & Dvir, D. (2002). One size does not fit all—true for projects, true for frameworks. *Proceedings of PMI Research Conference, Seattle, WA* 99-106. Newtown Square, PA: Project Management Institute.

Shenhar, A., Tishler, A., Dvir, D., Lipovetsky, S., & Lechler, T. (2002). Refining the search for project success factors: a multivariate, typological approach. *R and D Management, 32*(2), 111–126.

Teller, J., Unger, B. N., Kock, A., & Gemünden, H. G. (2012). Formalization of project portfolio management: The moderating role of project portfolio complexity. *International Journal of Project Management, 30*(5), 596–607.

Thaler, R. (1988). Anomalies: The winner's curse. *The Journal of Economic Perspectives, 2*(1), 191–202.

Tichy, N., Tushman, M., & Fombrun, C. (1979). Social network analysis for organizations. *Academy of Management Review, 4*(4), 507–519.

Turner, R. (1999). *The handbook of project-based management* (2nd ed.). London, UK: McGraw Hill.

Turner, R., Müller, R., & Dulewicz, V. (2009). Comparing the leadership styles of functional and project managers. *International Journal of Project Management*, *2*(2), 198–216.

Tushman, M. (1977). Special boundary roles in the innovation process. *Administrative Science Quarterly*, *22*(4), 587–605.

Voss, M., & Kock, A. (2013). Impact of relationship value on project portfolio success—Investigating the moderating effects of portfolio characteristics and external turbulence. *International Journal of Project Management*, *31*(6), 847–861.

White, D., & Fortune, J. (2002). Current practice in project management—An empirical study. *International Journal of Project Management*, *20*(1), 1–11.

16

When Stakeholders, Goals, and Strategy Conflict

JOHN WYZALEK, PfMP

Strategy is the unifying constant in portfolio management. It envelops a portfolio. It provides the portfolio's foundation, as well as the final goal that a portfolio manager strives to attain. A portfolio's strategy guides all portfolio activity, and portfolio managers always turn to it when determining how to most effectively manage their portfolios and derive benefits.

There is another constant that confronts portfolio managers. "The only constant is change" would have become a tired saw in business management if it were not such a powerful business force. In the current global, technology-drive economy, business managers more and more must reckon with it. The Project Management Institute's (PMI) *The Standard for Portfolio Management* (2013) recognizes change as a constant and reckons with it and its affects on strategy. The Standard describes many processes, tools and techniques, and artifacts that enable portfolio managers to deal with change. For example, it describes strategic change processes, risk management processes, performance management processes, and communications management processes, as well as tools, techniques, portfolio assets, and organizational assets, which all can be used to manage change.

Change is a constant happening all the time. But what does that mean for portfolio managers? When does a portfolio manager know when a change is occurring and how it will greatly affect the portfolio? To answer these questions, portfolio managers must pay attention to the other forces at work in a portfolio: stakeholders and their goals. In the perfect scenario, stakeholders and goals align with the strategy and support it. But this scenario is static and not capable of change. This chapter looks at scenarios when goals, stakeholders, and strategy conflict, which is when portfolio managers should pay attention because change is occurring.

Who Are the Stakeholders? What are Their Goals?

An easy way to answer these questions is to consult such portfolio assets as the portfolio charter, strategic management plan, and the communications plan. The stakeholder matrix used to analyze communications needs is useful in answering who are the stakeholders and what are their goals. Table 16.1 presents a sample matrix. It

Table 16.1 Sample Stakeholder Analysis Matrix

STAKEHOLDER GROUP	STAKEHOLDER ROLE	STAKEHOLDER INTERESTS
Sponsors	Funding and resources Strategic direction	Benefits in alignment of strategy
Governance	Oversight and authorization Resource management Benefit delivery management	Performance Benefit delivery Internal and external compliance
Component Managers and Teams	Completion of work to deliver benefits	Delivery of benefits Resource use

identifies stakeholder groups and lists their roles and interests, which are helpful in identifying their goals. This matrix identifies the sponsor as a stakeholder, and the goal of benefit delivery in alignment with strategy.

But does this matrix identify all stakeholders? The PMI Standard (2013) takes a broad view of what is a portfolio stakeholder and, as this chapter will explain, so should portfolio managers. It defines a stakeholder as any person or entity that is affected or believes itself to be affected by the portfolio. The matrix in Table 16.1 leaves out many common stakeholders that fit this definition. These include:

- Customers
- Investors
- Retailers
- Suppliers
- External regulatory bodies
- Marketing executives and managers
- Public relations executives and managers
- Organizational managers
- Competitors

This stakeholder list can be categorized in a way that some of these stakeholders fall into the groups listed in Table 16.1. Still, most cannot be fit into this matrix, and a portfolio manager should have a listing of all stakeholders, because identifying as many stakeholders as possible enables a portfolio manager to recognize as many stakeholder goals as possible, and subsequently, as many sources of changes as possible.

Stakeholders can be affected by a portfolio as well as affect it. Their goals represent the ways they can influence a portfolio both positively or negatively. In other words their goals represent opportunities or threats. Opportunities and threats are detailed in risk management documents, which are another source for identifying stakeholders and their goals.

For example, a portfolio management team has developed a probability and impact matrix, and on this matrix it identifies the following opportunity: "Delivery of services through mobile platforms." This benefit touches multiple stakeholders: customers, competitors, external regulatory agencies, retailers, investors, and organizational managers and departments.

Table 16.2 Portfolio Opportunity and Stakeholder Goal Matrix

PORTFOLIO OPPORTUNITY	STAKEHOLDER	STAKEHOLDER GOAL
Mobile Service Delivery	Consumers	Higher delivery speed Lower delivery cost Ubiquitous 24x7 service Secure transaction
	Investors	Higher return from higher profits because of larger market share
	External regulators	Consumer protections such as secure and private service delivery for consumers
	Customer service executives	High quality and efficiency of service delivery to consumers
	Customer service line managers and staff	Providing service delivery to consumers

Listing this opportunity in the matrix provides stakeholder goals: mobile platforms meet a demand of consumers for faster and cheaper services, represent an advantage over competitors, and possibly satisfy investor demand for more profits through greater market share.

A portfolio opportunity and stakeholder matrix can be developed to document stakeholders and their goals per opportunity, as shown in Table 16.2.

Just as portfolio components are prioritized to maximize efficiency and ensure alignment, stakeholder goals need to be prioritized for the same reasons. Table 16.3 shows the matrix augmented with stakeholder priorities, where 1 is the highest priority, and 5 is the lowest.

This rating shows that stakeholders value their goals and subsequently place a high priority on them. To derive more information about stakeholder goals and their importance to the portfolio, an additional criterion can be added to the matrix to make it more meaningful. This criterion is the degree of alignment of the goal with

Table 16.3 Portfolio Opportunity and Stakeholder Goal Priority Matrix

PORTFOLIO OPPORTUNITY	STAKEHOLDER	STAKEHOLDER GOAL	PRIORITY FOR STAKEHOLDER
Mobile Service Delivery	Consumers	Higher delivery speed	2
		Lower delivery cost	1
		Ubiquitous 24x7 service	2
		Secure transaction	1
	Investors	Higher return from higher profits because of larger market share	1
	External regulators	Consumer protections such as secure and private service delivery for consumers	1
	Customer service executives	High quality and efficiency of service delivery to consumers	1
	Customer service line managers and staff	Providing service delivery to consumers	1

Table 16.4 Opportunity and Stakeholder Goal Alignment Matrix

PORTFOLIO OPPORTUNITY	STAKEHOLDER	STAKEHOLDER GOAL	PRIORITY FOR STAKEHOLDER	ALIGNMENT BETWEEN OPPORTUNITY AND GOAL	OPPORTUNITY ALIGNMENT SCORE
Mobile Service Delivery	Consumers	Higher delivery speed	2	1	2
		Lower delivery cost	1	1	1
		Ubiquitous 24x7 service	2	1	2
		Secure transaction	1	2	2
	Investors	Higher return from higher profits because of larger market share	1	2	2
	External regulators	Consumer protections such as secure and private service delivery for consumers	1	2	2
	Customer service management	High quality and efficiency of service delivery to consumers	1	1	1
	Customer service line managers and staff	Providing service delivery to consumers	1	5	5

the opportunity. Table 16.4 presents the matrix with this added criterion, which is listed in the fifth column from the left.

The last column in Table 16.4's matrix is the opportunity alignment score, which is the product of multiplying the two previous columns. A low score indicates a high degree of alignment between the opportunity and stakeholder goals. These scores can be averaged, and a low average denotes an opportunity with a high level of stakeholder support. This matrix has an average score of 2.125, which denotes a high level of stakeholder support. A high average also represents an opportunity and set of goals that should be further explored and possibly added to the portfolio. For example, this matrix indicates that mobile service delivery should be a goal for the portfolio and indicates resistance to this goal from the customer service team managers and members whose jobs could be replaced by an automated, mobile service.

Alignment between Stakeholder and Portfolio Goals

The portfolio charter and roadmap identify the portfolio strategy and the goals being pursued to deliver the benefits to realize that strategy. For example assume the portfolio previously discussed has as a strategic goal: "Service delivery that increases the organization's market share." The portfolio's roadmap includes such goals as increasing the quality and efficiency of service delivery and increasing market penetration. The roadmap includes a lean initiative program and a program for branch location expansion.

So, how do these strategic goals square with the stakeholder goals previously discussed? The matrix in Table 16.4 can be easily modified by adding columns representing strategic goals and associated components. Table 16.5 presents this new matrix. Not surprisingly, stakeholder goals and strategic goals are not in complete alignment as shown in Table 16.5.

At a minimum Table 16.5 indicates risks to the portfolio because it is not delivering benefits fully in line with these stakeholders. This portfolio's manager and governance body could decide to initiate a new mobile platform component and to realign components. They could also decide to perform market benchmarks to analyze the appropriate response to this misalignment with stakeholder goals. Most importantly, they can view this misalignment as the effects of a strategic change in the marketplace, which market benchmarks could further corroborate. The risks can be escalated, and the portfolio may be reviewed to see how it can deliver benefits in line with this new strategy.

Table 16.5 Alignment Between Stakeholder and Strategic Goals

STAKEHOLDER	STAKEHOLDER GOAL	PRIORITY FOR STAKEHOLDER	ALIGNMENT BETWEEN STAKEHOLDER GOAL AND "SERVICE DELIVERY THAT INCREASES THE ORGANIZATION'S MARKET SHARE"	STAKEHOLDER ALIGNMENT WITH LEAN INITIATIVE	STAKEHOLDER ALIGNMENT WITH BRANCH EXPANSION DELIVERY
Consumers	Higher delivery speed	2	1	1	4
	Lower delivery cost	1	2	1	3
	Ubiquitous 24x7 service	2	1	2	4
	Secure transaction	1	2	2	1
Investors	Higher return from higher profits because of larger market share	1	2	2	3
External regulators	Consumer protections such as secure and private service delivery for consumers	1	3	3	2
Customer service management	High quality and efficiency of service delivery to consumers	1	1	1	1
Customer service line managers and staff	Providing service delivery to consumers	1	1	1	1

Conclusion

When stakeholder and portfolio strategic goals conflict, strategic change is under way. The example in this chapter used a group of stakeholders external to the portfolio organization. This group was chosen to highlight how changes in its goals signal strategic change for the portfolio organization. They are an important group of stakeholders because the organization cannot exist without their support (i.e., purchasing the organization's products or outside financial investment.) Portfolios must deliver benefits to this group in order to survive.

When goals conflict among stakeholders inside a portfolio organization, they may also signal a strategic change. More likely they may signal a conflict that is due to the allocating of resources, which could be resolved by performance management processes, more effective communications management, or a review of the portfolio component priorities. For example, assume the portfolio governance board does initiate a mobile platform delivery component, but it does not prioritize it properly, and conflicts occur with branch expansion initiative. In another scenario, conflicts occur because the branch stakeholder has more political clout within the organization.

Managing internal stakeholders is a must, but portfolio managers must also manage the needs of those external stakeholders, such as consumers or retailers. Where would Nokia and Blackberry be today if they had more quickly reacted to changes in consumer's goals? Microsoft? IBM? The music recording industry? Change is constant, and its effects are ever more powerful. Managing conflicts between portfolio strategic and stakeholder goals enable portfolio managers to effectively manage change.

Reference

Project Management Institute. (2013). *The standard for portfolio management,* Third Edition. Project Management Institute: Newtown Square, PA.

17

Portfolio Alignment

A Strategic Approach through the Predicative Value of Outliers' Data

ING. KARI DAKAKNI, MRes, CENG

Introduction

Project portfolio managers rely on data-driven approach to collect and analyze data based on a combination of metrics and methods for making strategic choices, to determine how the business will look like five years out, and particularly to balance resource allocation. The metrics used to select projects, range from quantitative (e.g., return on investment (ROI) for financial models and financial indices; probabilistic financial models; option pricing theory; etc.) to qualitative (e.g., alignment with company strategy; strategic approaches; scoring models and checklists; analytical hierarchy approaches; etc.). Despite of the use of a variety of methods and metrics, projects selection and project prioritization are two of the most cited project activities that are still presenting the weakest facets respect to all the other processes involved with project portfolio management as reported in (Cooper, Edgett, & Kleinschmidt, 1998). However according to recent surveys, companies still have difficulties in the selection criteria (Partners, 2013). Consequentially the poor performance in selection, prioritization, and thus in the decision-making process is reflected into the misalignment and in the general poor performance of the portfolio management. Is it possible that hidden or undiscovered predictive indicators within project portfolio data, with their distributions that have been overlooked within the standard methods, could address project issues such as selection and prioritization from misalignment?

Accurate data collection and management are one of the biggest challenges for a portfolio manager. Decision-making processes based on: missing data, misleading data, incomplete data, and at worst, data that are often voluntary ignored because of normative thinking behaviors or epistemic certainty,* are limiting the thinking of how to resolve persistent problems; these issues should be addressed either in data management and/or in organizational culture. In general, systemic problems often originate

* Overestimation of themselves

from a lack of accuracy in data analysis, data screening, misreporting, or from the so called "normalization of the deviance."* In addition rare events are perceived and portrayed by managers as either unique or unclassifiable (Christianson, Farkas, Sutcliffe, & Weick, 2009). Therefore, rare events or rare data points are often handled as 'accidental manifestations of underlying organizational processes' (Lampel, Shamsie, & Shapira, 2009, p. 835). When organizations are facing unusual events, whether costly or beneficial, they have the opportunity of learning from experience. Further information over systematic failures may contain outliers' data that are holding back the nature and the causes of uncertainties.

Outliers are usually omitted from standard analysis in many fields because they are deemed to lead to inflated errors or distortions of statistic estimates. However, according to the recent literature and recent empirical findings, outliers turned out to be more prevalent in project management than previously thought. *"Organizations that steer the performance of their projects on averages alone are blind to outliers and their impact"* (Budzier & Flyvbjerg, 2013, p. 21). They hold valuable information and can provide insight into previously unknown, unpredicted, and unexplained impacts to projects (components) at the portfolio level.

The existence of a mechanism, triggered by influencing factors, and generating outliers will be argued. Secondly, the outliers' analysis may reveal patterns that can lead to the opportunity to anticipate systems behavior. Outliers are not unique and random events to be simply rejected from analysis; outliers need to be evaluated through a quantitative and qualitative† approach.

Since a new theoretical approach is shedding new light over a predictive valence of certain class of outliers' data, this chapter will discuss the "potential application" of these findings and observations and how they can be exploited and adapted to portfolio management.

Outliers that exhibit a degree of probabilistic predictability, even though they cannot be predicted by a probabilistic approach, are named dragon kings (Sornette, 2009), and they seem to be the outliers of the outliers. The mechanism that produces dragon kings outliers is still in a research stage and not yet precisely detected for all the potential applicability fields. However the mechanism that produces outliers in portfolio management will be investigated within the interrelationships of the portfolio variables in the perspective of the influencing factors.

These observations, with the opportune distinctions, will be used to exploit the "predictive" valence of data at the portfolio level according to the results already reached in some of the investigated fields.

New metrics will be defined to analyze the interrelationships between data, with the aim of leveraging the value of the outliers data of the portfolio processes to

* It will be explained in the Jeopardizing Data Quality in the Communication Channel.

† For qualitative methods, it is a set of techniques used in disciplinary matters, primarily sociological research without the aid of formulas, mathematical models, and/or statistics.

monitoring, if any extra information can be reached, thus to lower the risk of unforeseen shocks resulting in a long-term improvement to portfolio performance. Other benefits may include improved governance through transparent accountability. A new model approach, with new metrics included, is proposed to stress any predictive value within the outliers and to check if any link between data accuracy, strategic information, and/or data interpretation and influencing factors exist. This model is embedded in a conceptual framework in which a quantitative analysis on the predictive valence of outliers should follow based on the qualitative observations that will be conducted. Data considered out-of-range or discarded are de facto holding the key to predicative large-scale events.

Strategic Alignment

Today's organizations heavily focus on projects and services, and departments are struggling to overcome the challenges related to the management of resources, costs, timing, and projects. The main difficulties arise at the consolidation level in the entire portfolio of projects. Many organizations have systems for reporting time and planning, but these tools often do not allow visibility of the whole as one entity. The use of technology and processing of information in the management field is growing in strategic importance for all types of organizations. However, it is mainly carried out with a poor consolidated view both of the entire process and in structuring to give priority to investment in the projects. In addition a set of methods and technologies that optimize transmitting, receiving systems workflow, and in information processing are often neglected within the organization. It follows that the lack of standards and methodologies impacts consistently the way of measuring and monitoring success. In other words, there is no inherent method to align the project activities to business priorities. The portfolio alignment then becomes a project in itself, emerging through its complexity. The embedded features of the portfolio are indices: continuously changing, killing/go* decisions, and communication items, which will be considered as performance metrics. To depict the interdependencies within the portfolio metrics, a different and opportune portfolio set of variables must be chosen as well (for instance: continuously changing culture versus employee's ergonomics, emotions, and engagement; killing/go decisions versus budget resources; and communication versus skills and roles) (Table 17.1).

What does project portfolio alignment mean and how does project portfolio alignment evolve over time? Project portfolio management defines the dynamics to undertaking organizational objectives throughout the development of a strategic plan, which responds to a specific idea, proposal, or vision. The portfolio includes projects, programs, and subportfolios, or better defined as components, and this collection must

---

* For commercialization products, operations, and projects e.g., when they fail during testing, market launches, wrong technology, etc.

Table 17.1 Metrics Proposed with Their Quantitative and Qualitative Analysis

SELECTED METRICS AND EXAMPLES OF THEIR APPLICABILITY	QUANTITATIVE (EVERYTHING CONCERNING THE METRICS THAT CAN BE EASILY COUNTED)	QUALITATIVE (TO INVESTIGATE DIVERSITIES, OUTLIERS, AND OTHER OBSERVATIONS AMONG ITS INTERRELATIONSHIPS)
Continuously Changing Culture	• New Technologies • Level of Engagement • Level of Confidence • Lack of Expertise • Competitive Research and Development (R&D)	• Feedback from Employees • Success Stories • Sentiment • Opinions • Emotions • 'Know How' • Ergonomics • Ethic Values
Killing/Go Decision	• Projects • Proposals • Ideas • Resource Distribution • Level of Confidence	• Perception of Resources Allocation • Perception of Success • Perception of Contribution • Impacts of the Elimination
Communication and Information	• Users • Feedback • Information Passed • Data Structure • High-Powered Computing	• Feedback • Comments • Meeting Feedback • Quality of Information • Level of Benefit • Level of Relevance • Time Sensitive Information Distribution

be managed simultaneously and holistically so it is viewed as a whole entity not as collection of parts. All the components must be continuously aligned with the strategic objectives. At a given time frame the composition of a portfolio represents a particular "time-view" of all the selected components reflecting the strategy as a sum of intent, direction, and progress driven by a strategic alignment at that particular time. Beside the components, the portfolio includes the activities for all the organizational priorities identification, which must reflect the initial planned investment to determine the prerogatives for the components selection according to the alignment plan. To assure alignment the portfolio must be managed in a coordinated way to meet, within time cost and scope of the single components, all the objectives defined by the strategy. The interdependencies between components are implemented by the alignment. The alignment represents a key process to focus on objectives and to validate strategic updates toward organizational governance occurring at any decision-making levels and inboard to the strategic planning process to leverage the components' selections for delivery value.

Therefore a continuous iterative updating process between alignment and the decision-making process (Figure 17.1) is a necessary prerogative to assure a constant control of the dynamics between environment, strategy, and objectives. The process of continuous alignment constantly impacts each level of the knowledge areas and strategy (Figure 17.2). Strategy, therefore, is not a static item.

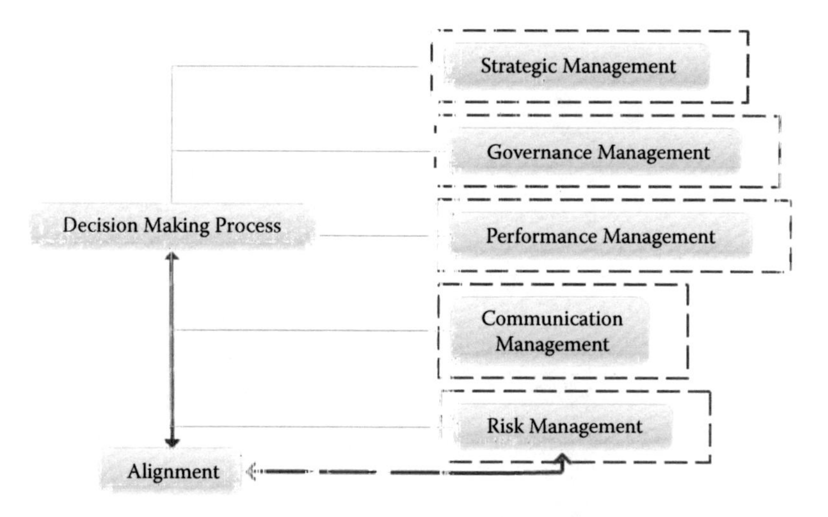

Figure 17.1 Mutual dependency between alignment and the decision-making process.

While a rigid strategic orientation, without the necessary flexibility, could be also inappropriate for some organizations, certain "flexibility" in revisiting and improving the strategy is always preferred in any case. For example, if there are foreseeable events, such as "maturation" of a product life cycle; presence of other products targeting the same market; or changes in technology, demands, environment, climate, etc. for which an immediate response in the sense of changing is required, then an update to strategic alignment should follow. Updates and revision processes should follow at each level of the development plan in particular for performance, communication, and risk management, since they continuously and unfailingly impact the strategic alignment. To pursue the vision plan, a continuous alignment between strategies and objectives must be implemented so as to ensure a robust framework and resilience from internal and external impacts (Project Management Institute (PMI), 2013).

Strategy is the technique of identifying the objectives and a large area of operations for the development of a broad framework in which actions provide the means to achieving goals with the least resources possible.

The alignment is a complex process, and it involves many other processes embedded within it, several areas of knowledge, and many variables. There is no inherent tool to inspect such complexity. The strategy varies because conditions change rapidly during development of the portfolio (Figure 17.3). The aligning process group defines the portfolio optimization, focused to omit, select, and improve all the portfolio components throughout the categorization and evaluation with their appropriate modification. Therefore *alignment* and *decision making* are mutually interdependent processes; if one is missing the other is ineffective. Their implementation responds and corresponds to a specific development of the strategic pattern. Hence the alignment results in a set of actions, guided by the strategy of selecting, prioritizing, and balancing the components, and leading throughout decision-making processes finally to success because of the attainment of the objectives. The achievement of the objectives requires a high level of

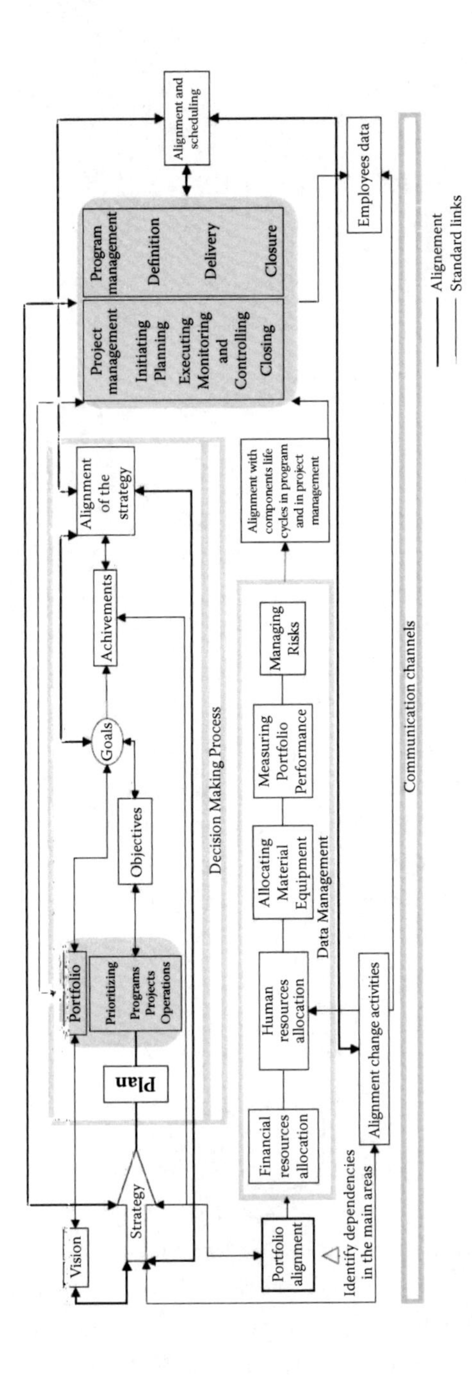

Figure 17.2 Continuous project portfolio alignment process.

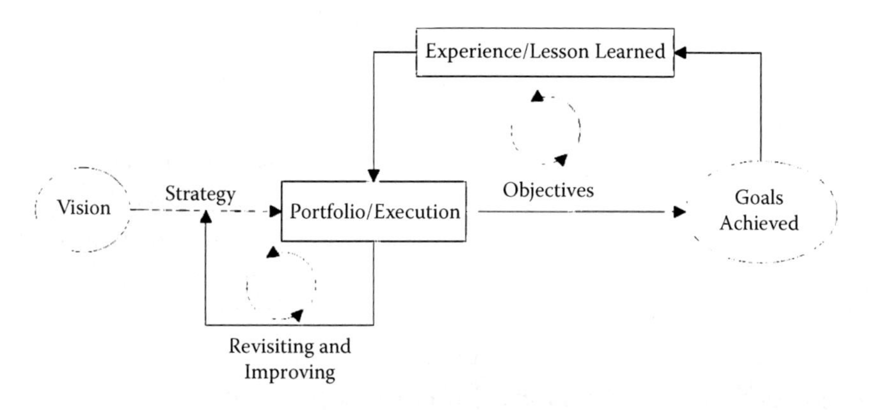

Figure 17.3 Strategy is not a fixed item in portfolio management.

conformity by all the members to comply and engage with the organizational objectives and establish trust among the members. Those objectives are obtainable by means of an appropriate controlling framework (performance management) and a *wide and wise* participation of all the members. When the objectives are achieved, then the alignment is updated toward the strategy. To obtain a *strategic alignment* the achievements need to be constantly enhanced and compared with feedback inputs, i.e., the strategies, must be clearly defined toward a common understanding (for all the members) of the components' purpose, inside the portfolio and toward the methods for their accomplishment.

The portfolio is monitored with the constant review and oversight on portfolio performance indicators and other available data. The execution of the portfolio management is linked also to the success rate resulting from the background experience of the organization: achieved or missed objectives collection, input data from the organization, and others. All processes are interacting with each other, and therefore, any change affects the whole system. Interconnections require tradeoffs into selection or variation in the roadmap, communication plan, and risk management.

Optimal alignment achievement depends on the effectiveness of framing the mechanism of the interdependencies between strategy, portfolio, and objectives and constantly taking into account the use of multiple portfolio methods. Despite the increase in the literature in managerial methodologies, effort to improve the rigorous treatment of organizational alignment is still a challenge. Official surveys underline chronicle misalignment in the linkage between strategies – objectives. More than 50% of the organizations, in fact, reveals that strategy is hardly aligned with objectives (Partners, 2013). Many questions thus remain concerning the use of methods; which approaches are better than others and what results organizations are realizing. Much progress has been made, but significant challenges remain.

In the panorama of portfolio methods, financial models are the most popular, because of the high expectancy in return of investment. Yet, not surprisingly in many cases the sole selection of higher-return and higher-profit projects turned out to generate the worst economic value (Cooper, Edgett, & Kleinschmidt, 1999). However, what is surprising is that despite their poor performance and *distorted perception** for achieving an effective control of the process, they are still the dominating decision processes without considering in tandem other qualitative approaches (for instance a strategic approach that is more than using scoring models, checklists, etc.).

Then, is this the right way to simply and routinely calculate the total value of all the components of a portfolio to allocate the available resources toward those components which have a greater return on investment? New approaches and multiple methods to promote a better overall assessment could be the preferred way. From the organizational perspective, the numbers behind several reports on the current status of portfolio management reveal also a need to go beyond the paradigm of structural and social organization. Systemic and synergistic phenomena of misreporting, obfuscation, and corruption caused by human factors† do depict indeed a need in the evolutionary progression within the organization, a deeper realistic awareness, and one oriented toward a more ecological and more cooperative way of managing the organization in the social and in the technical sides; it is not necessarily toward the maximization of profits, as shown by many theoretical economists, specialists in the field, and case histories of past events. Organizations with higher profits often are not the most profit oriented (Merck & Co., Walmart, etc.) (Kay, 2012).

Risks and the Interpretation of Certainty

Misalignment may be derived from wrong products into the market, failures, accidents, casualties, high losses, and also enduring obfuscation of information. Those circumstances contribute as a dangerous cocktail among a distortion of reality in which risks rise within the organizations. Moreover when external risks, such as the environmental instability in which most of the organization develops its own productiveness, the increase of socio-economic competitiveness, and political instability abound, and effective governance and sustainability are affected. Additionally factors such as the global demands for goods and services, worldwide competitiveness, the need of using more multifunctional and heterogeneous teams rather than individuals, cost base reduction, and competition from low cost economies require extra consideration. The manufacturing industry, for instance, has the ambition to increase in exploiting in highly automated production systems with the use robots even for the simplest packaging and positioning operations, etc. It reveals a strategic plan to attack inexpensive competitors on the cost side but also in the social "awakening" side. This explains the antithesis of the "anachronistic" need of continuous inexpensive labor demands, hardly specialized in such countries as China, in spite of complying with growing, fast development products in a low-cost and low-price segment and thus devoting workers to highly specialized control tasks of equipment, mechanisms, processes, logistics,

* It refers to managers that base their decisions often on incomplete or misleading data.

† Influential factors

engineering, etc. Obviously similar circumstances exist, which are not contingent, but are well defined through a slow evolution over time but with the defined characteristics of predictability and presenting a basic pattern to create a strategy. Such risks should not be left aside until a major event will shift uncontrolled changes impacting the eco-social stability. Is it a risk or a certainty? While an explanation sufficiently basic to highlight the importance of severe implications of social nature is required, at the TEDxReset* Conference in Istanbul, the futuristic, Thomas Frey, predicted that over two billion jobs will disappear by 2030, leaving fewer opportunities for individuals without specialty areas because factories will be more and more automated in the future. Whether risks can be transformed in opportunities or certainties it is a matter of how the organization excels in the sense of continuously changing parallel to its environment and how some peculiar aspects of the fragility of the human/organization complex will be portrayed (Figure 17.4). Major risks for portfolio management are inherent within the organization's complex.

In modelling complex decision problems it is often necessary to distinguish between uncertainty that is inherent to the phenomenon under investigation and uncertainty that is due to our lack of knowledge (or data omission, incomplete, etc.), which can be reduced by obtaining additional information. The awareness beyond this distinction should be kept in mind during the decision-making process.

On the other side, certainty defines a degree of "beliefs" or the 'blindness' toward an imminent certainty; similarly it is referred to epistemic certainty regarding an event that surprisingly turned out to be the opposite than previously thought. Probability is the possibility that any given event could occur where each event has the same probability of occurring than others.† There is a substantial conflict between maximum certainty and maximum likelihood. Certainty decreases with increasing stakes; therefore, the rate between the degree of certainty and degree of belief is a probability function. Absolute certainty does not exist because it responds to a high degree of complexity, but speculating on a no-certainty principle is likewise wrong. With the

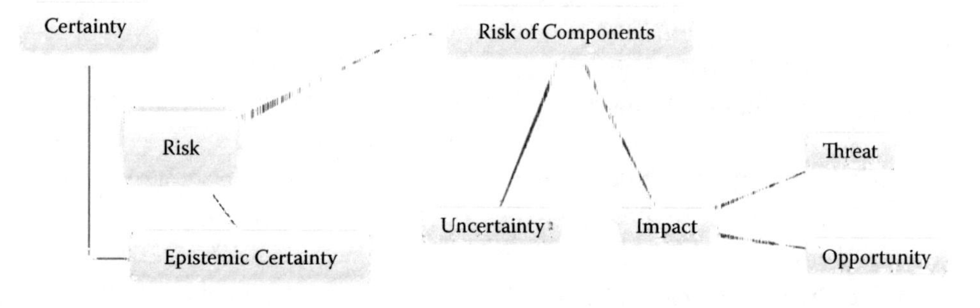

Figure 17.4 Risks threats or opportunities and epistemic certainty: A degree of beliefs.

* TED stands for Technology, Entertainment, and Design. It is an annual event where thinkers are invited to present.

† Laplace Theorem

knowledge ascriptions on which considerations support the decisions, it is evident that interest on the degree of certainty is related to the degree of knowledge. In fact, it would be similar to admitting that knowledge is not increased by any added value by experience. Nevertheless the conflict itself impacts the perception of the risks, both internal and external to the organization.

External risks have a higher degree of uncertainty with respect to organizational internal risks. What may represent a priori knowledge of an imminent risk of external type, if referred to an extreme case such as a *black swan** event, this could have unimaginable implications from the organization's point of view. But also there are a few things that claim to reverse an imminent catastrophic certainty already in place. These events, out of the reach of rational linear systems, posit complicated complex adaptive systems' theories and global level mechanisms (in the global sense of the world) applied to the stock market, to the economic crashes, and to distribution of earthquakes, etc.

Uncertainties are inherent part of the system. Endogenous stressors can be both addressed from a preventive point of view by acting in a robust way before the disruptive event occurs and considering the sole isolated sources of failure. This source most often is identified with the final stage in which the error occurs. Nevertheless internal risks can be monitored during their evolution, and not necessarily errors are generated in the phase where they exhibit themselves. Rising risks producing cascading failure can reveal foreseeable and controllable patterns of future misalignment in which often reside the root causes of the slow and inadequate reaction of the organization toward external risks or human factors issues. Internal risks correspond thus to a specific pattern caused by a fundamental organizational transformation. According to the punctuated equilibrium model, in other words periods of revolutionary changes alternated with periods of incremental adaptation (Gersick, 1991), in which the human factor has a significant effect, rising risks are unobserved. Retrospective analysis of past events may provide outstanding results for internal risks before uncontrollable severe escalation may occur. It represents a valid "territory" to explore and highlights in strategic retroactive analysis of the relevant data ('reengineering' the past events to improve the future) for unveiling patterns of important transformation changes constantly impacting and influencing decision making and alignment processes. To detect these important signals or symptoms of hidden information, positive or negative from which failure of projects or programs because of cost overruns, communication poorness or networking lethargy, and other errors can be derived, the organization's internal risks should be suppressed with analysis of outliers. Outliers characterize the occurrence of extreme value distributions and how they produce interdependencies variances into other portfolio variables. Outliers must be framed therefore into a qualitative observation perspective.

* Black Swan, from Nicholas Taleb "is a symbolic metaphor to define an event, with the following attributes: unpredictability, consequences, and retrospective explain ability".

Emerging Properties of Portfolio Complex System

Qualitative Metrics for a Strategic Alignment

Emerging complex properties are inherent properties of portfolio management. Portfolio management is a complex system exhibiting interconnected sub-assemblies or subsystems which behave as a whole. As in the social, biological, and technological fields, networks display common topological features. In project portfolio management they measure the level of interaction between the various functional groups at multiple levels through the knowledge areas. Complex systems are not able to exist as a whole entity without coordinated control of the communication processes and without affecting the information quality after a certain period of time. Data represent the raw material to structure the information from which representative organizational magnitudes will be extrapolated. For example, organizations often struggle because of the numerous disconnections of the data caused by the proliferation of millions of tools (Excel, Power Point, etc.) without fitting them into a big picture. This disconnection is a clear symptom describing how communication has not caught yet a *cybernetic* perspective of self regulation. Transmitting and receiving processes at the organizational level need to be controlled to confirm the information distribution and coordination at each level into the organization. Second, they need to be self regulated against the influential factors that may impact the veracity of the original transmitted information. As emerged from the observation of the dynamics of obfuscation,* portfolio managers and managers in general, base their own decision on the quality of the information and rely on the veracity of information. The quality of the information improves, and it is directly proportional to the quality of the *decision-making process*. It uncovers in fact the interrelationships among the portfolio variables.

Financial measures used as a decision methodology do not yield the best results although they are the most used and most popular. In order to unpack the interconnections mechanism between projects and components among the way they are selected is due to underlining that its complexity is an increasingly emergent property in a holistic sense. It is also proportional to the environment in which the organization is located. This awareness should endorse trans-disciplinary approaches toward a dynamic project portfolio management. A qualitative model approach thus will be preferred for inquiring into these paths where knowledge areas variables are cross correlated. The environment and the organization are embedded in a circular relationship so the need to chase the multiple interdependencies through their own development in the portfolio plan leads to developing a set of metrics from which interdependencies themselves become evident.

* The dynamics of obfuscation investigate the motives behind certain behaviours. Obfuscation and distortion of information, among the subordinates or among the executives, are observed before, during, and after severe accidents.

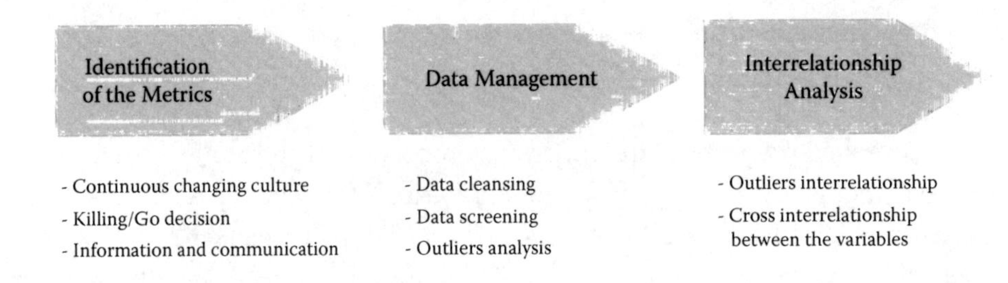

Figure 17.5 Overview of the approach.

The organization is in constantly interfaced with the environment through the people and their families, the place with its local features, and the events or mega events (natural catastrophes, Olympics, sport competitions, and other various events) from which it is daily affected according to characteristics such as being slow, adaptive, or ongoing. It reveals endogenous key factors that unveil a hidden influence for steering the organization toward a certain level of performance. Figure 17.5 describes a broad overview of the approach used.

The ability to identify the most appropriate path within the many alternatives, that an environment can offer, is a crucial stage at each time frame of the portfolio decision-making process. The continuous changing culture is the deepest expression of an organizational soul in which its mentality is reflected through its actions. The killing/go decision is the watchdog of portfolio management; its balance is crucial. Communication and information distribution are the supporting skeleton of the company on which to base decisions. Therefore performance metrics* such as: continuous changing culture, killing/go decisions, information, and communication, etc. embedded with each other will provide insight about the mechanism of interdependencies. When priorities change, then misalignments or even opportunities can be created.

Causes of misalignment often occur because of the intrinsic limitations that are part of the organizational structure or in the "deep structure"† (Gersick, 1991). Old paradigms in organizational and strategic thinking, such as organizational ecology or agency theory, according to which organizations are static societal structures (Levy, 2000) embedded in a slow-moving environment contrast greatly with the new conceptual portrait of a modern firm in a fast-paced context. The paradigm is a reflection of a societal structure in which it was created and closely related to the "informal constraints" that shape human behavior particularly such as customs, traditions, sanctions, and codes of conduct, etc., whereby they affect local enduring features, which silently form the organization. So if the environment is highly competitive, then the ability to change continuously within a critical time frame is a core capability of success. Organizations that do not present a cascading adaptation within their environment and produce resistance to change usually are seeking to maintain complex networks

* The metrics need to be numeric and conveniently quantified.

† This deep structure is what persists and limits change during equilibrium periods, and it is what disassembles, reconfigures, and enforces wholesale transformation during revolutionary punctuations.

and commitments (with buyers, suppliers, etc.) ending with a stasis and without yielding to a substantial transformation (Romanelli, 1994). The ability to continuously change is an organizational metric that reflects itself into the portfolio performance (as expressed before in the previous footnote) the number of probes, experimental products, strategic alliances, futurists' shows, pavilions or research would be the better fit to quantify this metric for further analysis. From the literature many pessimistic orientations glean the fact that some organizations have mediocre capabilities to directly reach the objectives that may be due to listless reactions, human factors, etc. Failures or extreme events trigger changes. The "organizational listless" does not trigger revolutionary changes but is the systematically failures or impediments that do create the conditions for altering the strategy for organizational survival purposes. When unforeseen events or underestimated shocks occur into an organization, the dynamics of the alignment undergoes the pressure impacting the strategy and altering the conditions at a particular time frame of the portfolio alignment. Therefore, the ability to react based on reliable information comprising risks and opportunities is an essential property of a given organizational system. To avoid a complete process alteration, with consequential changes to the objectives, the organizational strategy must focus on the ability to mobilize additional resources to recover (resilience, resource allocation management, etc.), invert, or take advantage of a new situation and increase when possible the performance of the portfolio to go through a new period of adaptation. Most delays and cost overruns in fact arise from unexpected elements that did not enter into the plan (Taleb, 2010).

As shown in Figure 17.6 after a shock the system responses present different behaviors such as a robust behavior that maintains the performance unaltered, an under performance behavior creating misalignment with the previous conditions, or a threat; or an adaptive behavior in which performance is enhanced results. Misalignment for underperformance occurs because of a lack in the awareness about the differentiation

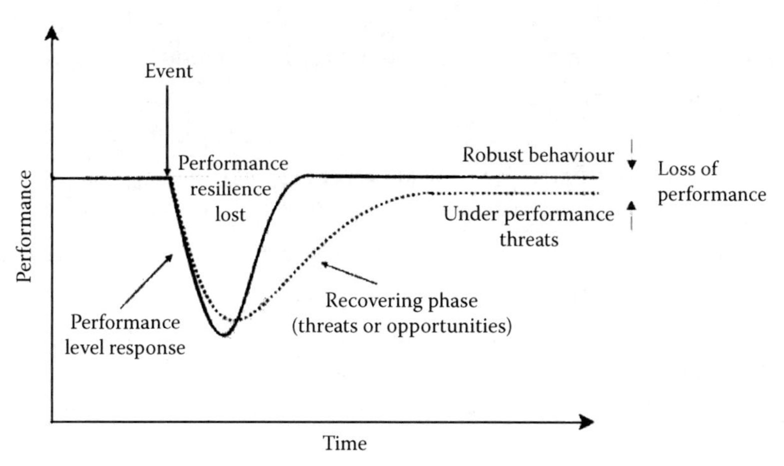

Figure 17.6 The performance is having a shock, which impacts the resilience and the robustness (see also Figure 17.7). In this case, the event is a threat.

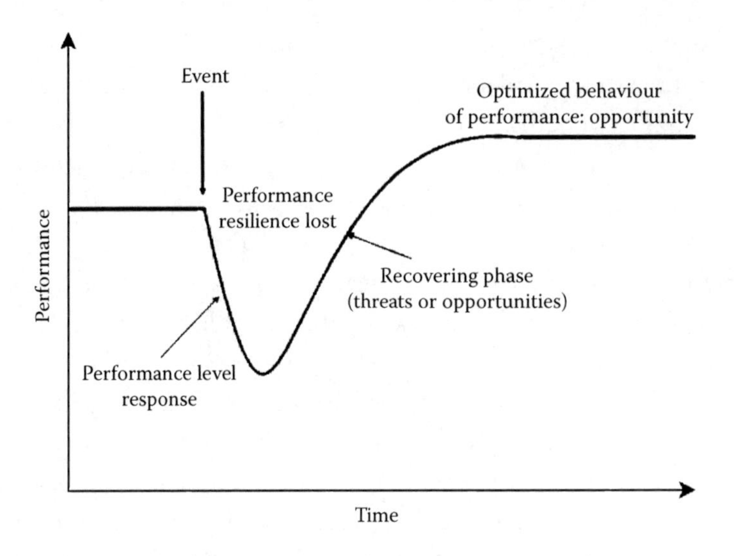

Figure 17.7 In this case, the event represents an opportunity.

between the intrinsic robustness of the system, which aims to maintain the alignment between strategy and objectives and therefore ensures a certain adaptability of the organizational system, rather than robustness toward the external environment. Risks then may represent opportunities. There are indeed other situations in which environmental stresses imply major deviations from the original objectives. Then it is vital for the company to try to convert external stimuli into new opportunities. Precisely protecting the organization from the environmental threats implies, both integration or merger of the threats into the organization, without distribution of the power with the aim to broaden the consensus, and a redistribution of the power at the executive level induced though a variation over the original organizational objectives (Selznick, 1948).

In a constantly mutuality* of information exchange, the success criteria are strictly linked to the ability to change rapidly and with resilience. An organization must be focused on a dynamic structure to have the necessary resilience coming from unforeseeable events and being prepared to process large amounts of data with people engaged in the process (Figure 17.7).

Often to reach such level of competitiveness the organization must processes data sets in a methodical and structured way peculiar of a certain maturity model level† (Partners, 2013). A certain maturity model is an a priori condition that determines the measurability of the performance metrics. Level 4 and 5, according to the maturity model, are organizations focused on better structured data management and on innovative practices involving both social issues and technological aspects (both processes and tools).

Other challenges, around disconnections and misalignment, frequently are rooted in the organizational culture as an expression of the influencing factors (e.g., human

* Mutuality, general meaning: condition or state of reciprocity or sharing

† Planview developed the Innovation Management Maturity Model. The model encompasses a five-tier grading system from 1 (low) to 5 (high) for organizations.

factors). Communication management is the most impacted by it.* In emergency situation often executives had truncated data regarding the real conditions of the situation to be reported. Executives seem to cope comfortably with this large informal behavior. There is evidently an inherent deficiency of alignment between the organizational/corporate paradigm, still shaped over an old-fashioned semi-static model, and the paradigm determined by the social, economic, and technological pace where growing market opportunities represent a vital source for any given organization. This view described, however, is facing the necessity of revitalizing the approach into business with transformation mentality, less risk aversion, and business flexibility over data sets management in a modern context where often there is a shortage of time to process the incredible amount of information.

Since 1948, important features for organizational surviving skills have been observed such as the stability of the communication lines, stability in the informal relationships in the organization, continuity of policy and its sources, and homogeneity of views with respect to the meaning and role of the organization (Selznick, 1948). Beyond the great efforts made since 1948, with implementation of efficient communication channels outlined in modern organizations, major challenges still persist. It follows that transparent management of information and clear communication become indispensable prerogatives to determine the way of collecting information and managing data repositories such that the quality of decisions endures along with the strategic alignment. The way of managing communication and information together with their results in terms of amount, quality, and heterogeneity of their distribution is considered as a performance metric.

Another metric is represented by the derived benefits from the elimination process defined by the killing/go decision process with a consequent determination of the range of time or time limits within each decision, for component selection and elimination, needs to take place to avoid misallocation of organizational resources.

Two analogies are presented to empirically explain the concept. One is taken from natural science and the other from the engineering field. The killing/go decision process 'embodies' the organisms' mortality of the evolutionary theory according to which "organisms produce more offspring than actually can survive.† This is reflected also in engineering, where the mortality failures are higher during the initial stage of the product's life cycle. As in the natural science the mortality measures the fitness level of each organism in respect to its environmental adaptability i.e., its attitude to perform relatively to a required function. In parallel, the elimination of one or more components in the early stages during the development of portfolio is a symptomatic outcome of the many generated ideas/proposals, which need time to adapt, evolve, or

* Some survey reports state executive decision makers are often uninformed or provided with misleading information, regarding a "possible threat" for the organization. A majority of the surveyed companies is unstructured, but only the 29% of the executives think that poor decision making leads to poor choices (The Economist, 2007).

† Inference of Darwinian Theory

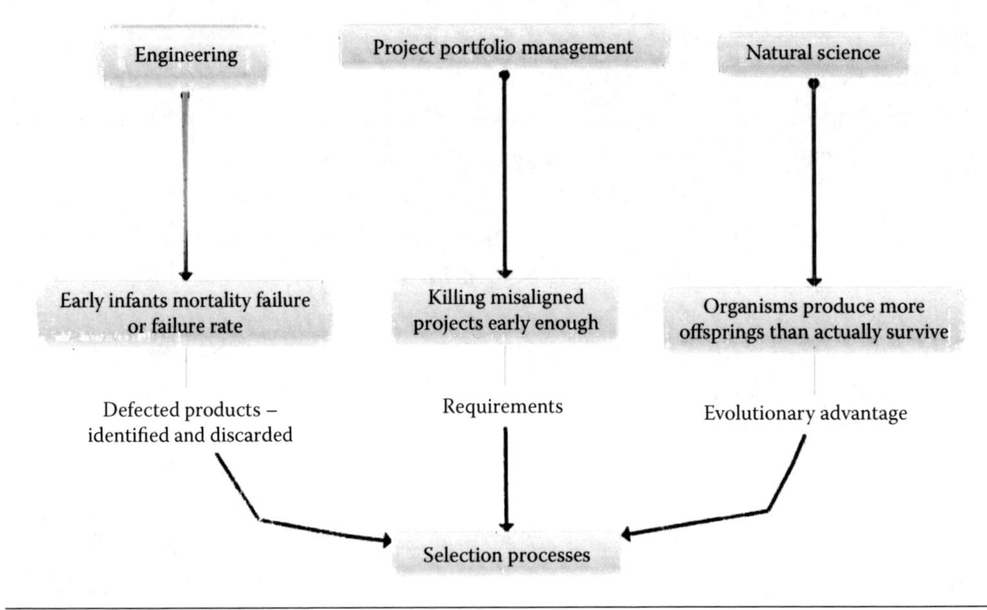

Figure 17.8 Similar behavior.

become obsolescent when interfaced with environmental factors or the information process (Figure 17.8). In fact a new know-how input into the organization may discard a previous decision.

Conversely the delay in 'killing the components' may have negative and considerable impacts on revenue, cost overruns, or other components (perception of unlimited resources). Therefore, killing/go decision actions benefit portfolio optimization and reinforce strategic alignment. These observations define the first stage of the approach (Figure 17.9). Killing/go decision is interpreted here as a portfolio metric performance or the health state of an organization in respect of the amount of crowding ideas, projects, and proposals in which innovations may be concealed.

However, delays in killing/go decision of projects are a bias created by a perception of unlimited resources.

The portfolio is considered a complex system, exhibiting strong emergent properties so the metrics need to be well portrayed to represent it. These metrics have been chosen for their representativeness and strategic importance in the main areas of the organization: the culture, the information channels structure, and selection/decision process flows with the relative allocation of resources (to give priorities to components).

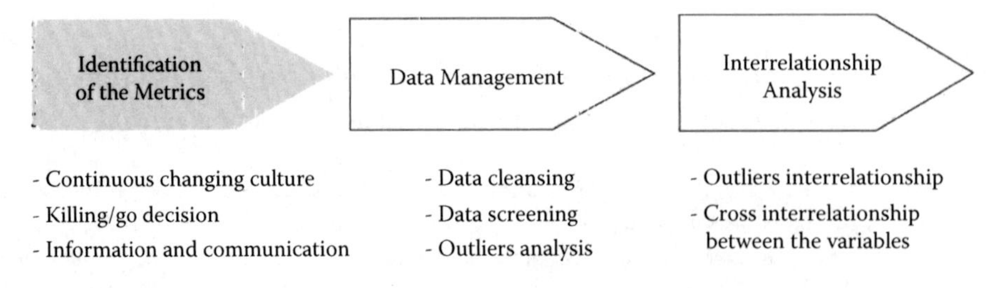

Figure 17.9 Metrics.

Each of these properties of portfolio performances is not independent of one another. Their interconnections/interdependencies may present circular feedback between the components revealing important relationships as typical features of all complex systems. Also, the presence of extreme values in particular data sets, such as outliers, may have significance for highlighting connections and links of cause and effect, whether overt or hidden, among the various organizational variables. Outliers thus should be checked beyond the cross-projects functions such as cross-project scheduling, time tracking, resource loading, and analysis along with cross-project consolidation and reporting. The purpose of the proposed approach is to appraise how to locate a type of predictive valences out of the outliers' data thus improving the monitoring aspect of the portfolio management or enhancing scenario planning, project prioritization, and ROI assessments.

Internal Risks and Potential Failures

Organizational failures, especially when they present extreme features, attract and fascinate many government senior officials, academics, and politicians because of the implication in the likely socio-economic impacts into the environment in respect to safety, ecological, political implications, and in the societal structures. Organizational failures frequently keep occurring presenting systemic symptoms with similar causes such as biases, misinterpretations, cultural, self-interest etc., which may find the root causes in managerial behavior. These circumstances suggest limited consideration about internal risks (Figure 17.10) such as human factors or an inadequate response to the internal stressors. They (internal stressors) develop into the system amplifying the errors already in place through data omission, data interpretation, or data misrepresentation and then collapse into unseen mechanisms that generate poor performance and hidden rifts into the portfolio in general. Omitting data points, especially extremes and influential points

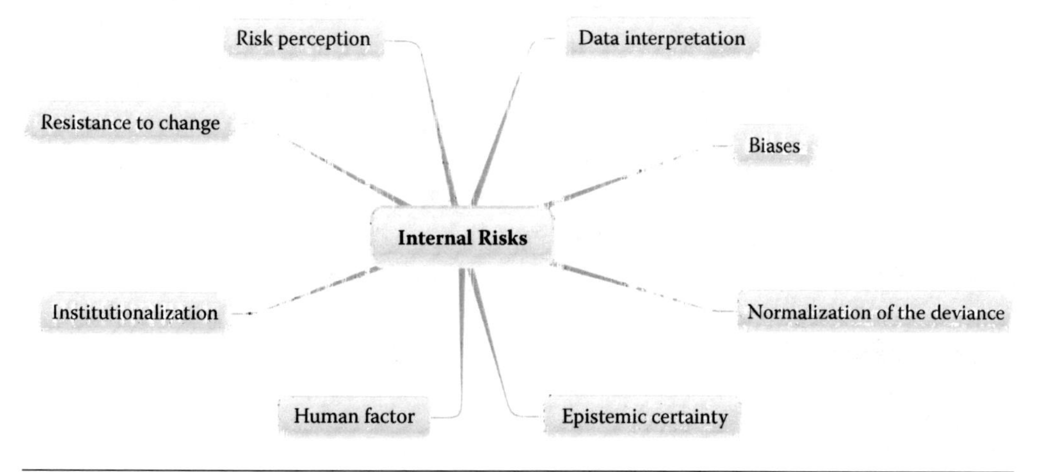

Figure 17.10 Variety of Internal Risks—*Institutionalization* is a concept related to process improvement. It is not just a generic practice description, but it implies the process is well established as the work is performed, and there is consistency in performing the process.

simply erased from data sets, is a common practice. At other times data can be discarded because there is an ongoing questioning over their representativeness.

Decontextualizing data sets from one system to another, for instance, also leads to incorrect outcomes and distortion of the reality, i.e., extension of binary properties to system variables and interpreting them in a complex system world. Exposures in the real world are not captured by binaries, and "*Decision makers and researchers often confuse the variable for the binary*", because often the terminology between bets, predictions, and exposures is confused (Taleb & Tetlock, 2013, p. 1).

To establishing an effective decision-making process and recover the misalignment caused by internal risks, a major review of data structuring is needed. In addition, one of the major internal risks is dismissing outliers root causes analysis without appropriate investigation while they may generate "credible" risk patterns and therefore make predictable what before was unpredictable. From the literature, prevention and preventive actions on research of related topics, few people questioned these assumptions and paradigms. It is noted that checking for the presence of outliers is done only 8% of the time in many disciplines (Osborne & Overbay, 2004).

Measuring performance and managing risks without considering outliers is one way organizations hide and accept large risks with lack of information about them (Budzier & Flyvbjerg, 2013). The Challenger disaster (Vaughan, 2009) showed, for example, that hidden processes slowly enlarge the risk accepted by organizations.

Beside the a priori rejection of outliers, another important observation is putting forward the dilemma of processing information faster and the limitations of time to make decisions in a recursive environment. Global networks and computer enhancement of the information create scarcity for attending and digesting the information in real time. Communication systems in an organized society need to distinguish the relevant information to address them to the right premises and to exploit them to contribute to speed up the information systems involved with the decision-making process. How can these two aspects be combined? Observations create opportunities. The widespread practice of simply discarding data without having a full understanding of their value should be questioned. Outliers can rise from different mechanisms, from errors, and from an inherent meaningful variability of a given data set or from different mechanisms; therefore, it is important to point out a range of causes that have generated them. These points can substantially affect the results and the subsequent interpretation (Stevens, 1984). Therefore appealing to a valid discriminant method for categorizing the nature of the outliers is an unavoidable choice. In addition an important aspect is the management of the communication channels in different information distribution systems.

Jeopardizing Data Quality in the Communication Channels

To have a better understanding of each portfolio organization's time frame conditions, data need to be well structured and accurate in order to be accessible at any level and with the same amount and contribution of knowledge and information between the executives, their subordinates, and employees. The ability to systematically generate information and build knowledge based on background and experiences theoretically secures competitive advantages, and therefore, assures the characteristics of continuous changing transformations into the organizational culture. Many firms compete and change continuously. Nevertheless many organizations suffer from poor and inaccurate information structuring because of many factors, such as the use of financial measures, misplaced resource allocation, or the poor choice of KPIs as indicators of a steady state of a portfolio at a given time frame. The choice and the representativeness of the performance indicators need to be estimated from time to time according to the goal of the analysis. Depicting the right ones is therefore a question of thinking and sense making. The evaluation of the selection, prioritization, and termination in portfolio management is supplied by different sources of the process groups such as the communication plan, costs, success rates, customer satisfaction, and commercial data bases (e.g., standardized cost estimating data, industry risk studies, and information and risk data bases), and budgeting and historical data projects are continuously updated. Data are reviewed for resource allocation (availability and capability), aligning process groups, process assets, business value, historical data (updating the portfolio management plan and trend analysis), portfolio performance variances, market data, communication management, and elicitation collection techniques, etc. Communication is by definition the set of phenomena that are aimed at information distribution. It is in itself complex, which seems odd that organizations still do not deal with it from a cybernetic perspective to improve its accessibility. Nevertheless, because the management levels have continuous overlaps, synergy, and interconnections throughout all the portfolio management processes, it is difficult to keep the same quality level of information distribution when data are coming from different knowledge areas; in this case the risk is to present a partial picture without considering the entire portfolio of components.

With the increasingly recursive environment, different factors need to be taken into consideration, such as the diverse nature of the risks and also outliers data. There is a clear distinction to do for commensurability of risks in different environments, internal and external to the organization. Other internal risks may be triggered by organizational behavior, such as voluntary exposure to danger, possible manipulation of the information, misreporting, social desirability, and self-presentation (DePaulo, 1992). The risk, whose event represents its realization in certainty terms, makes tracing the root causes of technical errors too elaborate. Therefore the human decision-making process intersected with unpredictability and dangerous technology can culminate with terrible facets. Data relay both information and economic value. Information and communication management become the key metric for structuring information distribution to detect the interdependencies between self-interest, caused by human factors and organizational priorities and interests. This approach should lead

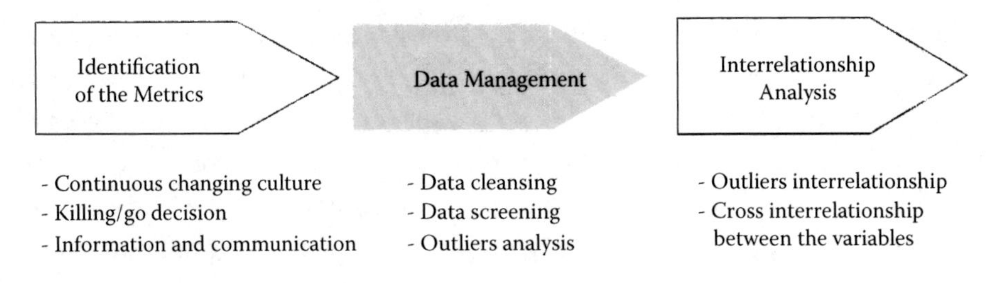

Figure 17.11 Data cleansing.

to uncover those (influential) factors constantly influencing the portfolio alignment and its performance.

The second stage of the model involves the process of data cleansing (see Figure 17.11) throughout a self-regulating process based on cybernetic dynamics. The purpose of this task is to select and trace relevant data, included outliers, from a previous spurious collection because misreported or just errors and discriminating them from outliers can be generated from *hidden mechanisms.*

To identify these risks, and reduce communication systems vulnerability, a fundamental practice is to abstain from the *normalization of the deviance** and to investigate the existence of points or special deployments which highlight misalignment between the various levels of the information process. Furthermore the introduction of the cybernetics self-regulating information process outline irregularities caused by the optimal or suboptimal time allocation, tasks, queuing, etc., by the side of the communication channels users, as information is perceived and exploited. The optimization of the transmitting, receiving phase thus would improve the systems workflow. To follow the feedback control will be carried out with the aim of highlighting the trends and thus preventing same patterns of behavioral irregularities in the future, as shown in the following, Figure 17.12.

The final part of this conceptual approach will be devoted to the connections and interrelationships of the performance metrics, within their power law distributions inter-correlated with the portfolio variables opportunely chosen by the managers.

* Diane Vaughan, sociologist at Columbia University, studied the Challenger accident. She explains that: *"Normalization of the deviance in social field means that people within the organization become so much accustomed to a deviant behaviour that they don't consider it as deviant despite the fact that they far exceed their own rules for the elementary safety. But it is a complex process with some kind of organizational acceptance. The people outside see the situation as deviant whereas the people inside get accustomed to it and do not. The more they do it, the more they get accustomed. For instance in the Challenger case there were design flaws in the famous "O-rings", although they considered that by design the O-rings would not be damaged. In fact it happened that they suffered some recurrent damage. The first time the O-rings were damaged the engineers found a solution and decided the space transportation system to be flying with "acceptable risk". The second time damage occurred, they thought the trouble came from something else. Because in their mind they believed they fixed the newest trouble, they again defined it as an acceptable risk and just kept monitoring the problem. And as they recurrently observed the problem with no consequence they got to the point that flying with the flaw was normal and acceptable. Of course, after the accident, they were shocked and horrified as they saw what they had done"* from the Interview: Diane Vaughan Sociologist, Columbia University—by Bertrand Villeret interview—Editor in chief—ConsultingNewsLine—http://www.consultingnewsline.com/

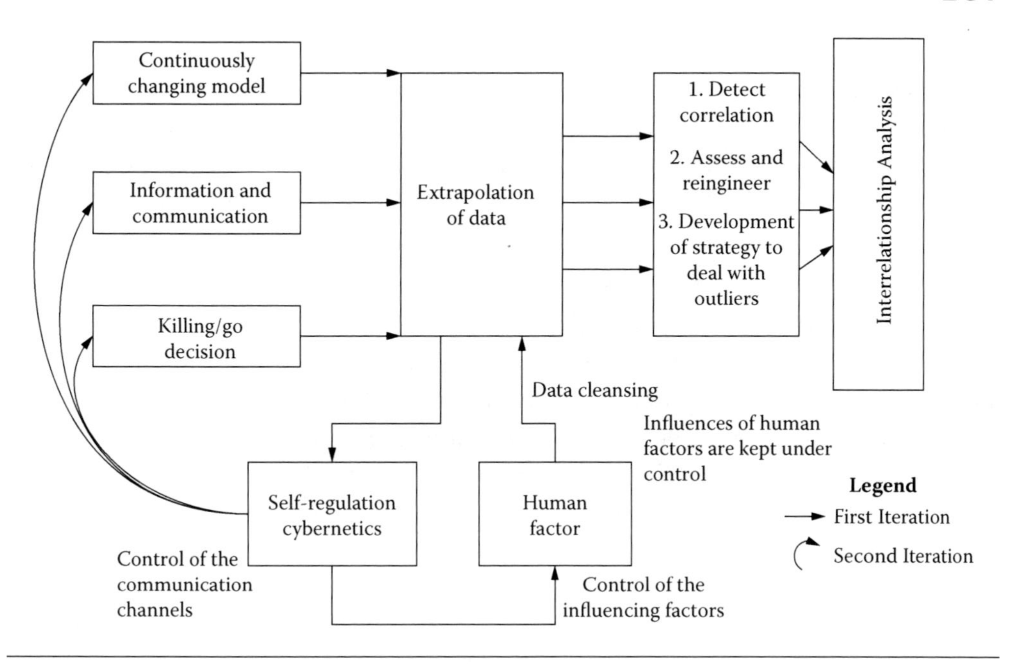

Figure 17.12 The qualitative model for cybernetic dynamics.

Outliers in a Project Portfolio Recursive Environment Outliers are extreme data points far outside the norm of a variable or a population. They are differentiated from the inlier data points because of a consistent deviation within the inspected sample and from other observations. An outlier is an observation that lies an abnormal distance from other values in a random sample from a population; extreme points in the data cloud. Nevertheless, the use of the term "Outlier" almost reveals the intent to instill an idea of mistaken points, which must be rejected. An outlier is a data point whose response does not follow the trend showed by the rest of points in the data set. It is called an influential point if it unduly impacts any results in the prediction, estimation, or in the hypothesis aspect of the analysis. The problem of outliers is one of the oldest in statistics, because their presence can lead to error rates, distortion of parameters, or estimates, etc. These points can substantially affect the results and their subsequent interpretation. In many field such as statistics, medical sciences, and finance, outliers are removed since they represent unexpected results during observation. Outliers have also been defined as values that are "dubious in the eyes of the analyst" (Dixon, 1950, p. 488). Hawkins described an outlier as an observation that "deviates so much from other observations as to arouse suspicions that it was generated by a different mechanism" (Hawkins, 1980, p. 2). Such behavior seems to be the "rule" for different reasons, for example, risk appraisal relying on the strategic interpretation of sensitive and meaningful available data, as reflected in the long-established engineering tradition with well-established concepts to characterize risks. Under these conditions, therefore, an observation on outliers in general is worthy of a deeper insight for an optimal plan development phase in project portfolio management. Do meaningful outliers exist that can be interpreted as predictive signals pointing out instability of the system or anticipating cost and schedule overruns? From project performance literature the outliers that are often associated with quantitative measures are defined as unique and idiosyncratic. However from empirical findings overruns shows regular and stable patterns, and they are not idiosyncratic events and follow a power law (Budzier & Flyvbjerg, 2013). With the growing intricacy in human organizational systems the sole quantitative approach is not a complete tool for investigating the complexity of the rising of intersecting cross-disciplinary findings. Consequentially it is necessary to embed the quantitative with a qualitative angle over the meaning of outliers generated from the data distributions of the portfolio items. To serve as inspiration for an inquiry, the qualitative remarks will be engaged, not only from a phenomenological point of view, but more importantly in correlation to their generating causes. The traditional way of identifying interdependencies and root causes tend to be in a limited and oriented to small systems Outliers' screening may constitute a method of observation that goes beyond the mistaken or misrepresentation to point out hidden information lurking in data sets. From project management literature that major causes of outliers are internal and controllable by organizations, there is the possibility that control of the outliers could have a positive impact on the decision-making and risk management practices.

In the light of these observations about the quality of the data and the obviousness of how later decisions and the results achieved may be affected, it is necessary to consider the long-sought "hidden mechanism" that lurks behind the outliers (interpreted as dragon kings class) can be triggered by the interference of the influencing factors (e.g., human factors).

Brief Summary of Power Law Distribution In reference to the events distribution, Gaussian distribution is the most important distribution precisely because it finds numerous applications in many fields. Gaussian distribution is a realistic representation not only of the physical phenomena, but because it can be implemented over different disciplines in the social sciences, in which they have been observed and even denoting common behavioral patterns. Mostly they are linear phenomena with independent variables. Probabilistic methods have been the main approach used to depict the occurrence of extreme events in terms of prediction and characterization. However, there are situations in which the nature of the structures is complex thus it is difficult to find a suitable mathematical representation of it. For instance, in the case of a small scale, a curve whose points are touched by a tangent can be merged with a tangent itself and therefore may lose the pattern of its original structure. With the introduction of the fractals' concept, described by a power law distribution and scale invariant, emerges the importance of scale laws and power laws. Power laws can be generated in natural and man-made system found in various applications, i.e., Gutenberg Richter for the earthquake distribution. Although fractals and earthquake distributions are related, they express a different concept apart from the overlap of the points of the tangent, in fact the exponent, in the case of earthquake distribution, refers to an event size distribution, rather than to a simple metric property of the phenomenon; extreme events are completely ignored by Gaussian distribution. Nature tends not to use Euclidean geometry in achieving its forms but uses a far more complex geometry to describe itself. In conclusion, the Gaussian distribution is not always exploitable to describe phenomena that occur in nature in the social and economic environment. Nevertheless power law distributions are characterized by certain criticality. In case of extreme events occur they alter the mean value of the distribution, creating unstable Pareto distributions. Power law distributions quantify the dependence of the frequency of the item, such as the size, or in general other measures as inverse power. Furthermore, if the variance is potentially infinite and the confidence interval large, it means the larger are the values of extreme events, and the frequency of these events result unpredictably, making some models unusable from a predictive point of view. Extreme rare, disastrous events happen without notification and cannot be predicted. This was highlighted by Mandelbrot when he worked on fractal properties of the curves that describe the financial markets, coming to assume that the stock market crash occurs much more frequently than assumed by a log-normal distribution (Remorov, 2014).

In this chapter, the stock market crash will not be discussed further except to add it to the list of those external factors that can have a substantial impact into the organization. Unfortunately the unpredictability of the stock market is an intrinsic property governing this complex system, because it responds to its own balancing physiology between slacks and opulence, and if anything could be done, what should be done is to revisit worldwide the role of economy.

Distribution Features Power law distributions were introduced into organizational studies to explain the size distribution of business firms (Simon & Bonini, 1958). A power law distribution of a variable has a certain exponent α ($\alpha > 1$*) representing the fatness of the tails of the distribution. Per α values that are very small impact the outliers in the center as the distribution increases (Taleb, 2010). In the normal distribution, "tail" is meant as the bottom shape, left and right, of the probability distribution. When the average is greater than the median (the difference between the mean and the median is positive), the distribution is positively skewed. If the median is greater than the average (the difference between the average and the median is negative), the distribution is negatively skewed. The kurtosis is instead an indicator of how the scores are concentrated toward the mean or thicken toward the extremes of the distribution. It depicts how "heavy" the tail is.

In correlation and linear regression analysis it is assumed that there is a linear relationship between the two variables, even is not, so it is comparable as trying to fit a

* The normalization of the power law distribution:

$$p(x) = C\ x^{-\alpha}\ ;\ 1 = \int_{x_{min}}^{\infty} p(x)dx = C\int_{x_{min}}^{\infty} x^{-\alpha}dx = \frac{C}{1-\alpha}[x^{-\alpha+1}]_{x_{min}}^{\infty}$$

straight line into a non-linear equation. Standard statistical methods would fail trying to analyze non-linear data; i.e., data transformation somehow constrains the data into a straight line. But this is the best way to fit data into statistical methods (even other advanced statistical methods that reject transformation). The transformation—truncation—robust method may produce outliers in the same way it is supposed to reduce them.* A fat tail is a property of some probability distributions, which exhibit extremely large kurtosis, skewness (asymmetric property of the distribution). In the fat tails lie the outliers.

Dragon Kings Outliers

The term King was coined for an analogy to those kings,† (Sornette, 2009) for which their cumulative wealth in their own country is distributed following a Pareto‡ law distribution. According to it all the wealth lies in the tail of the distribution. Dragon Kings are exceptional events, derived from the definition of outliers but with a diverse interpretative need in which resides the importance beyond the extrapolation of the fat tail distribution of the rest of the population. They coexist with power law distributions in large variety of systems. The term "outlier" emphasizes the spurious nature of these anomalous events, suggesting to discard them as errors (Sornette, 2009). For this reason the name Dragon Kings has been coined. Sornette found mainly six areas§ of the statistical distribution in which the presence of Dragon Kings has been observed. He hypothesized that there is class of dynamics which produces extreme events, such disruption, material rupture, or failures, beyond the extreme value of the distribution, precisely the "Dragon Kings". They can be recognized with precursory signs. In 1987, the concept of "self-organized criticality"(Bak & Tang, 1987) was introduced to unfold that most significant events in a distribution are generated by the same mechanism as the smaller siblings. But an important message is that there is no unique methodology to diagnose Dragon Kings (Sornette, 2009). However in the Dragon King theory, it is assumed that the distribution of large events in a system are generated by different mechanisms: "*dragon-kings exhibit a degree of predictability, because they are associated with mechanisms expressed differently than for the other events*" (Sornette, 2009, p. 1). There is a class of outliers therefore that clearly is generated by different mechanisms (Hawkins, 1980). The extreme events of the Dragon King

* Transformation, truncation, and the robust method are some statistical procedures used by researchers to protect data from being distorted by the presence of outliers.

† King Buhimol Adulyadej (Thailand), Sheikh Khalifa bin Zayed al-Nahayan (United Arab Emirates), Sultan Hassanal Bolkiah (Brunei), Sheikh Mohammed Bin Rashid al-Maktoum (Dubai), Prince Hans Adam II (Liechtenstein) , Sheikh Hamad bin Khalifa al-Thani (Qatar), King Mohammed VI (Morocco), Prince Albert II (Monaco) and so on.

‡ The Pareto index is still a measure of inequality of income distribution.

§ City sizes; acoustic emissions associated with material failure; velocity increments in hydrodynamic turbulence; financial drawdowns; energies of epileptic seizures in humans and in modelling animals; or earthquake energies.

class occur in some specific fields* discovered recently. It is assumed to occur with an increase of the dynamic mechanism depending on the application field; it could be due to a coupling strength, positive feedback loops processes emerging transiently. But the mechanism is generic. These empirical evidences are found in different fields but always exhibit the same behavioral pattern.†

Applicability of the Approach at the Portfolio Management Level

Systematically investigations, regarding conditions and circumstances under which breakdowns and dysfunctions cascade into a "Dragon King" event, involve many disciplines in a trans-disciplinary framework. As far as the purpose of this chapter is concerned, outliers as Dragon Kings will be referred to only as extreme data points detected from portfolio variables distributions, which may have been originated from "hidden mechanisms" of interrelationships between portfolio variables and influential factors, not necessary to the same one that generates the population of the given data or events.

As Dragon Kings are generated by a different mechanism, the outliers in portfolio management can be attributed to a hidden mechanism inherent within the interconnections between the components. This mechanism is triggered by the presence of the factors of influence. These factors impact different items in portfolio management, including variables, risks, alignment, etc. Then it is assumed that this influence leads to a percentage of impact for each variable, which influences and distorts the original value that a given variable previously had. This percentage of impact, an influencing part, is responsible for the production of most of the extreme values, outliers (except those produced by errors in the reports, see Figure 17.13). There is almost a necessity to assess the classes of outliers but not from the statistical or phenomenological point of view, as already exists (influential data points, leverage points, etc.), but from an *etiological* perspective. This influencing part then is reflected creating a distortion mechanism into the interrelationships between the portfolio variables.

To summarize the influential part of human factor and feedback loops generating from portfolio variables will be considered as the hidden mechanism for detecting contingent asynchronies impacting cross-projects scheduling and alignment.

Internal risks may lead as well to "disruption or failure" in an uncertainty regime that not necessarily originated in the same phase in which they exhibit themselves. This uncertainty should be taken in consideration during the decision-making process. For example regarding the NASA disaster there were not only merely engineering issues

* Coupling strength that increases within a system – may apply to earthquakes and financial crashes; Condensation in a type Bose-Einstein dynamic – explains the agglomeration of the urban areas; Heterogeneity decreases – describes the material rupture; Transient positive feedbacks (in the sense of system dynamics) emerge in the system - material rupture.

† To underpin the mechanism of generating Dragon Kings it is fundamental to observe these phenomena: low heterogeneity coexisting with strong interaction, producing therefore poor synchronization; correlated percolation processes; and positive feedback mechanism (referred to system dynamics).

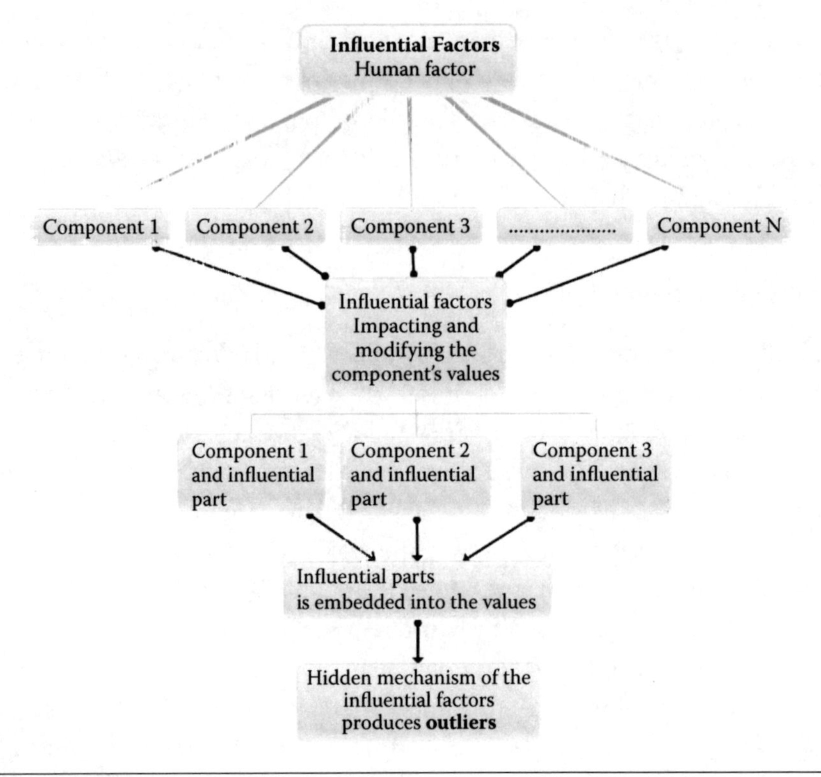

Figure 17.13 The influential part is the percentage of influence determined by the influential factors (human factor, normalization of deviance, etc.).

involved but also cultural impacts on the decision-making processes. There is a sociological explanation that explores much deeper the root causes concerning the engineering design failure of the O-rings, which has been attributable the NASA Challenger accident (Leveson, 1986,Vaughan, 2009). To capture the slow and constant process in which system vulnerability exhibits itself, a synchronized reinforcing processes regarding the interaction with human factors, groups, technology, culture, economic climate, etc., must be considered. However complex systems develop behavioral patterns in ways that are not recognized in the statistical distributions, and there is a hopeful pervasiveness to intercept predictive signals through experimenting different paths.

Predictions beyond the Data and Interrelationships

Is there any hidden information beyond outliers' data?

Dragon Kings may have a significant predictive value. Nevertheless analysis on historical data can be conducted to detect any type of outliers by reverse engineering and root-cause analysis of the issues. Outliers are viewed as sensors to refer to, when identified, the hidden mechanisms within the interrelationships that are impacting portfolio performance metrics. The hidden mechanisms have been ascribed to influential factors; i.e., human factors. Subsequently a number of outliers impacts all the portfolio items directly and indirectly; for example, cost, schedule, and benefit are influenced by it. Cost, schedule, and risk distributions have fat tails and are detected by large amount of variability or discrepancies between median and means or both, and variability and discrepancy are indicators for the presence of outliers. At the same way discrepancy between the mean and the median indicates skewness and the length of the tails of distributions (Agresti & Finlay, 2008). Depending on the exponent "α", of the distribution, for example, it also does not converge for a large value of kurtosis, skewness, and variance; hence, the distribution exhibits a fat tail, where the concentrated outliers exist. In projects and other components (for which crude measures prove the findings), cost, schedule, and risk have fat tails implying that outliers are in the center of the distribution, and this means that then in the "normal values" there are plenty of outliers. Consequently the strategic importance of information has not been adequately taken into account; i.e., organizations are blind while they are steering the performance of their projects, as in the case of the Challenger disaster (Budzier & Flyvbjerg, 2013). This would sound sobering considering that firms on average manage $200 million in projects each year. During that time these organizations will realize that 74 million of their projects are at risk of failing" ("Strategies for Project Recovery," 2011, p. 2).

Undoubtedly the theory of Dragon Kings, for which its applicability in other disciplines is still in a developing stage, suggests that most of the extreme events are not ascribable to the concept of "*black swan*", because they are preceded by warning signals, exhibiting a degree of probabilistic predictability where small overruns foreshadow the large one. Dragon Kings, in one of the application fields of the theory, the financial market, financial bubbles were revealing certain maturation toward instability in which a series of positive feedbacks loops (in system dynamics sense) between investment strategy and traders led to an exponential growth that may culminate in a crash. That could be a simple disruption for a generic system or an epileptic attack in the case of epileptic seizures observations. As already pointed out, the financial crises are considered only as examples, in particular in this case where the implementation of the theory has interesting implications but outside the current view of this chapter. Dragon Kings, hunting, and related theories can be successful mainly in detecting approaches in the area of composite material. The theory showed valid predictions on reliability of the whole composite-based structures, for example, when material rupture occurs. Various components reach their maturity to plunging/going toward obsolescence, for which it is empirical to imagine how a constant state of regime may affect the original characteristics of the component or rather than the material as in this case. From the engineering perspective it is possible to refer to several expository situations: for instance the Gaussian curve can describe the various stages of the entire cycle of material performance, based on picking samples at various stages and therefore is able to predict if a replacement or if a maintenance schedule is needed. The main objective of this analysis is to provide different temporal and operation phases for the observation of the samples and consequentially collect the relevant data regarding the status of their conditions when ruptures occurs gradually. Yet there are other situations in which rupture is unpredictable because of changed operating conditions. It recurs to altering operating parameters, for temperature, pressure, or manufacturing as in the case of the "Stress Corrosion Cracking" for material rupture. Power law distributions have a historical bound with engineering practice: to detect and predict ex ante when a disruption would occur or estimate distribution over expected losses. They are a long and traditional approach in risk characterization. Some engineering aspects, however, do not provide large dependencies to the human factor impacts unless they involve managerial decisions. The greatest danger is represented instead by cutting costs. If cost reduction is excluded, however, the inherent linearity of the engineering and design promotes more issues related to reengineering of events and thus the ability to understand the hidden causes behind the failures.

Undoubtedly the main aspect of Dragon Kings theory lies in the versatility of its assumptions that can really be applied on a large scale and in a variety of fields and disciplines and applications as well as with some extra assumptions and precisions. Beyond this new conceptual angle there is also the attractiveness of integrating dragon theory for detecting those parameters that cause asynchrony or positive feedback loops (in system dynamics sense) in the system (portfolio management in this case) and assuming that those parameters are greatly influenced by the human factors or more in general by the influential factors discussed in the last paragraphs. While the relationships in the case of engineering are of binary nature, in all cases in which the human factor is involved, high complexity exists among the interdependencies in all the items of the portfolio. The qualitative model is framed, therefore, in a dynamic structure in which every single agent acts iteratively toward the alignment improvement, analyzing first, the predictive signals coming from the detection of the outliers, and afterward by cross comparing the performance metrics with the data from which outliers have been extracted or generated, after being filtered from the human factor disturbance. The human factor and its influence impact on the former data vulnerability of the organization uses self-regulated data feedback to serve the purpose of filtering and cleansing data. After this procedure it will be possible to analyze data distribution with the outliers that may be present (Figure 17.14). The dragon theory's usefulness for this purpose is represented by considering it as a conceptual tool of a cognitive survey to investigate the interrelationship among the portfolio items. Clearly a general framework exists from this approach, and it is the desire to increase the resilience of

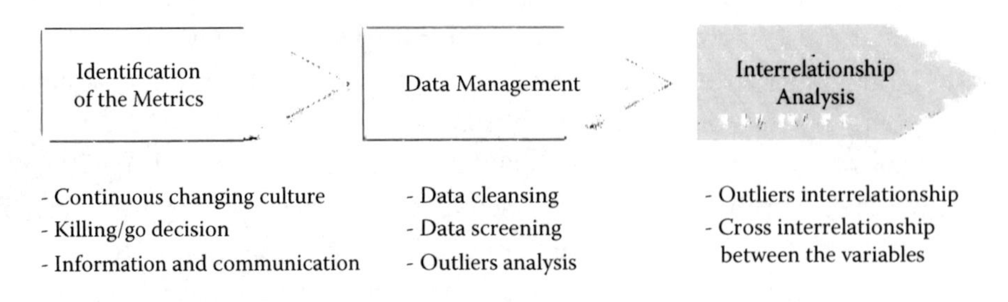

Figure 17.14 Outliers interrelationships analysis.

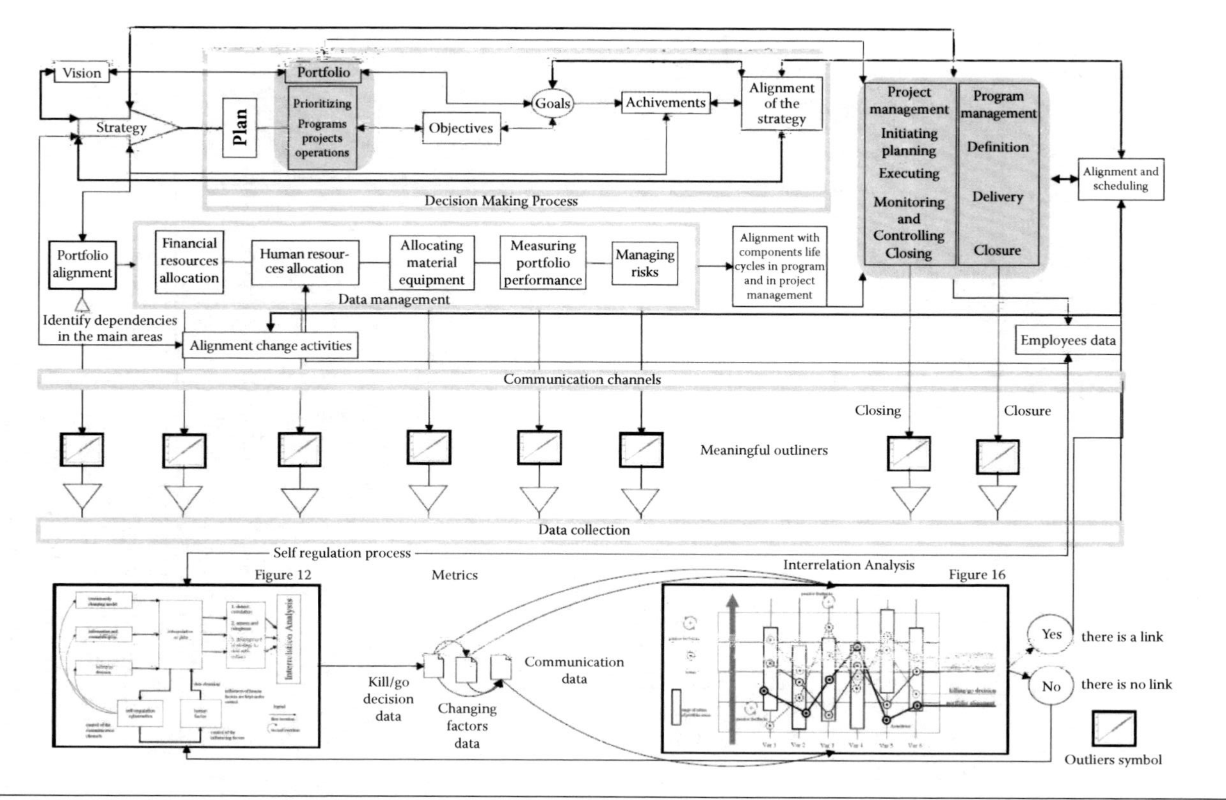

Figure 17.15 Portfolio framework of the qualitative model.*

* In the qualitative model map as shown in Figure 17.15, three different types of alignment are described respectively for scheduling, for strategy, and for change activities. The first area is the components scheduling in which the queuing time and priorities of the various components' tasks are time-dependent and must be aligned; alignment is constantly updated with the strategy, with the alignment to be identified during the change activities. In addition the alignment interacts with the six main areas (areas defined in PMI, 2013, p. 9). For each area data are collected and required information are extracted based on the explanations in the paragraphs regarding outliers analysis. See the procedure described in Figure 17.12, the self-regulation of the communication. Ultimately the choice of the metrics allows the analysis of the correlation, as shown in Figure 17.16.

organizations and systems, tightening efforts to recover an obvious lack of immediate responses and reactions by the governance structure if the same accountability should increase, see Figure 17.15.

If actually behind the misalignment, the product's failure or disruption of a system or of the portfolio components, reside helpful symptoms to emphasize endogenous nature of the vulnerability of the system, it is worthwhile to investigate how to prevent these maturation effects toward instability affecting performance.

However, it is possible to capsize these findings with retroactive observations. If in fact, malfunctions or poor performance of the past were hiding distribution of events in which the presence of the Dragon Kings, not yet known, had revealed sources of an imminent harm, then this type of knowledge would have allowed changing the course of events that brought the system to collapse. It then follows with the same determination with which the system goes to its "ruin or death point" state, it sends strong signals that need to be interpreted according to the circumstances and the different levels of application. The theory partially explains that in the case of detection of these Dragon Kings "signals" what could be done is to act against the symptoms they present or determine whether there is more insight to bring concerning root-cause analysis. For example, in the case of the distribution of epileptic seizures, the patient would receive the maximum indication on when and how to take the necessary drugs to alleviate the symptom (which sometimes is a positive sign of an internal unease) instead of searching its etiology.* Statistical distributions are valid instruments and tools, but they need to be understood in tandem with many other considerations and observations on qualitative aspects.

Observations on the Approach

Outliers are stable and not random phenomena following power laws distribution (Budzier & Flyvbjerg, 2013) triggered by a different mechanism as for the Dragon Kings.

There are two hypotheses at the base of this approach; the first is that meaningful outliers are generated by a mechanism depending on the mutual inter-correlation between the portfolio variables. The second is that the human factor is an influencing element that continuously impacts this mechanism thus producing extreme data points.

If large events are attributed to a cascade of small and continuous changes through the time, then some evidence before "maturation" into a collapse or a system failure should be recognized before, thanks to the behavioral patterns. Then the meaningful data points will serve as a necessary tool to re-engineer a given problem. The new angle consists in deselecting the theory of the "gradual maturation". The sequence of events is constantly impacted by the influencing factors, and its dynamics are therefore distorted.

* The term derives from the Greek language etiology (cause = aitia and logos = word / speech) and is used in medicine, law, philosophy, physics, theology, biology, and psychology in relation to the causes that produce phenomena.

This new perspective suggests that impacts of the influencing factors (human factor, normalization of deviance, etc.) should be first suppressed, as explained in Figure 17.12, before starting the analysis of data according to the interrelationship analysis phase (Figure 17.14).

After the suppression of the influencing factors, an analysis over the issue's phenomenology should start. Therefore paths that indicate a mutation state in place into another, toward instability, should be better recognized within the different functions and knowledge areas, thus acting to obviate any damage. There are also premises to drive the development toward a given direction and strategies toward the targets/objectives, as well as reinforcement where there is a lack of cohesion, performance, and communication in project portfolio management. The three metrics represent the general performance of the portfolio in the sense that their variability is associated with the variability of increased or decreased values of the quantitative variables linked to the portfolio (cost overruns, schedules, time, etc.).

The major obstacle is the amount of data required from many different managerial areas: information on the financial results, resource requirements, timing, probabilities of completion and success for all projects and other components, and data regarding employees, etc. Much of this information simply is not available, and, when it is, its reliability is suspect (or incomplete). Further, these mathematical portfolio approaches historically have provided inadequate treatment of risk and uncertainty; therefore, a high maturity level is a priority for data management.

In addition it can be assumed that behind the mechanisms that led to a malfunctioning, overruns, disruptors, etc. of a given system, may be found within historical data. In this case the information has strategic importance and needs to be extrapolated for lessons learned. Anomalies in data sets among the historical data will be analyzed to check out the existence of patterns, which can be applied to events that have not yet occurred. This property should then be exploited to understand recurrent behaviors that can be used to prevent systemic issues and therefore reduce the risk of misalignment.

There is a strong belief that root causes that generate the relative misalignment in the portfolio may be discovered through the depiction of these interdependencies. Of course this theory needs to be experimented with crude measures as well.

The occurrence and impacts of outliers depends and differentiates the portfolio at the knowledge areas and at any stage of the development processes. It has been highlighted how the failures exhibit themselves at a certain phase of the delivery path of the process; nevertheless this point is not necessary the point in which the issue originated and then causes the disruption. So the phenomenology of the failures can be expressed and represented by:

- Components "failure"
- Impacts from projects or programs selection/elimination on the interdependent components
- Impacts of indirect factors on overall portfolio components

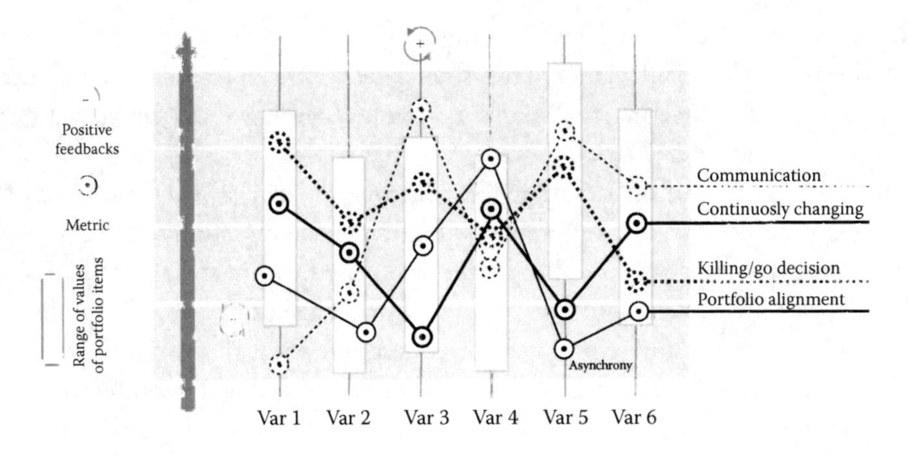

Figure 17.16 Global view of the dynamic of the interrelationships between related portfolio items and performance metrics.

The interrelationship analysis between the various agents involved with the portfolio management is the key process to investigate and to unveil the hidden mechanism behind the causes of misalignment or failures. It is the last part of the approach.

In the global view example, Figure 17.16, the three metrics chosen, continuously changing, killing/go decision, and communication, are constantly impacted by the system variables (that can be multiples, here just six for example purposes). Their variability affects the portfolio performance. The variables are different within their range of values, and they may exhibit during a chosen time frame of observation (e.g., a quarter, a month, etc.). But they can also exhibit extreme values, outside the range of values. When some of these outside or outstanding values reflect a set of outliers values of the metrics, then they need to be analyzed with the cross dependencies between all the portfolio components to trace eventually, similar behavior, proportionalities, root causes, and significant variances, etc.

Potential Application Areas

Variable distributions in portfolio management for further cross-interrelationship analysis research may be taken from the following items:

From decision-making processes:

- Market data
- Customer satisfaction
- New trends
- New products, new technologies
- Mega events (natural disasters, Olympics, conventions, etc.)
- Emergency management

Data from alignment:

- Costs
- Time to finish
- Time scheduled
- Competitiveness
- Iterative processes detecting new objectives
- Re-alignment of strategy and tactics

Data from employees:

- Distributions of skills
- Level of confidence
- Employee historical performance data
- Ergonomic – from design to usability
- Cybernetics (formal and informal communication)
- Study "people at work"
- Anthropometric data
- Mental efforts/errors
- Skilled, rules, and knowledge-based behaviors issues

Conclusions

This chapter has addressed the problem of alignment within complexity. Often the alignment is a result of random variables control and sequences of, decision-making process interacting with each other without a clear vision of the current status. It has highlighted and emphasized aspects of intentional misleading issues, linked to the human factor; technical aspects of incompatibility of the enterprise architecture unable to sustain and effectively analyses the amount of data ever increasing; analyze philosophical aspects linked to the value, ethics, and participatory accountability of public life; and environmental aspects related to the locus of the organization. The approach shown introduces an element of control and an element of interpretability of data. Decisions are based on data that represent the grounds of strategic choices and actions. Any alteration (intentional or casual) of the data implies chain related consequences traceable with presence of outliers. Observations on these extreme values in general, discarded from analysis a priori or even ignored, are here ordered for capturing predictive signals and anomalies. They may reveal the existence of hidden mechanisms that links variables of the same system or subsystems. The intention to uncover interdependencies between metrics and portfolio variables is presented to observe the mechanism that generates outliers. Definitely the innovative angle consists in a unique approach (a method should follow up) that can detect the existence of cause-and-effects linkages between the portfolio items, events, and influencing factors that may alter the alignment from fine grained to the executive level. These events may represent opportunities as well as risks.

Neglecting outliers may cause impacts to the organization, in terms of losses in profit, lack in performance, missed opportunities, cost overruns, and failures, etc. It has been highlighted as well, that reengineering historical data to depict behavioral patterns may represent a significant contribution for reducing risks and capturing predictability.

Major issues at the portfolio level are still perpetrating although new technologies and data base management will be further developed. The sole financial metrics are not sufficient to establish well-defined parameters for making the right decisions in the selection process. This approach adds as well a strong qualitative value to sustain the standard financial methods in portfolio decision making. Therefore to better achieve a strategic alignment and enhance the representativeness of the metrics chosen, a method for cross-relating the variables has been presented. Undoubtedly the chapter presents a new theory that leaves room for further developments creating new streams of research and applications for which its theoretical foundations must be highlighted empirically with organizational data. At the portfolio level, data flows in rivers, and the only prerequisite is an organization with a medium-high innovation maturity level.

References

Agresti, A., and Finlay, B. (2014). *Statistical methods for the social sciences* (4th Edition). Harlow, Essex, UK: Pearson Education Limited.

Appleseed Partners and OpenSky Research. (2013). *Fourth product portfolio management benchmark study*. Planview. Retrieved from: http://www.planview.com/company/press-releases/introduces-innovation-management-maturity-model/

Bak, P. & Tang, C. (1987). Self-organized criticality: an explanation of 1/f noise. *The American Physical Society*, *59*(4), 381–384.

Budzier, A. & Flyvbjerg, B. (2013). Making-sense of the impact and importance of outliers in project management through the use of power laws. In *IRNOP Proceedings* (pp. 1–28). Oslo, Norway.

Christianson, M. K., Farkas, M. T., Sutcliffe, K. M., & Weick, K. E. (2009). Learning through rare events: Significant interruptions at the Baltimore & Ohio Railroad Museum. *Organization Science*, *20*(5), 846–860. Retrieved from http://pubsonline.informs.org/doi/abs/10.1287/orsc.1080.0389

Cooper, R. G., Edgett, S. J., & Kleinschmidt, E. J. (1998). Best practices for managing R&D portfolios. *Research-Technology Management*, *1*(1), 20–33.

Cooper, R. G., Edgett, S. J., & Kleinschmidt, E. J. (1999). New product portfolio management: practices and performance. *Journal of Product Innovation Management*, *16*(4), 333–351.

DePaulo, B. M. (1992). Nonverbal behavior and self-presentation. *Psychological Bulletin*, *111*(2), 203–243.

Dixon, W. J. (1950). Analysis of extreme values. *The Annals of Mathematical Statistics*, *21*(4), 488–506.

Gersick, C. J. G. (1991). Revolutionary change theories: A multilevel exploration of the punctuated equilibrium paradigm. *Academy of Management Review*, *16*(1), 10–36.

Hawkins, D. M. (1980). *Identification of outliers*. London: Chapman and Hall.

Kay, J. (2012). Obliquity. *Capitalism and Society*, *7*(1). Retrieved from http://journals.psychiatryonline.org/article.aspx?articleid = 181052

Lampel, J., Shamsie, J., & Shapira, Z. (2009). Experiencing the improbable: Rare events and organizational learning. *Organization Science*, *20*(5), 835–845. Retrieved from http://pubsonline.informs.org/doi/abs/10.1287/orsc.1090.0479

Leveson, N. G. (2008). Technical and managerial factors in the NASA Challenger and Columbia losses: Looking forward to the future. In Kleinman, D.L., Cloud-Hansen, K.A., Matta, C., Handelsman, J. (Eds.), *Controversies in science & technology: From climate to chromosomes.* New York: Mary Ann Liebert, Inc.

Levy, D. L. (2000). Applications and limitations of complexity theory in organization theory and strategy. In Rabin, J., Miller, G.J., Hildreth, W.B. (Eds.), *Handbook of strategic management.* Second Edition. Boca Raton, FL: CRC Press.

Osborne, J. W. & Overbay, A. (2004). The power of outliers (and why researchers should always check for them). *Practical Assessment, Research & Evaluation*, *9*(6), 1–12.

PM Solutions. (2011). *Strategies for project recovery*. Retrieved from http://www.pmsolutions.com/collateral/research/Strategies for Project Recovery 2011.pdf

Project Management Institute. (2013). *The standard for portfolio management.* Third Edition. Newtown Square, PA: Project Management Institute.

Remorov, R. (2014). Stock price and trading volume during market crashes. *International Journal of Marketing Studies*, *6*(1), 21–30. doi:10.5539/ijms.v6n1p21

Romanelli, E. & Tushman, M. L. (1994). Organizational transformation as punctuated equilibrium: An empirical test. *The Journal, Management*, *37*(5), 1141–1166.

Selznick, P. (1948). Foundations of the theory of organization. *American Sociological Review*, *13*(1), 25–35.

Simon, H. A. & Bonini, C. P. (1958). The size distribution of business firms. *The American Economic Review*, *48*(4), 607– 617.

Sornette, D. (2009). Dragon-kings, black swans and the prediction of crises. *SSRN Electronic Journal.* doi:10.2139/ssrn.1470006

Stevens, J. P. (1984). Outliers and influential data points in regression analysis. *Psychological Bulletin*, *95*(2), 334–344.

Taleb, N. N. (2010). *The black swan: The impact of the highly improbable fragility.* New York, NY: Random House.

Taleb, N. N., & Tetlock, P. E. (2013). On the difference between binary prediction and true exposure with implications for forecasting tournaments and decision making research. *0*, 2–7. Retrieved from http://ssrn.com/abstract = 2284964

The Economist. (2007). In search of clarity Unravelling the complexities of executive decision-making. *Economist Intelligent Unit.*

Vaughan, D. (2009). *The Challenger launch decision: Risky technology, culture, and deviance at NASA.* Chicago: University of Chicago Press.

18

Delivering Organizational Value in the Zone of Uncertainty

DR. LYNDA BOURNE, DPM, FACS, FAIM

The dynamics of the global economy and the increased complexity of delivering value to stakeholders have had consequences for organizations. Uncertainty about complex and ever-changing roles, and uncertainty about what success means has increased for individuals. Until recently there was a view that managing an organization's work could be made relatively straightforward—if only we could get the processes and controls right! The formidable mixture of multiple developments and operational activities at differing stages of delivery were recognized as obstacles that could be overcome with more and better planning and controls. In reality, processes and controls are only part of the story: without an understanding of the nature and sources of the complexity we all encounter, there will be much wasted effort.

This complexity, borne of uncertainty and unpredictability, affects the ability to develop, implement, and control an organization's portfolios, programs, and projects to deliver value to the organization: its source is a combination of technology, organization processes introduced to 'balance' the organization's work, and relationships between stakeholders. This chapter will focus on the third of these factors: management of the relationships between the organization and the many stakeholders who can affect or who are affected by this work.

The more the organization 'balances' the portfolio through selecting work to provide a broader range of opportunities for organizational value, the more complicated the mix of knowledge, action, and relationships becomes. These are all in the domain of the *stakeholder factor*. People apply knowledge (and misconceptions), people act (and make decisions), and people must work with others to achieve this value (relate). It is people—*stakeholders*—who make the design decisions and purchase decisions; people are responsible for the approvals and prioritizations. It is people who may or may not comply with, or who misinterpret, the processes; and it is people who implement the solutions. At any step in the course of following the organization's prescribed procedures and processes, the intention of the strategy may be diverted (or subverted) through conflicting goals, lack of understanding of the importance of following a process, or just plain apathy. People are the major risk: managers who do not properly support the implementation of decisions they make, or managers who do not allow their program or project managers sufficient time to monitor, modify, and manage to the strategies and plans developed in the portfolio management process.

The principles and processes, plans, and measures that are put in place in organizations to support effective management of portfolios, programs, and projects are important not only for consistency of approach but also to help build a culture within the portfolios through common symbols—language. (Hofstede, Hofstede, & Minkov, 2010). The effectiveness of the application of these fundamental building blocks is enhanced through focusing on the stakeholder relationships.

The chapter is organized as follows: first a description of the 'Zone of Uncertainty'—the optimistic view and the reality. The next section will describe potential remedies for the disruption and disappointment that occurs in the Zone, principally related to improved stakeholder relationship management. Any discussions of portfolio are based on this simple but succinct definition (Microsoft, 2008) that portfolio management is ***'doing the right things'***—effectiveness; ***'doing things right'*** is execution enabled by project management/work management (efficiency).

The Zone of Uncertainty

The 'Zone of Uncertainty' is the 'murky' area between the level of governance and strategy that defines the executives' view of how to achieve business success, and the projects or other work that deliver value to the organization. In the Zone the hopes and dreams of the organization's strategists collapse under the weight of the complications and complexity of turning vision into value, characterized by the perception of lack of control, lack of certainty, and lack of direction.

The Optimistic Approach

As part of the regular business planning cycle and in response to current environments or expected future environments, an organization defines its business strategy for a particular interval. From that strategy emerges clear strategic objectives, which will then be converted into portfolios, programs, and projects. The senior stakeholders' expectations, and those of all but the most cynical, are that going from A to B will be straightforward—all that is needed is the perfect combination of the right approvals, structures, and processes in place (command) and then measures and reporting against those measures (controls). The transition from the business strategy to the output of the project occurs in the Zone.

The Zone is the highly complex and dynamic region set between an organization's strategic vision and the projects created to deliver that vision. It includes all the initiatives introduced by senior (or middle) management to deliver value. Zone activities include the development of portfolio and program management to ensure that all work done by projects:

- Delivers the organization's strategy in a balanced way,
- Is appropriately funded and resourced, and
- Is tracked to deliver the benefits outlined in the business case.

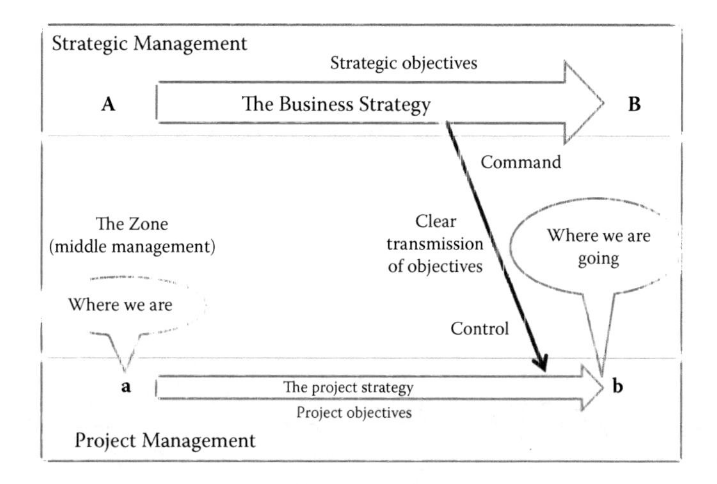

Figure 18.1 Idealized view of how projects will deliver the organization's business strategy.

Figure 18.1 describes the idealized (linear and predictive) view of the Zone. It shows how, theoretically, management's vision has clear direction and clear outcomes, unclouded by their own culture or perceptions, and without any divergence or conflict; and how, through clear transmission of strategic (organizational) visions into tactical (project) objectives, projects will be delivered to the required time, cost, and quality. The assumption built into this concept is that it is possible to develop the perfect schedule, resource it effectively, and implement the plan without deviation. A further assumption is that there will be a universal agreement from all stakeholders that the chosen path is the best and only path for successful provision of value to the organization. A third and more dangerous assumption is that the 'truth', depicted in the schedule will only vary slightly, and is an accurate representation of what will occur.*

The 'Zone of Uncertainty'—The Reality

Figure 18.2 shows the more likely picture of what will actually happen in the Zone. Management's expectations remain unwavering,† but the outcomes are not so predictable. The impact of the change that the project is to deliver and other events within the organization and outside of it may result in the expectations of stakeholders not being delivered. The simple direct line of assumption between the executive's articulation of strategic objectives, and the strategic alignment between these objectives and the project that is approved and resourced will most likely deviate. The certainty of the 'truth' of the schedule is affected by unexpected events, and senior stakeholders and the project team will react to try to regain control of the delivery of the objectives.

* This is the myth of the 'truth' of the project schedule and the assumption that it is possible to predict the future and build stable detailed plans of how the outcomes of the project will deliver the business strategy.

† Their attention is now on other visions and strategies, believing that the process will automatically proceed without any more senior management intervention.

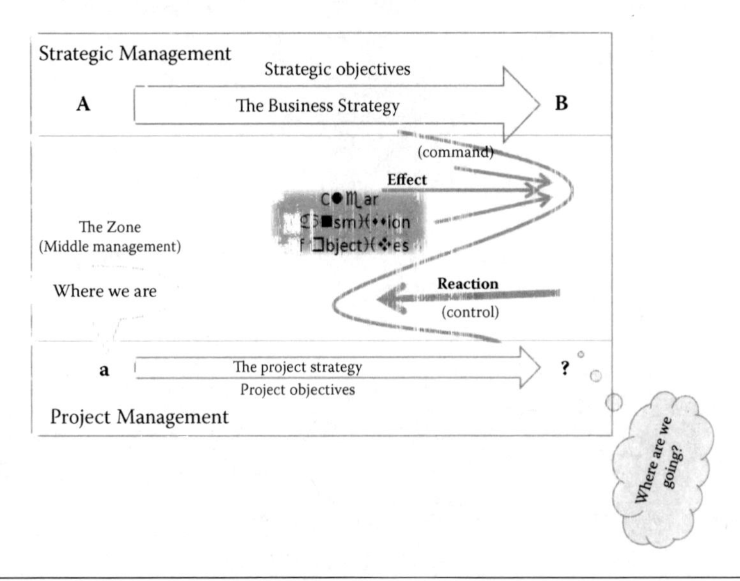

Figure 18.2 A more realistic view of the transition from business strategy to project.

Often these adjustments cause more disruption within the project and its relationships with other projects as well as portfolios and programs.

The Project Management Institute's (PMI) *A Guide to the Project Management Body of Knowledge (PMBOK® Guide)* (2013, p. 7) defines these relationships among portfolios, programs, and projects:

> Portfolio management aligns organizational strategies by selecting the right programs or projects, prioritizing the work, and providing the needed resources; whereas program management harmonizes its projects and program components, and controls interdependencies in order to realize specified benefits. Project management develops and implements plans to achieve a specific scope that is driven by the objectives of the program or portfolio it is subjected to, and ultimately, to organizational strategy.

As stated earlier in this chapter the processes and practices that define how best to do portfolios, programs, and projects from the perspective of the project management professional bodies are well defined and provide guidelines on the 'what must be done'. The process of delivering project outputs or delivering value to the organization through portfolio management assumes adherence to these processes. Success could not be assured without taking into account the *stakeholder factor*. Value can be delivered to organizations through the application of the discipline of compliance to process: application of and compliance with, consistent sets of processes have delivered value in the form of reduction in rework within software engineering projects.* This leads to improvement of customer satisfaction and return business. It is still, however, important to read into the words 'application' and 'compliance' the notion that the practitioners need to be trained,

* This information was obtained in conversation with a Capability Maturity Model Integration (CMMI) assessor, supported by a view of a confidential report, prepared by a software development organization and shown to me with approval of the client.

and managers need to recognize the importance of investing in the implementation, training, and measurement of the implementation of these processes sustainably.

Complexity

There is no doubt that the work of portfolio, program, or project management is complicated:* and often complex.† An organization's work will always be complicated—what will make it complex is the combination of technical complexity, the specific selection and management of work in order to balance portfolios, and the web of relationships with the stakeholder community in the environment of unpredictability.

We live with unpredictability, often unconsciously, every day and every minute. We do not know what is going to happen next; we take it on faith that the sun will rise tomorrow and that our plans for tomorrow will be able to be executed. The best example of this facility we have of ignoring the reality of not being able to predict the future is the project, program, or portfolio *schedule*. The schedule is developed by the team, including experts, but is at best just a 'guesstimate' of the best way to deliver the outcomes within the time frame, budget, and the available resources to achieve the greatest 'value' for the stakeholders or the organization. Despite the best intentions of all concerned, and even with the experience of others who have done similar work before, the plan assumes that there will be only a few complications, and the contingencies and other risk responses planned in will be sufficient. But we know that unexpected events will occur, and the carefully developed plan will need to be re-thought time and time again. This is normal. What is not normal is when stakeholders 'believe' the plan: believe that it is TRUTH. Managing the unrealistic expectations of the stakeholders based on this belief is a key part of stakeholder relationship management. The area between the expectations of how the organization must plan, operate, and control all the work and each project, program, and operational activity is a minefield of unpredictability and complexity—so expectations that the planned actions and changes will occur in a linear fashion will inevitably lead to failure (in the eyes of someone).

The complexity of human relationships is seen in situations where:

- People with different interests, loyalties, cultures, and interactions with one another are put together to deliver something:
  - Teams come together causing unpredictable behaviour particularly when there are differences in national organizational or professional culture
- Differences magnified by different perspectives of the people involved:
  - Those who implement, those who use the products, those who benefit from the outcomes of the project, or those who establish a regulatory environment

* Complicated = a large number of interconnected and interdependent parts.

† Project work is complex if it consists of many interdependent parts each of which can change in ways that are not totally predictable and which can then have unpredictable impacts on other elements that are themselves capable of change (Cooke-Davies, Crawford, Patton, Stevens, & Williams, 2011).

- Clients cause scope changes or delay important decisions
- Managers react inappropriately to cost, schedule, scope, or quality pressures (Cooke-Davies, 2011).

Stakeholders

Anyone who has ever had to engage the stakeholder community of a project, program, or portfolio will be aware of the unpredictability of each individual, and the additional complexity of understanding which groups or individuals are important and supportive are complicated. This section will discuss the fundamentals of understanding and engaging stakeholders.

The concept of 'stakeholder' existed long before management writers (in the West) took up the cause and adapted the word and concept for an organizational purpose. www.dictionary.com defines 'stakeholder' in the following ways:

1. The holder of the stakes of a wager (this was the original meaning of the word).
2. A person or group that has an investment, share, or interest in something, as a business or industry (this is now the generally accepted usage in the English-speaking business world).
3. In Law: a person holding money or property to which two or more persons make rival claims.

The meaning and concept of 'stakeholder' in the English-speaking world is hazy, with no common agreement on who are stakeholders. How something is defined and how it is expressed in language gives a very good indication of what the word or concept actually IS in that culture. What stands out to me when considering the various translations of 'stakeholder' is that in each culture the translation points to quite specific ideas about stakeholders. When the idea or need of managing stakeholder relationships is imported into the non-English speaking world, the translation of 'stakeholder' magnifies the haze.* From discussions with colleagues from different language backgrounds:

- In Spanish: tenedor de apuestas (holding the wager) and partes interesadas (interested parties)
- In French: *Des parties prenantes* (parties who are taking) and *intervenantes* (intervening)
- In German: *Beteiligten* (involved) and *Anspruchsgruppen* (who have a claim)
- In Dutch: *belanghebbenden* (having a stake)

* In interviews with managers in organizations in the Spanish-speaking countries in South America, I discovered that 'stakeholder' translated has many meanings, often focussing on just one attribute rather that the more inclusive definition becoming more accepted in the English-speaking world. This led to my informal enquiry of my international colleagues regarding how 'stakeholder' was translated.

- Paul Dinsmore (Dinsmore, 1999) referred to 'stakeholders' as the '*ones who have the beef*'. This seems to be a consistent view of 'stakeholder' in Brasil.*
- In Japan: 'stakeholders' are 'related people' or 'people sharing risk and profits.'†
- In China: 'stakeholders' are 'participants with related interest'.‡

From this brief analysis it is possible to conclude that there is no innate or generic meaning for stakeholders, and so any discussions about stakeholders in organizations needs to begin with a definition. This definition comes from PMI (2013, p. 563).

> An individual, group, or organization who may affect, be affected by, or perceive itself to be affected by a decision, activity, or outcome of a project.

The strength of this definition of stakeholders is that it not only takes into account the ubiquity of stakeholders but also the diverse functions that they may have in the organization, making it almost impossible to ignore any individual or group that fits the definition of stakeholder. An additional strength is the focus on perceptions as a characteristic of 'stakeholder'. Perceptions and expectations of stakeholders are one of the most important factors in successful stakeholder engagement and therefore to the organization's outcomes. Understanding what each key stakeholder expects to gain (or lose) from the outcomes of the work (either its success or failure) is essential to the success of the work and the perceptions of the stakeholder community.

A stakeholder has a 'stake' in the activity, portfolio, program, or project. It is important to consider the nature of a stakeholder's stake when defining a stakeholder's needs, requirements, or how the individual or group can impact the organization's activities. This stake may be:

- An interest: a circumstance where a person or group is affected by a decision, action, or outcome.
- Rights: legal—as enshrined in legislation or moral—environmental, heritage, or social issues.
- Ownership: legal title to real property, intellectual property, or a worker's reward for his or her experience or knowledge.
- Contribution in the form of knowledge or support.

A Methodology

It is not possible to measure the expectations or perceptions of people objectively and foolish to guess or assume. It is only possible to measure trends or changes in satisfaction or support. A consistent approach—a methodology—for tracking any changes is

---

* I presented at conferences in Brazil in 2009 and 2011—this is a consistent theme of all discussions with project managers and Project Management Office (PMO) practitioners in that country.

† I had the opportunity to speak to a group of project managers in Japan in January 2013, and asked the question of how 'stakeholders' were translated in Japanese at that time.

‡ From e-mail correspondence with Bob Youker, previously working for the World Bank, now retired.

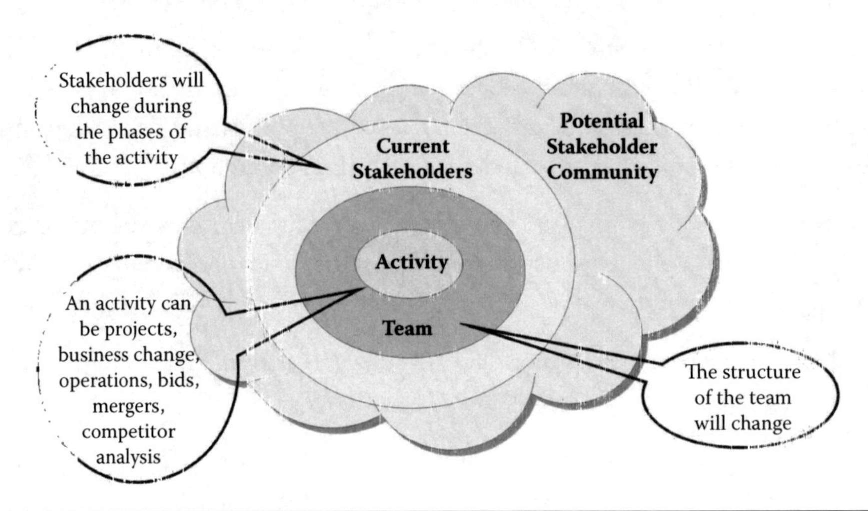

Figure 18.3 Stakeholder relationships.

critical. This methodology will enable recording of significant elements, such as support to provide a baseline, and then, through data gathering in a consistent manner, it is possible to see what has changed in these elements—improvement, or otherwise, in support for example will be a good indicator of whether the team's efforts at engagement have been successful.

The ***Stakeholder Circle***® methodology is based on such a concept. Figure 18.3 shows the relationships between the activity* and its stakeholders. All decisions or understanding of the relationships are made from the perspective of the manager and the team. Surrounding the activity itself is the team, often overlooked in many stakeholder engagement processes. Surrounding the team is the community of stakeholders that has been identified as being important to the success of the activity at the current time—'*time now*' in the life of the activity. The outermost circle references potential stakeholders: those who may, or will, be important to the success of the work at a later stage. By differentiating current stakeholders and potential stakeholders in this way, confusion about which stakeholders are important at any particular time and how best to manage the current relationships will be minimized, while ensuring that planning for future relationships is managed effectively. The stakeholders in the outer circle must also be considered in risk management planning because they may cause the activity to be at risk of failure in the future. Alternatively, these stakeholders may need to be considered in an organization's marketing plans, as potential customers.

Managing Stakeholder Relationships

The ***Stakeholder Circle*** is a five-step methodology based on the concept that any activity can only succeed with the informed consent of its stakeholder community, and managing

* The concepts defined in this paper and the methodology applies to ALL activities that an organization approves, resources, and funds to achieve its strategies and goals.

the relationships between this community and the team will increase the chances of success. The team must develop knowledge about this community and appreciation of the right level of engagement. This information will help define the appropriate level and content of communication needed to influence stakeholder's perceptions, expectations, and actions. Although consistency in approach to analysis of the stakeholder community is helpful, it is also necessary to emphasize that stakeholder relationship management is complex and cannot be reduced to a formula: each person is unique, and the relationships between people reflect that uniqueness and complexity.

The methodology consists of five *steps*:

- *Step 1*: identification of all stakeholders;
- *Step 2*: prioritization to determine who is important;
- *Step 3*: visualization (mapping) to understand the overall stakeholder community;
- *Step 4*: engagement through effective communications; and
- *Step 5:* monitoring the effect of the engagement.

Step 1: Identify This consists of three activities:

- Developing a list of stakeholders;
- Identifying *mutuality*:
  - How each stakeholder is important to the work, and
  - What each stakeholder expects from success (or failure) of the project or its outcomes (his or her expectations)
- Categorize: document each stakeholder's *Influence Category*:
  - *upward* (senior stakeholders), *downward* (the team), *outward* (public, government, suppliers, shareholders, etc) and *sideward* (peers of the manager of the activity)

The output of this step will be a list of all stakeholders that fit the definition of stakeholder.

How many stakeholders? Beware of 'STAKEHOLDER MYOPIA'!

Some organizational activities are large and complex and may affect many stakeholders. For example, construction of public facilities or national infrastructure projects will affect private citizens, landowners, and the natural and historical environment. For such projects, it is essential to recognize and accept that there will be large numbers of stakeholders. There is often an unconscious boundary on what a 'good number' of stakeholders can be—this is *stakeholder myopia.* It is important for the team and for their managers to understand that while the initial number of stakeholders identified may appear unwieldy or overwhelming, *Step 2: Prioritize* provides a structured and logical means to prioritize the key stakeholders for the current time.

Step 2: Prioritize Most stakeholder management methodologies rely on an individual's (or the team's) subjective assessment of who is important. The approach adopted in the ***Stakeholder Circle*** methodology attempts to provide consistency in decision making about stakeholders. It does this through a structured decision-making process where team members agree and rate the characteristics of stakeholders to assess their relative importance. *Step 2: Prioritize* provides a system for rating and therefore ranking stakeholders. The ratings are based on three aspects:

- *Power:* the power an individual or group may have to permanently change or stop the project or other work;*
- *Proximity:* the degree of involvement that the individual or group has in the work of the team; and
- *Urgency:* the importance of the work or its outcomes, whether positive or negative, to certain stakeholders (their stake), and how prepared they are to act to achieve these outcomes (stake).

The team applies ratings to each stakeholder, for 1–4 for *power*, and *proximity*, (where 4 is the highest rating) and 1–5 for each of the two parts of *urgency*—*value* and *action* (where 5 is highest). By adding the ratings it is possible to develop a ranked list of stakeholders important to the success of the work at this time in its life cycle.

Why choose these prioritization attributes? The three attributes of *power, proximity, and urgency* are the essential elements for understanding which stakeholders are more important than others. The definition of *power* seeks to identify those who have power over the continuation of the work itself or no power at all. *Proximity* provides a second way of identifying how a stakeholder may influence the work or its outcomes, through regular, close, and often face-to-face relationships in influencing the outcomes of the work.† The immediacy of this relationship contributes to trust between members of the team, and more effective work relationships as the team members understand the strengths and weaknesses of those they work with on a regular basis. *Urgency* is based on the concept described in (Mitchell, Agle, & Wood, 1997) whose theory described two conditions that may contribute to the notion of urgency:

1. Time sensitivity: work that must be completed in a fixed time, such as a facility for the Olympic Games
2. Criticality: an individual or group feels strongly enough about an issue to act, such as environmental or heritage protection activists.

In the ***Stakeholder Circle***, *urgency* is rated through analysis of two sub-categories: the *value* that a stakeholder places on an outcome of the work, and the *action* that he or she is prepared to take as a consequence of this stake. The inclusion of *urgency* in the prioritization ratings balances the potential distortion of an organizational culture

---

* The definition of *power* will only depend on the joint agreement of the team making the assessment of the ability to cause change; it does not depend on any other definition.

† *Proximity* simply defines how stakeholders are involved in the work of the project or activity.

that identifies stakeholders with a high level of hierarchical power as most important. If *power* and *proximity* are the only measures, stakeholders such as the 'lone powerless voice with a mission', who can cause significant damage to successful outcomes if ignored, will not be acknowledged.

Step 3: Visualize—Mapping Complex Data The objective of every stakeholder mapping process is to:

- Develop a useful list of current stakeholders and assess some of their key characteristics;
- Present data to assist the team's planning for engaging these stakeholders;
- Reduce subjectivity;
- Make the assessment process transparent;
- Make the complex data collected about the stakeholders easier to understand; and
- Provide a sound basis for analysis and discussion.

Presenting complex data effectively will be directly useful to two important stakeholder groups: the organization's management generally requires information in the form of lists, tables, pictures, or graphics, whereas the project team responsible will need charts and graphics for analysis of the community to highlight potential issues. The mapping from the ***Stakeholder Circle*** fulfils all these requirements.*

The ***Stakeholder Circle*** Figure 18.4 is an example of how data gathered during *steps 1 and 2* of the ***Stakeholder Circle*** methodology is shown. In addition to this graphic the names of the stakeholders represented is printed alongside. Key elements of the ***Stakeholder Circle*** are:

- Concentric circles that indicate distance of stakeholders from the work of the activity;
- The size of the block represented by its relative length on the outer circumference, which indicates the scale and scope of influence of the stakeholder; and
- The radial depth of the segment indicates the stakeholder's degree of power.

Colours† help the interpretation and indicate the stakeholder's *influence category* relative to the activity: orange indicates an *upward* direction, green indicates a *downward* direction, purple indicates a *sideward* direction, and blue indicates *outward.*

Step 4: Engage The fourth part of the ***Stakeholder Circle*** methodology is centred on identifying engagement approaches tailored to the expectations and needs of these

* Reading the ***Stakeholder Circle*** map of stakeholders is best in colour. For examples of the outputs of ***Stakeholder Circle*** go to www.stakeholdermapping.com and http://www.stakeholdermapping.com/stakeholder-management-resources/papers/#P021 .

† Colours are not shown here: for a full colour representation refer to www.stakeholder-management.com or Bourne (2009).

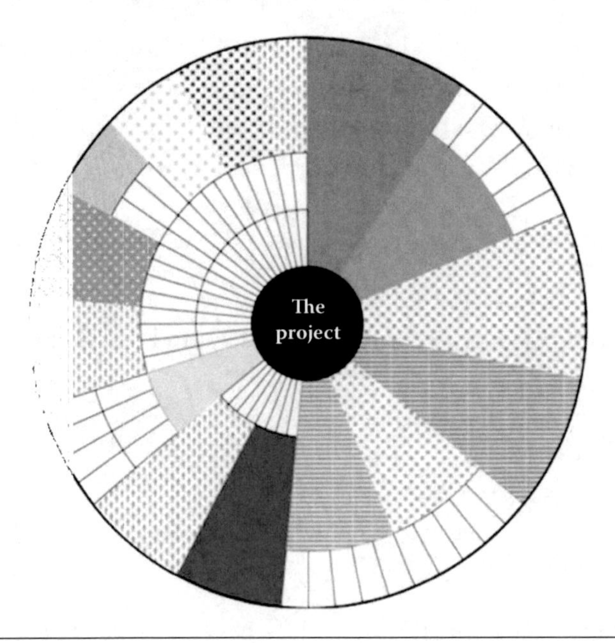

Figure 18.4 The Stakeholder Circle.

individuals or groups. The first step of this analysis involves identifying the level of interest of the stakeholder(s) at five levels: from committed (5), through ambivalent (3), to antagonistic (1). The next step is to analyse the receptiveness of each stakeholder to messages about the project: on a scale of 5, where 5 is – direct personal contacts encouraged, through 3 – ambivalent, and to 1 – completely uninterested. The third step is to identify the target attitude: the level of support and receptiveness to messages that would best meet the mutual needs of the project and the stakeholder. Figure 18.5 illustrates two stakeholders' engagement levels.

Step 5: Monitor Effectiveness of Communication The matrix illustrated in Figure 18.5 becomes the engagement baseline and starting point for measuring communication

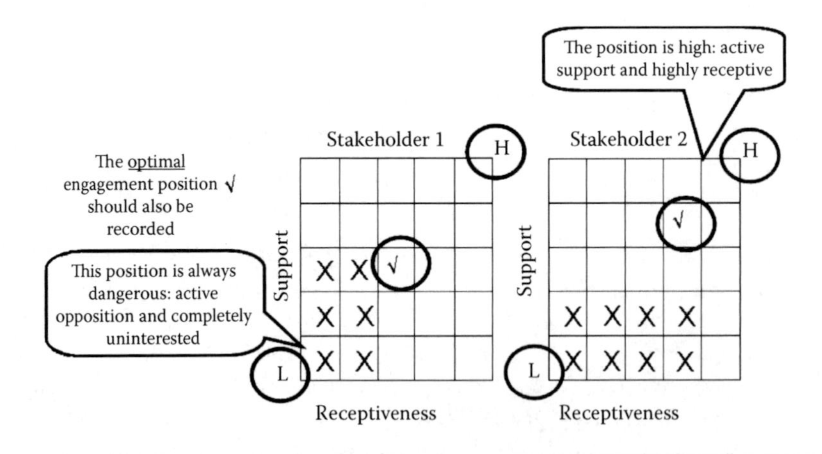

Figure 18.5 Stakeholder engagement profile.

effectiveness. A stakeholder's *attitude* toward an organization or any of its activities can be driven by many factors including: whether involvement is voluntary or involuntary; whether involvement is beneficial personally or organizationally; or whether the level of a stakeholder's investment is either financial or emotional in the activity. If the individual's or group's stake in the activity is perceived to be beneficial, or potentially beneficial to them, they are more likely to have a positive attitude to the activity and be prepared to contribute to the work to deliver it. If on the other hand, they see themselves as victims, they will be more likely to hold a negative attitude to that activity. Any assessment of *attitude* will need to take into account the following elements:*

- Culture of the organization doing the work;
- Identification with the work and its outcomes or purpose;
- Perceived importance of the activity and its outcomes; and
- Personal attributes, such as cultural background, personality, or position in the organization.

Any relationship requires constant work to maintain; this applies to family relationships, friendships, management of staff, and maintenance of professional networks. Relationships between an organization and its stakeholders are no different. The team must understand the expectations of all of the important stakeholders and how can they be managed through targeted communication to maintain supportive relationships and to mitigate the consequences of unsupportive stakeholders for the benefit of the organization and its activities.

Typologies of Stakeholders: An Aid to Engagement

The ***Stakeholder Circle*** methodology has attempted to provide a way to identify and otherwise define groups of stakeholders to assist with developing appropriate communication strategies and implementation plans to improve the attitude of important stakeholders. (Fassin, 2012) has developed a useful typology based on four categories:

- Stakeowners: 'legitimate' (traditional) claim on the firm;
- Stakewatchers: pressure groups—possess only an indirect claim;
- Statekeepers: regulators who impose external control and regulations on the firm; and
- Stakeseekers: seek to have a voice in the public debate and 'pretend' to have a claim on the firm.

* These elements are also aspects of any individual's cultural background, which will be discussed later in this chapter.

By combining the analytics of the ***Stakeholder Circle***[*] with these categories is a useful way to develop communication strategies and practical implementation of that communication can be developed.[†]

The Communication Plan

The basis for an effective communication plan is defining for each stakeholder:

- The *purpose* of the communication: what does the team need to achieve through the communication;
- The most *appropriate* information: most effective *message*—its format and delivery method;
- *Targeted* communications to meet the expectations and requirements of the stakeholder and the capacity and capability of the team.

Based on each stakeholder's unique engagement profile, a communication plan can be developed. The communication plan should contain:

- *Mutuality*: How the stakeholder is important to the activity AND the stakeholder's *stake* and expectations;
- Categorization of influence (*upward, downward, outward, sideward, internal, and external*) and/or the categories of (Fassin, 2012). Stakeholders from each category will require messages presented in different content and format;[‡]
- Engagement profile preferably in graphical form: level of support for the activity and receptiveness to information about the work AND target engagement: necessary levels of support and receptiveness;
- Strategies for delivering the message:
  - *Who* will deliver the message?
  - *What* the message will be: regular activity reports or special messages?
  - *How* it will be delivered: formal and/or informal, written and/or oral, technology of communication – e-mails, written memos, meetings?
  - *When:* how frequently it will be delivered, and over what time frame (where applicable)?
  - *Why:* the purpose for the communication: this is a function of mutuality—why the stakeholder is important for activity success, and what the stakeholder requires from the activity?

---

[*] Go to www.stakeholdermapping.com for details of the ***Stakeholder Circle*** methodology and access to the software that supports it

[†] These ideas are explored in more detail in my forthcoming book: *Making Projects Work: Effective Stakeholder Management and Communication* scheduled for publication in 2014.

[‡] For example, a stakeholder from the *upward* category will probably only need a summary in the format that is most familiar—financial professions probably require spreadsheet information, and human resource managers may prefer graphics. Communication to the team, however, will be more effective in detail with clear instructions on what has to be done and how will do it.

- *Communication item:* the information that will be distributed – the content of the report or message.

Effective Communication

Regardless of how well the communication strategy and plan are crafted, other factors must be considered, such as the role of the stakeholder.

- The sponsor or other *upward* stakeholder may require exception reports, with briefing data sufficient to be able to defend the activity and *no surprises*;
- Middle managers who supply resources need time frames, resource data, and reports on adherence to resource plans and effectiveness of resources provided; therefore, more comprehensive information;
- Staff working on the activity and other team members need detailed but focused information that will enable them to perform their activity roles effectively;
- Other staff need updates on progress of the work, particularly information on how it will affect their own work roles; and
- External stakeholders will also require regular planned and managed updates on the activity, its deliverables, its impact, and its progress.

Factors that Affect Communication Effectiveness

Other factors may act as barriers to effective communication. Awareness of these factors and their consequences may drive the timing and context of the communication activity. They will be described in more detail in the next section of the chapter.

- Personal reality: conscious and unconscious thought processes will influence how individuals receive and process any information they receive;
- Cultural differences: differences in communication requirements may be caused by cultural norms influencing the preferred style of presentation, content, and delivery of information. These differences may be national, generational, professional, and organizational;
- Personality: personality differences may also dictate the how and what of effective communication. A senior manager with limited available time and a preference for summary information will have no patience for information delivered as a story, whereas a team member or a stakeholder with a different personality style may find the delivery of facts not interesting enough; and
- Environmental and personal distractions will include noise, lack of interest, fatigue, or emotions—if either the sender or the receiver is known to 'have a bad day', or is feeling unhappy, it is better to postpone any face-to-face communication until another occasion.

The answer to that question—*What makes us who we are and how do we operate in our social world?*—lies in a complex web of our own 'reality' formed by our own experiences (and how our brain makes sense of those experiences), our culture, whether national, professional, generational, and our gender. This web influences how we live and work and relate to others. And within the work environment, the culture of organizations affects the project and its stakeholders.

Perception and 'Reality'

Researchers have long taken an interest in the different ways that individuals make sense of their surroundings—'their world'. (Weick, 1995) developed the concept of 'sensemaking' by which people make sense of their organizational environment. Sensemaking refers to how we interpret an unfamiliar situation within the framework of previous experience, then incorporating the new data, or resolving the current issue, or adapting to a new environment.[*] Weick (1995) has interpreted this process in one way, while neuroscientists have taken a completely different approach for how we construct our reality and how we learn and make sense of new situations.[†]

When new information is presented to us, the new data are compared with our existing mental maps to find connections between new data and existing frameworks. If there are no connections the brain will try to make the connections fit into the existing framework. New information or stimulation bombard the brain continually so the brain will take shortcuts. We have expectations about what we are going to read or experience, and therefore, we 'see' it in that frame—not necessarily what actually happened. Such approximation means that often we misunderstand or misinterpret what we observe.

There is no reality 'out there'; only the reality we discern through the filters of our experiences, our knowledge, and interests. Each person will therefore have constructed a different reality so that he or she may describe the same scene in totally different ways.[‡] The brain truly sees the world according to its own wiring, selecting and ignoring information depending on its filters.

[*] To illustrate this concept, Weick (1995, p. 55) relates the story of a small military unit sent on a training mission into the Swiss Alps and who became lost in a snow storm. One of them had a map, and with the assistance of that map they planned their journey back to their base. When the storm subsided they began their journey back to base. On that journey, they did not always find the landmarks that the map showed, but with the help of residents of the villages they passed through they eventually found their way back to base, tired, hungry, and cold. That was when they discovered that the map was a map of the Pyrenees and not the Alps! The map was not the blueprint, it was only the artefact that helped them get started; the rest of the journey was facilitated by cues from the environment, incorporating new information, and acting with purpose.

[†] There are obviously many other theories contributing to an understanding of how we construct reality. It has long been a question that philosophers have grappled with—reality and the relationships between the mind and reality through the means of language and culture.

[‡] (Horowitz, 2013) describes what happens when she turns a daily 'walking around the block' with her dog into an exercise of perception by inviting people from different professions to walk with her and describe what they 'saw'. Each one of them drew her attention to different aspects of the same pathways she had walked on many times before: psychiatrist, economist, her 19 month-old son, an architect, and eight others. They all 'saw' aspects of that block that she could never have imagined.

Personality

Personality is a second factor to consider in communicating to engage stakeholders. The term 'personality' is derived from *persona* meaning *mask,* and refers to an individual's distinct pattern of thoughts, motives, values, attitudes, and behaviours. There are many typologies for categorizing personality. The best known of these typologies is the Myers-Briggs Indicator (MBTI).

(MBTI) measures psychological preferences in how people perceive the world and make decisions (Kroeger & Thuesen, 1988). Often people will act in ways that cause an individual confusion or anger, because it seems wrong or unreasonable, but in fact is just different from how another individual might act. (Kroeger & Thuesen, 1988) state that when this happens it is not the problem of the person who seems to be wrong or unreasonable, it is the person who is feeling the anger! Their view is that it is important to try to understand why these others are acting the way they do and also why it is causing such a reaction. They turn to a typology such as MBTI to assist in this understanding process.

The MBTI uses four pairs of alternative preferences which when combined provide 16 possible types of personality* :

- Introversion (I) or Extraversion (E) – 'attitudes';
- Sensing (S) and Intuition (N) – 'functions';
- Thinking (T) and Feeling (F) – 'functions';
- Judging (J) and Perception (P) – 'lifestyle'.

Using information about the personality attributes or behaviour preferences about others may be useful in understanding why people act and react how they do. As with any method or schema to understand others (and perhaps oneself) it can only give an indication.

Culture

Personality and perception of reality define characteristics of individuals and is not necessarily dependant on the individual's nationality, age, race, or gender. (Hofstede et al., 2010) has defined culture as: 'software of the mind'. A person's culture (national, professional, and organizational) influences how messages will be sent and received (their communication style), which in turn influences how people from different backgrounds can work together harmoniously. Understanding how cultural background affects communication style reduces misunderstandings and helps build empathy.

Culture is learned from parents, teachers, peers, and 'heroes' throughout childhood and well into adult life. (Hofstede et al., 2010) have defined four ways to describe

* For most effective results it makes sense to use the MBTI as used by groups who have received the appropriate training in applying the instrument and analysing results. However for a quick assessment to get a feel for the MBTI process go to: http://www.personalitypathways.com/type_inventory.html.

how culture manifests itself—symbols, heroes, rituals, and values. From an analysis of these four it is possible to develop an understanding of any type of culture and also to provide a means to compare cultures.

- *Symbols are words, gestures, pictures, or objects that carry a particular meaning that is recognised as such only by those who share the culture.* In the context of delivering organizational value that is the central concern of this chapter, the symbols will relate to the language, processes, and practices developed to formulate the structure and the discipline of portfolios, programs, and projects within the organization.
- *Heroes are persons, alive or dead, real or imaginary, who possess characteristics that are highly prized in a culture and thus serve as models for behaviour.* In the context of this subject heroes will be the CEO who delivers organizational success through increasing shareholder value, the MBA, or the accountant (Hofstede et al., 2010).
- *Rituals are collective activities that are technically superfluous to reach the desired end but that within a culture are considered socially essential.* They are carried out for their own sake. Rituals can range from how and whom we pay respects to others, religious ceremonies, or business conferences (a way of reinforcing group identity). Rituals include 'discourse', the way that language is used in text and talk, in daily interaction, and in communicating beliefs. Many meetings and progress reports are ritualistic observances that perform no real business function.
- *Values are broad tendencies to prefer certain states of affairs over others.* Because they are acquired early, many values remain unconscious to those who hold them. The value system is central to culture and is best understood through understanding pairings such as: good/evil; dirty/clean; dangerous/safe; or abnormal/normal; paradoxical/logical; irrational/rational (Hofstede et al., 2010).

In the organizational environment 'values' are often expressed as 'on time, within budget, and to scope'. This approach may not necessarily be the best approach for delivering value to the organization, and its imposition may impose undue stress on those whose role it is to deliver the outcomes within the constraints of scarce resources and funding and minimal interest of sponsors and other important stakeholders.

Cultural diversity within a project team may take the following forms:

- Generational and gender—a team may contain representatives from as many as four different generational groups: baby boomers; Gen X, Y, or Z. Generational differences may cause misunderstandings based on communication preferences, attitudes to work, and even language.
- Industrial or professional—Managers; professionals (engineers, accountants, and teachers); and blue collar workers. Once again they will have different communication styles, language, and approaches to work.

- National—consider a mix of Asian, Anglo-American, and Latino cultures – here also there will be different communication styles, language, and approaches to work.
- Organizational—Corporations, Government departments, and Universities will all have different structures and focus.

(Hofstede et al., 2010) developed typologies of culture from research he carried out for IBM in the 1980s and updated in 2010. He has defined five dimensions and with recent collaborations with other researchers added a sixth*:

- Power distance (weak/strong) (PDI): an indicator of dependence relationships in a country.
- Collectivism/individualism (IDV): Individualism defines societies in which the ties between individuals are loose: everyone is expected to look after himself or herself and his or her immediate family. Collectivism defines societies in which people from birth onward are integrated into strong, cohesive in-groups, which continue to protect people throughout their life time in exchange for unquestioning loyalty.
- Femininity/masculinity (MAS): masculine society defines gender roles as distinct: men are supposed to be focused on material success, and women are supposed to be more concerned with quality of life. In a feminine society gender roles overlap. And both men and women are supposed to be modest, tender, and concerned with the quality of life.
- Uncertainty avoidance (UAI): members of a strong UAI culture feel threatened by ambiguous in unknown situations. There is a need for predictability in the form of written and unwritten rules.
- Long-term/short-term orientation (LTO): LTO stands for the fostering of virtues oriented toward future rewards in particular perseverance and thrift; short-term orientation (STO) stands for the fostering of virtues related to the past and present in particular respect for tradition, preservation of 'face', and fulfilling social obligations.
- Indulgent/restrained: This dimension has as its focus 'happiness' – a universally cherished goal. There are two main aspects: evaluation of one's life and description of one's feelings.

Generational Culture

From time immemorial, older generations have complained about the transgressions and lack of respect of the younger generations: complaints about the loss of respect of 'younger generation' were found in Egyptian scrolls 2000 BC. (Hofstede et al., 2010).

* The website: http://geert-hofstede.com/national-culture.html has profiles for most countries, and the ability to compare two countries on all five dimensions.

Zemke, Raines, and Filipczak (2013) have categorised the four potential generations that operate in the workplace today as:

- *Traditionalists:* born before 1943;
- *Baby Boomers:* born 1943 to 1960;
- *Gen Xers:* born 1960 to 1980;
- *Gen Yers:* born 1980 to 2000.

There can be many areas of potential conflict and misunderstanding in the form of different:

- Symbols, heroes, and rituals;
- Differences in values and points of view;
- Ways of working and thinking; and
- Talking and communicating.

To build strong relationships between the portfolio, program, or project and its stakeholders, it is important to understand these generational differences in order to bridge the inevitable gaps.

The focus in generational studies seems to be focused only on the differences and influences of the cohorts that have been identified in current studies. It is also important to recognise that each age group will have distinguishing characteristics that may also have an influence on how the different generational groups operate and how the other generational groups view them*.

Professional Culture

In discussions about culture and cultural differences that may exist within teams from the same generations and national groupings, there may still be different communities within the team or within organizations; these are professional cultures that exist within any project or organization today. (Schein,1996) identified three distinct cultures in manufacturing organizations:

- Operators—Exist within the part of the organization that builds the product or delivers the service. Their structure and values are unique to the organization or at least to their industry. Their culture and norms are built on trust and teamwork.

* For example *twenty-somethings* are ideological and believe that the older generations are cynical and complacent; at 30 most individuals will have met a life-partner and may even have begun to raise a family—this will change their world view. At 40, individuals will begin to recognise that many of their dreams may now never be fulfilled—this can lead to the phenomena of changing jobs of making life style changes. At 50 people have more money and fewer expenses and can indulge in things that they could not afford in their youth, such as fast cars or motor bikes. These opinions are based on my own observations of Western individuals and groups and some conversations with people in these age groups.

- Technical specialists—Designers and implementers of technology. These categories include project managers, engineers, and hardware specialists.
- Executives—Fiscal responsibility. They favour command and control systems and management techniques based on command and control.

There are now far more occupations that will contribute to success of the project or the organization in its portfolio, but these three categories still serve the purpose of understanding the different types of professional cultures. This understanding will assist communication by reducing the possibility of misunderstanding. Communication between these different groups can be improved through the effort of each group understanding the values, symbols, and rituals of the other groups (Schein, 1996).

Gender

The previous sections on perception, personality, and culture have focused on factors that cause individuals to think and act in the way that they do: it also illustrates how each individual is unique: they are not all the same; they cannot be treated all the same. There is one more point of difference that needs to be considered within this chapter—that is gender.

We think of ourselves in terms of gender—even if we don't realize it! (Fine, 2010). The social context we grow up in influences who we are, how we think, and what we do. All our social expectations and stereotypes are formed at an early age. In the Western world the gender stereotypes fit within the framework of the following—to a greater or lesser extent:

Female traits:

- Communal personality traits;
- Compassionate, loves children, dependent, interpersonally sensitive, and nurturing;
- To serve the needs of others.

The male traits:

- Agentic personality traits[*]
- Aggressive, leader, ambitious, analytical, competitive, dominant, independent, and individualistic;
- To bend the world to your command and earn a wage for it.[†]

[*] The capacity to exercise control over the nature and quality of one's life: the capacity to act in the world (National Centre for Biotechnology Information)

[†] This is the case for white, middle-class heterosexual men (Fine, 2010)

A person grows and changes in response to social environment. Each person develops a 'Wardrobe of Self' (Fine, 2010) to match all the social identities one person can assume.*

The gender stereotypes that provide the clothing for the 'Wardrobe of Self' are reinforced by society. In masculine countries, such as United States, Australia and the United Kingdom, boys are socialised toward assertiveness, ambition, and competition. They are expected to aspire to career advancement. In these same countries, girls are polarised between some who want a career and many who do not. In feminine countries, such as the Netherlands and Denmark, children should not be aggressive. Men and women many or may not be ambitious, and may or may not want a career.

Discourse—The Sharing of Information

In dealing with communication to stakeholders it is essential to consider all the differences in culture that we have discussed so far. The reason that it is important to understand gender differences is the way the men and women transmit and choose to interpret information. (Tannen, 2013) describes these differences a:

- 'Report talk'—the way that men communicate both formally and informally, transferring information to establish and maintain status that displays their abilities and knowledge.
- 'Rapport talk'—the way that women communicate both formally and informally to build and maintain connections, first validating the relationship to build rapport and then dealing with any business.

Neither of these ways of communicating is necessarily superior to the other—this is just how men and women have been socialized. But it also explains why there can be misunderstandings in both formal and informal conversations, where men try to 'fix' the problem by giving advice, and women want to talk about the problem without necessary needing the advice the men are seeking to provide. These misunderstandings will also explain the impression that many male managers have of the linguistic styles of their female colleagues. For example, women ask more questions, usually for clarification or deeper understanding: this has been interpreted by male managers as not knowing enough (Tannen, 1995).†

---

* My wardrobe is: Melbourne resident, teacher, grandmother, woman, university professor, writer, Baby Boomer. Depending on which identity I need to 'wear' I will have different approaches, perhaps use different language and tone, I will socialize in different ways. Who I am is sensitive to the social context at that moment.

† And of course there is the story of how women are willing to ask for directions, whereas men are reluctant to do so.

Organizational Culture

The final aspect of culture is the specific culture of an organization. When companies are part of international corporations, their planning and control systems will be influenced by the national culture specific to the country in which this branch of the company practices, even though headquarters will attempt to influence decision making, processes, and controls. Different organizations will display different characteristics, depending on their structure and mission: Corporations (for profit); Not for Profit, such as charities; and Government Departments or agencies. Within these characteristics will be other distinguishing features based on:

- Risk tolerance—are they risk avoiding or risk seeking?
- Charter—are they entrepreneurial or public-service oriented?
- Who benefits—shareholders? Selected groups of society? Or the public at large?
- Product orientation—manufacturing, product sales, service providers, or a mixture?
- National, regional, or multinational?

The culture of the organization will be formed from the mix of features: in turn the culture of the organization will influence how management is 'done' within the organization. Part of peoples' mental software consists of their ideas about what an organization should be like, with *power distance* and *uncertainty avoidance* affecting our thinking about organizations (Hofstede et al., 2010). Understanding these dimensions requires answering two questions:

- Who has the power to decide what? (power distance)?
- What rules or procedures will be followed to attain the desired ends? (uncertainty avoidance)?

Individualism and masculinity affect our thinking about people in organizations and not the processes, practices, and symbols of the organizations.

Conclusion

Only when the needs (expectations) of each key stakeholder and the stake or stakes he or she may have in the outcome are known and understood is it really possible to begin to understand their drivers or business needs. These expectations may be overly ambitious or unrealistic. With an understanding of the cultural background of key stakeholders, messages can be crafted and information offered to 'educate' stakeholders on what can be achieved in the context of expectations. The sponsor's needs cannot be assumed and may not even be related to the work itself. But this understanding of needs and expectations is crucial. It is a fundamental starting point for any campaign to reach agreement on alternative objectives and the management of the perception of the sponsor *but also* the perceptions of the stakeholders of the sponsor.

The focus of this chapter has been on the reality of what happens in the Zone of Uncertainty, the area between an organization's strategic vision and the projects created to deliver that vision. It describes the murky, complicated, and complex activity to achieve the vision. It also describes the difference between the optimism of the neat, linear processes, plans, and controls and the messy reality of its implementation. There are three major causes of the complexity that creates the Zone: technological complexity, the complexity of the work that results from seeking to achieve balance in the portfolio, and the *stakeholder factor*. The *stakeholder factor* describes the complexity of human relationships and human activity as the basic building block in the work of organizations. Improvement in the area of stakeholder relationships will reduce the messiness and confusion that everyone experiences within the Zone.

Stakeholder relationship management is not easy; it is also very time consuming and cannot be reduced to templates or standard e-mails. To deliver value to the organization effectively through portfolios, programs, and projects, analysis of the stakeholder community and its engagement requires exchange of information – communication. This is the only tool we have to develop and sustain robust relationships. Successful communication requires not only careful planning and thoughtful implementation, it also requires understanding more about each individual or group that makes up the stakeholder community, through an understanding of their cultural background and their mental maps—they make sense of their world and how they learn and act.

References

Bourne, L. (2012). *Stakeholder relationship management: A maturity model for organisational implementation* (Revised Edition). Farnham, UK: Gower.

Bourne, L. (for publication 2014). *Making projects work: Effective stakeholder management and communication.* Boca Raton, FL: CRC Press.

Cooke-Davies, T., Crawford, L., Patton, J., Stevens, C., & Williams, T. (Eds.). (2011). *Aspects of complexity: Managing projects in a complex world.* Newtown Square, PA: Project Management Institute.

Dinsmore, P. C. (1999). *Winning in business with enterprise project management.* New York: AMA Publications.

Fassin, Y. (2012). Stakeholder management, reciprocity, and stakeholder responsibility. *Journal of Business Ethics, 109*, 83–96.

Fine, C. (2010). *Delusions of gender: The real science behind sex differences.* London: Icon Books.

Hofstede, G., Hofstede, G. J., & Minkov, M. (2010). *Cultures and organizations: Software of the mind. Intercultural cooperation and its importance for survival.* New York: McGraw-Hill.

Horowitz, A. (2013). *On looking: About everything there is to see.* London: Simon & Schuster.

Kroeger, O., & Thuesen, J. (1988). *Type talk: The 16 personality types that determine how we live, love, and work.* New York: Dell Publishing.

Microsoft. (2008). *Project portfolio management: Doing the right things right.* downloaded March, 2014 from www.epmconnect.com

Mitchell, R. K., Agle, B. R., & Wood, D. J. (1997). Toward a theory of stakeholder identification and salience: Defining the principle of who and what really counts. *Academy of Management Review, 22*(4), 853–888.

Schein, E. H. (1996). Three cultures of management: The key to organizational learning. *Sloan Management Review, Fall, 1996*, 9–20.

Tannen, D. (1995). The power of talk: Who gets heard and why. *Harvard Business Review, 'on communicating effectively'*. September.

Tannen, D. (2013). *You just don't understand: Women and men in conversation.* New York: HarperCollins.

Weick, K. E. (1995). *Sensemaking in organizations.* Thousand Oaks, CA: Sage.

Zemke, R., Raines, C., & Filipczak. (2013). *Generations at work: Managing the clash of boomers, gen xers, and gen yers in the workplace.* New York: Amacom - American Management Association.

Index

T

U

W

Z

PGIL2021USA